# All-Digital RF Transmitters for Next Generation Wireless Communications

This book introduces the foundations of All-Digital Transmitters (ADTs), mainly based on Delta-Sigma and Pulse-Width Modulators, and then focuses on the main research challenges and achievements that have been reported. Architectures are presented for specific cases, along with low-level diagrams and implementation figures of merit.

*All-Digital RF Transmitters for Next Generation Wireless Communications* presents a broad, in-depth view of the technology and the challenges it faces. The authors present a set of ideas and architectures that enable the implementation of fully digital transmitters, bringing the digital domain closer to the antenna. In each chapter, the reader is presented with a set of challenges, accompanied by theoretical explanations and experimental validations of how the problems were addressed. In particular, the authors begin with the analog filtering requirements from the reconstruction filter, carrier agility, architectural techniques for designing wideband and high-speed delta-sigma modulators, and concurrent multi-band capability. The book also deals with various applications in which All-Digital Transmitters were integrated and have inherently surpassed state-of-the-art techniques, in terms of innovation, Figures of Merit (FoM), or scalability. The book concludes by summarizing the major findings and offering insights into future research directions to explore.

Written for RF system design engineers as well as digital signal design engineers involved in designing radios for digital communication systems, this book could be used by professionals and postgraduate students.

# All-Digital RF Transmitters for Next Generation Wireless Communications

Samuel Santos Pereira, Daniel Da Costa Dinis, Rui Fiel Cordeiro, Nelson José Valente da Silva, Luís Filipe Almeida, José Manuel Neto Vieira, Nuno Miguel Gonçalves Borges de Carvalho, and Arnaldo Silva Rodrigues de Oliveira

CRC Press
Taylor & Francis Group
Boca Raton London New York

CRC Press is an imprint of the
Taylor & Francis Group, an **informa** business

First edition published 2027
by CRC Press
2385 NW Executive Center Drive, Suite 320, Boca Raton FL 33431

and by CRC Press
4 Park Square, Milton Park, Abingdon, Oxon, OX14 4RN

*CRC Press is an imprint of Taylor & Francis Group, LLC*



ISBN: 978-1-032-12682-1 (hbk)
ISBN: 978-1-032-12683-8 (pbk)
ISBN: 978-1-003-22571-3 (ebk)

DOI: 10.1201/9781003225713

Typeset in CMR10
by KnowledgeWorks Global Ltd.

# Contents

# Preface

Since the start of this decade, we have been watching to a global spread of handsets, such as smartphones, tablets, as well as new network-dependent devices boosted by the omnipresence of wireless networks in our daily lives. This trend has led to successive generations of standards that provide higher throughput and support mobility. In just over three decades, humanity has witnessed four generations of mobile communications, which have significantly modified cellular infrastructure, and the fifth generation is now in full production. Massive number of RF front-ends, peak data rates of 10 Gbps (everywhere and every time), latencies lower than 10 msec (to support hard real-time applications), and huge device densities (in the so-called IoT scenarios), are some of the expected disruptive capabilities. Increasingly, this implies that the next generation of radios must be carefully designed to ensure a smooth transition to subsequent standards and technologies, while previous generations should not be jeopardized. Thus, it makes sense that features such as multi-standard support, flexibility, agility, real-time reconfigurability, and upgradeability shall be valued.

At the same time, as analog RF components do not follow Moore's Law, their integration and mass production impose major constraints that have induced a radical shift in mindset: moving from analog to digital-centric design of RF transceivers. Interestingly, this trend toward the digitalization of conventional analog front-ends has motivated radio research toward a concept idealized in the late '90s: the SDR. This idea envisioned a novel architecture for radio systems in which waveforms are completely synthesized in the digital domain, thereby achieving greater flexibility, reconfigurability, and efficiency. To accomplish this, the RF signal generation and analysis need to move closer to the antenna.

In view of the foregoing, if this change of mindset from the analog to the digital design has all the inherent advantages that the next generation of radios should have, to enable multi-standard, multi-band, agile, flexible, and reconfigurable radio front-ends, so one can foresee that the digitalization of the radios will underlie the future generation of wireless and mobile communication systems.

In our view, the ultimate target for these techniques is the concept of All-Digital Transceivers, in which everything is digital from end to end. In them lies the possibility of designing radio data paths that are completely digital from the baseband to the RF stage. The underlying idea explores the quantization of an $m$-bit digital signal into a 2-level binary representation, yielding constant-envelope output signals that can be directly amplified by highly efficient PAs, such as SMPAs. After amplification, a reconstruction filter is required to recover the signal before it is fed to the antenna. With such a concept, the analog front-end will be highly minimized, and therefore, unprecedented integration and massification of transceivers is leveraged.

The main purpose of this book is to present a broad and in-depth view of the All-Digital RF Transmitters. The text presents systematic design methods for fully digital RF transmitters, focusing on the most pressing research challenges. It was written for RF system design engineers and digital signal design engineers involved in the design of radios for digital communication systems.

This book adopts a practical, pragmatic approach, defining research challenges, proposing innovative solutions, and experimentally validating them. FPGAs were used as prototyping and evaluation platforms due to their inherent cost- and time-effectiveness. Notwithstanding the foregoing, all the concepts and architectures can be directly transferred and integrated into any other specialized application-specific circuit-level techniques.

In terms of document organization, Chapter 1 starts by laying out the background and motivation for this different reality of RF transceivers. Chapter 2 aims to provide a full overview on Digital Transmitter Architectures, spanning from the Classical RF Transmitter Architectures, to the All-Digital Transmitter Architectures, passing through the foundations on Pulse Encoding Techniques and the main research challenges associated with these types of techniques. After providing a stable comprehension level to the readers, Chapters 3 up to 7 propose and validate different architectures that have been proposed to excel the state-of-the-art literature in specific and well-known research challenges. In particular, Chapter 3 deals with the analog filtering requirements from the reconstruction filter, Chapter 4 with the Carrier Agility, Chapter 5 with the architectural techniques to design wideband and high-speed delta-sigma modulators, and Chapter 6 deals with the concurrent multi-band capability. Concluding the research challenges, Chapter 7 presents some of the different applications where All-Digital Transmitters were integrated and inherently surpassed the state-of-the-art techniques either in terms of innovation, FoM, or scalability.

Finally, Chapter 8 concludes this book by summarizing the major conclusions and future research lines that may be explored in the coming years.

We could not finish this Preface without emphasizing that this book reflects the most significant advances in All-Digital RF Transmitters achieved over the last decade, a period marked by rapid technological evolution and renewed research momentum in digital-centric radio architectures. Alongside the presentation of SoA techniques and experimentally validated solutions, this work also deliberately exposes the current limitations, trade-offs, and open challenges that still hinder the widespread adoption of fully digital transmitter paradigms. By highlighting these constraints and discussing emerging trends and future development paths, we aim to provide not only a comprehensive reference of what has been accomplished but also a clear roadmap for researchers and engineers to further push the boundaries of All-Digital Transceivers in the years to come.

# *About the authors*

**Samuel Santos Pereira** is a Researcher at the Instituto de Telecomunicações in Aveiro, Portugal. He earned his M.Sc. in Electrical and Telecommunications Engineering from the University of Aveiro in 2018 and completed his Ph.D. in Electrical Engineering at the same institution in 2024. His research focuses on all-digital transceivers, reconfigurable systems, software-defined radio, delta—sigma-based transmitters, and RF jamming techniques. Additionally, he has contributed to the development of C-RAN solutions for 5G networks and, more recently, has worked on satellite communication systems using both microcontroller and FPGA-based platforms.

**Daniel Da Costa Dinis** is a Satcom Space Segment Engineer contracted by Starion at the European Space Agency in Harwell, UK, where he oversees technology and product development initiatives. He holds an M.Sc. and Ph.D. in Telecommunications from the University of Aveiro (2013, 2019) and an Executive MBA from the Quantic School of Business and Technology (2021). With a diverse career spanning academia, research laboratories, industry, and institutional agencies, Dr. Dinis has held key roles at Instituto de Telecomunicações, Mitsubishi Electric Research Labs, Thales, Airbus Defence and Space, and All Space.

An accomplished professional, Dr. Dinis has co-authored over 20 publications, is a co-inventor of six U.S. patents, and has earned multiple awards for his contributions. He actively participates in several IEEE committees, serves on the ESA-funded SatNEx-V management team, and specializes in All-Digital Transceivers, Satcom systems, phased arrays, RF systems, and Software-Defined Radios.

**Rui Fiel Cordeiro** is a Senior FPGA Developer at EVS Broadcast, specializing in high-performance FPGA-based video broadcast systems. He holds a Ph.D. in Electronics and Telecommunications Engineering from the University of Aveiro, Portugal. His doctoral research at the Instituto de Telecomunicações focused on flexible, wideband all-digital radio transmitter architectures, with an emphasis on reconfigurable hardware, digital signal processing, RF system design, and embedded systems. Dr. Cordeiro's career includes a role as a Development Engineer at Bosch Car Multimedia, where he contributed to multi-technology radio solutions for automotive and vehicular communications. He later served as Principal Investigator at CEiiA, leading research on connectivity and positioning for sustainable mobility. Currently, at EVS Broadcast, he continues to advance cutting-edge FPGA solutions for the video broadcast industry.

**Nelson José Valente da Silva** is a Lead Wireless Engineer at Altice Labs, where he focuses on developing Smart Wi-Fi solutions for Tier 1 cable and telecom operators. He earned his M.Sc. and Ph.D. degrees in Electronics and Telecommunications Engineering from the University of Aveiro, Portugal, in 2008 and 2013, respectively. Dr. Silva has authored or co-authored three patents and over 20 papers in journals and conference proceedings. His previous research experience includes work at the Instituto de Telecomunicações in Aveiro, Portugal, on reconfigurable transmitters for software-defined radios, and at the IEETA Institute in Aveiro, Portugal, on reconfigurable embedded systems. His research interests span

Wi-Fi and smart home technologies, software-defined radios, all-digital transceivers, reconfigurable hardware, embedded systems, and signal processing algorithms and architectures.

**Luís Filipe Almeida** is a Researcher at the Instituto de Telecomunicações in Aveiro, Portugal, where he contributes to projects focused on FPGA digital design, Software-Defined Radios, and Signal Processing. He earned his Ph.D. in Electrical Engineering from the University of Aveiro in 2025, building on a career closely tied to academia and research institutions since completing his M.Sc. at the same university and Instituto de Telecomunicações. Dr. Almeida has co-authored several papers in international journals and conference proceedings and is a co-inventor of a national Portuguese patent. His research interests include Software-Defined Radios, All-Digital RF Transceivers, runtime FPGA reprogramming, digital system design flexibility and reliability, and satellite communications.

**José Manuel Neto Vieira** is an Assistant Professor at the University of Aveiro, Portugal, where he also earned his Ph.D. in Electrical Engineering. In 2004, he founded the Portuguese section of the Audio Engineering Society (AES) and served as its President from 2005 to 2011. In 2010, he received the prestigious Plug Award from APRITEL for his work on the "Bio-inspired Cochlear Radio." His primary research interests include digital audio signal processing, ultrasonic location, software-defined radio, and radar systems. In May 2018, he was appointed Pro-Rector for Information Systems at the University of Aveiro.

**Nuno Miguel Gonçalves Borges de Carvalho** is a Full Professor and Senior Research Scientist at the University of Aveiro and the Institute of Telecommunications, Portugal, where he also serves as the Director of the Department of Electronics, Telecommunications, and Informatics. He has coauthored several books on microwave systems, measurements, white-space communications, and wireless power, and has published over 400 research papers. Dr. Carvalho has held prominent editorial roles, including Editor-in-Chief of the Cambridge Wireless Power Transfer Journal and Associate Editor of IEEE Microwave Magazine. His research interests include software-defined radio (SDR) front-ends, backscatter communications, wireless power transfer, nonlinear distortion, and microwave measurement systems. He holds six patents and has actively contributed to IEEE MTT-S and URSI committees, serving as the 2023 IEEE MTT-S President and a Distinguished Lecturer.

**Arnaldo Silva Rodrigues de Oliveira** is a Senior Researcher at the Telecommunications Institute and an Associate Professor at the University of Aveiro, Portugal, specializing in computer architecture, digital systems design, software-defined radio, and embedded systems. He earned his Ph.D. in Electrical Engineering from the University of Aveiro in 2007. Dr. Oliveira actively contributes to several nationally and internationally funded research projects and has authored or co-authored 160 publications in journals and international conference proceedings. He is a member of IEEE, serves on multiple scientific committees for conferences, and is a regular reviewer for indexed scientific journals and magazines. His research interests focus on reconfigurable digital systems, software-defined radio, and next-generation radio access networks.

# *Acronyms*

$C_{eff}$ Coding Efficiency.

$\Delta\Sigma$ Delta-Sigma.

**$\Delta\Sigma$M** Delta-Sigma Modulator.

**$\Delta\Sigma$oF** Delta-Sigma over Fiber.

**3G** Third Generation.

**3GPP** third Generation Partnership Project.

**4G** Fourth Generation.

**5G** Fifth Generation.

**6G** Sixth Generation.

**ACLR** Adjacent Channel Leakage Ratio.

**ACPR** Adjacent-Channel Power Ratio.

**ACPR-L** Adjacent-Channel Power Ratio Lower.

**ACPR-U** Adjacent-Channel Power Ratio Upper.

**ADAAT** All-Digital Antenna Array Transmitter.

**ADC** Analog-to-Digital Converter.

**ADT** All-Digital Transmitter.

**AI** Artificial Intelligence.

**ASIC** Application-Specific Integration Circuit.

**ASK** Amplitude-Shift Keying.

**ASSP** Application-Specific Standard Product.

**AWG** Arbitrary Waveform Generator.

**BB** Baseband.

**BBU** Base Band Unit.

**BN** Beamforming Network.

**BP-$\Delta\Sigma$M** Band-Pass Delta-Sigma Modulator.

**BPF** BandPass Filter.

**BRAM** Block Random-Access Memory.

**BW** Bandwidth.

**C-RAN** Cloud Radio Access Network.

**CA** Carrier-Aggregation.

**CDM** Code-Division Multiplexing.

**CIC** Cascaded Integrator-Comb.

**CIFB** Cascaded-of-Integrators, Feedback Form.

**CIFB/F** Cascaded-of-Integrators, FeedBack/FeedForward.

**CIFF** Cascaded-of-Integrators, Feedforward Form.

**CMCD** Current-Mode Class-D.

**CMOS** Complementary Metal–Oxide–Semiconductor.

**CORDIC** COordinate Rotation DIgital Computer.

**CRFB** Cascaded-of-Resonators, Feedback Form.

**CRFB/F** Cascade-of-Resonators, FeedBack/FeedForward.

**CRFF** Cascaded-of-Resonators, Feedforward Form.

**CU** Central Unit.

**DAC** Digital-to-Analog Converter.

**DC** Direct Current.

**DDC** Digital Down-Convert.

**DDS** Digital Direct Synthesis.

**DFT** Discrete Fourier Transform.

**DSP** Digital Signal Processor.

**DUC** Digital Up-Conversion.

**E-O** Electrical-Optical.

**EDFA** Erbium-Doped Fiber Amplifier.

**EDGE** Enhanced Data Rates for GSM Evolution.

**EF** Error-Feedback.

**EVM** Error-Vector Magnitude.

**FFT** Fast Fourier Transform.

**FIFO** First-In First-Out.

**FIR** Finite Impulse Response.

**FoM** Figures of Merit.

**FPGA** Field-Programmable Gate Array.

**GSM** Global System for Mobile Communications.

**HP-ΔΣM** High-Pass Delta-Sigma Modulator.

**HSB** Higher SideBand.

**I/O** Input/Output.

**I/Q** In-phase/Quadrature.

**IC** Integrated Circuit.

**IF** Intermediate Frequency.

**IFFT** Inverse Fast Fourier Transform.

**IIR** Infinite Impulse Response.

**IoT** Internet-of-Things.

**IP** Intellectual Property.

**LEO** Low-Earth Orbit.

**LFSR** Linear Feedback Shift Register.

**LO** Local Oscillator.

**LP-ΔΣM** Low-Pass Delta-Sigma Modulator.

**LPWM** Limited Pulse-Width Modulation.

**LSB** Lower SideBand.

**LTE** Long-Term Evolution.

**LUT** Look-up table.

**MASH** Multistage Noise Shaping Modulator.

**MER** Modulation Error Ratio.

**MGS** Multi-Gigabit Serializer.

**MGT** Multi-Gigabit Transceiver.

**MIMO** Multiple-Input Multiple-Output.

**MZM** Mach-Zehnder Modulator.

**NB** Narrow Band.

**NMSE** Normalized Mean Square Error.

**NRZ** Non-Return-to-Zero.

**NTF** Noise-Transfer Function.

**NZ** Nyquist Zone.

**O-E** Optical-Electrical.

**OFDM** Orthogonal Frequency-Division Multiplexing.

**OOK** On-Off Shift Keying.

**OSR** OverSampling Ratio.

**PA** Power Amplifier.

**PAM** Pulse Amplitude Modulation.

**PAPR** Peak-to-Average-Power Ratio.

**PFM** Pulse-Frequency Modulation.

**PLL** Phase Locked Loop.

**PLM** Pulsed Load Modulation.

**PWM** Pulse-Width Modulation.

**QAM** Quadrature Amplitude Modulation.

**QPSK** Quadrature Phase-Shifting Keying.

**QPWM** Quantized Pulse-Width Modulation.

**RAM** Random-Access Memory.

**RAN** Radio Access Network.

**RBW** Resolution Bandwidth.

**RF** Radio-Frequency.

**RoF** Radio-over-Fibre.

**ROM** Read-Only Memory.

**RRH** Remote Radio Head.

**RU** Remote Unit.

**SDN** Software Defined Network.

**SDR** Software-Defined Radio.

**SFDR** Spurious-Free Dynamic Range.

**SFP** Small Form-factor Pluggable.

**SISO** Single Input Single Output.

**SMPA** Switched-Mode Power Amplifier.

**SNR** Signal-to-Noise Ratio.

**SoA** State-of-the-Art.

**SoC** System on Chip.

**SQNR** Signal-to-Quantization Noise Ratio.

**SR** Symbol Rate.

**SSB** Single-Side Band.

**SSMF** Standard Single-Mode Fiber.

**STF** Signal-Transfer Function.

**TDM** Time-Division Multiplexing.

**TIDS** Time-interleaved Delta-sigma.

**TTD** True Time Delay.

**TX** Transmitter.

**USB** Universal Serial Bus.

**VBW** Video Bandwidth.

**VCO** Voltage-Controlled Oscillator.

**VHDL** Very High Speed Integrated Circuit Hardware Description Language.

**VLSI** Very-large-scale Integration.

**VMCD** Voltage-Mode Class-D.

**VSA** Vector Signal Analyzer.

**ZOH** Zero-Order Hold.

# Chapter 1

# Introduction

## 1.1 Scope and Motivation

In recent years, the rapid growth of wireless communications has led to the development of numerous new standards operating at different frequencies, using dissimilar coding and modulation schemes, and targeting various applications. Such vast proliferation drives research efforts towards the development of new, flexible radios capable of adapting to various communication scenarios. Moreover, the exponential growth in the use of wireless devices leads to further utilization of the electromagnetic spectrum. In this sense, improving RF spectrum management, as well as using it efficiently, is also increasingly vital.

From the utilization of Orthogonal Frequency-Division Multiplexing (OFDM) in Long-Term Evolution (LTE) to the increase of carrier frequencies and the Multiple-Input Multiple-Output (MIMO) application in Fifth Generation (5G), several steps have been made towards that objective. However, some of these techniques, such as MIMO, which includes Phased Array Beamforming, Spatial Multiplexing, and Diversity Coding, face several implementation challenges. For instance, in MIMO, increasing the number of transmitters in an antenna array can be too bulky and too expensive using conventional transmitter architectures.

The concept of Software-Defined Radio (SDR), which was introduced in [130], presents a fully digital solution for a transmitter. In this concept, all signal processing is implemented in the digital domain, and then the signal is converted into the analog domain, with only power amplification, filtering, and radiation needed to transmit the signal. The ideal SDR concept still eludes us; however, All-Digital Transmitter (ADT) presents an attempt to implement this concept. In an ADT, the baseband signal processing, upsample, and upconversion are implemented in the digital domain. Then either Delta-Sigma Modulator ($\Delta\Sigma$M) or Pulse-Width Modulation (PWM) is used to reduce the number of bits to just one per sample. This allows ADTs to use one-bit Digital-to-Analog Converters (DACs) in order to send the signal to the analog domain. The use of a single-bit DAC can reduce the footprint and power consumption of the transmitter when compared to typical multi-bit or Nyquist Radio-Frequency (RF)-DACs. With this, an ADT approach can take advantage of the high integration level of modern digital circuits to create a large number of transmitters in a small footprint and at a relatively low cost. These characteristics make ADTs one of the most promising choices to implement MIMO-based transmitters.

Furthermore, massive MIMO is being proposed [5], where it relies on the law of averages, powered by the extreme number of antennas, to create a system resilient to noise, fading, and hardware imperfections [113]. Massive MIMO can use (near) constant envelope signals, which are perfect to be generated by an ADT approach, eliminating the need for high Peak-to-Average-Power Ratio (PAPR) signals, which in consequence eliminate the need for high linearity and high dynamic range constraints in the RF front-end. Instead, Switched-Mode Power Amplifier (SMPA) could be used in the front-end, reducing not only cost but also power consumption.

DOI: 10.1201/9781003225713-1

Currently, a significant portion of the energy is wasted in the transmission path of the RF front-end. In fact, considering the total power consumption of a cellular network, over 60% is consumed by the base stations [198], which have an overall power efficiency of approximately 5 to 10% [28, 106]. Additionally, considering the total power budget of a typical base station, over 50% is used in the transmission path [60, 106, 198], thus making a clear point that improving the power efficiency of the RF front-end is fundamental. Moreover, due to its significant contribution to overall power consumption in a cellular network, it is also clear that even small power efficiency improvements in the RF transmission path will have a substantial impact on the overall network's power consumption, allowing for a significant reduction in the environmental footprint.

Additionally, a transmission chain based on a Field-Programmable Gate Array (FPGA)-implemented ADT presents a higher degree of flexibility and graceful degradation when compared to a traditional chain. This is due to factors such as frequency agility, which allows the online reconfiguration of the carrier frequency to compensate for a failed subsystem, thereby increasing the overall system's lifespan. The ability of online reconfiguration may also be a crucial step towards network automation. Furthermore, the digitalization of the RF chain powered by FPGAs will also facilitate a closer integration of the transmission chain with Artificial Intelligence (AI). This integration has the potential to accelerate one of the primary objectives of the upcoming Sixth Generation (6G) network, which is the comprehensive integration of AI across all layers of the network [166]. Focusing on the physical layer, AI has already been demonstrated to be useful in channel estimation under scenarios affected by In-phase/Quadrature (I/Q) imbalance, non-linearities, and quantization errors [40] and in digital predistortion [218], while deep learning approaches have been presented as a way of reducing the design complexity while increasing accuracy and robustness in symbol detection [225].

Recently, ADTs have also been pointed out as a possible solution to optimize size, weight, power, and cost in satellite payloads, especially when Low-Earth Orbits (LEOs) are concerned. Their ability to implement MIMO transmission paired with their frequency agility makes them ideal for this type of implementation. On the other hand, their relatively small size and cost can help reduce both the manufacturing and launch costs of these satellites.

### 1.1.1 5G and Beyond—Changes and Challenges

With 5G already being deployed all around the world, data rates higher than one gigabit per second are being delivered to some users, and, due to the rise of the Internet-of-Things (IoT), thousands of devices are being connected to a single macrocell, with many more expected in the near future [11]. The increase in data rate per user, accompanied by the rise in the total number of connected devices, could represent a thousand-fold increase in the so-called aggregated data rate, which refers to the total data rate in bits per second per unit of area [11]. Another target for 5G is the low latency, which is expected to be less than 1 ms. However, at the time of writing this book, some limitations of 5G are already being unveiled. Furthermore, some researchers around the world have already started to draft what 6G might be [5]. With target peak data rates of 1 Tbps and expected spectral efficiencies ten times higher than 5G, it is clear that the current technology is not suitable for such high demands. Furthermore, 6G is expected to serve new types of user equipment which require a different approach to the way base stations are designed [227].

A solution to these problems could arise from the combined benefits of extreme densification of base stations and increased bandwidth (primarily by migrating into the millimeter wave and terahertz bands), together with enhanced spectral efficiency provided by MIMO and massive MIMO. The use of multiple frequency bands will also ensure that all scenarios are covered. On the one hand, low frequencies can provide large coverage, which is important for rural areas. The high-frequency band, on the other hand, will provide most of the

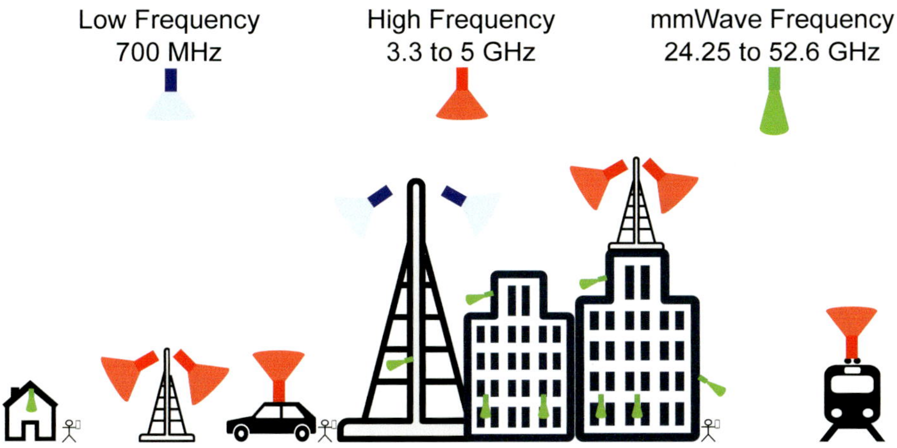


**FIGURE 1.1:** Different 5G bands and their applications.

bandwidth needed in cities, especially inside cars and other forms of transportation. On the other hand, the millimeter wave band will enable coverage of other specific use cases where a large bandwidth is the primary requirement. A depiction of the various frequency bands used in 5G and their corresponding use cases is shown in Fig. 1.1.

To meet all these requirements, cooperation between access points and network terminals using different heterogeneous technologies will be necessary to support the needed flexibility and spectral efficiency for 6G networks. Therefore, 6G radios must efficiently answer the new communication challenges while, at the same time, being capable of incorporating previous standards. An increase in the use of new wireless standards with different frequencies, diverse coding and modulation schemes, and targeting various applications is also common nowadays. In these scenarios, highly flexible, power-efficient, and wideband radio transceivers are required to adapt to changes in the communication environment. In fact, for such an approach, the physical layer of the radio must be able to support the simultaneous transmission of multi-band, multi-rate, and multi-standard signals, which is impractical or inefficient with conventional radio architectures [15]. Furthermore, the integration of more and more RF chains, or equivalently, the massification of the number of RF front-ends, is one of the main flagship characteristics of 5G and will be even more important in 6G. This has been a consequence of the renewed interest in the spatial domain, which is being highlighted as a key enabler to improve radio resource efficiency while increasing capacity. The use of multiple antennas has shown improvements in this domain [13]. As the conventional RF architectures are too bulky and too costly to provide an effective answer to these new trends, the use of ADTs techniques in this field in the next few years may also be a promising research track. Moreover, for true massive-MIMO implementations, the number of antenna sites needs to increase as well. This puts the current Radio Access Network (RAN) architecture under pressure due to its lack of scalability. To overcome this limitation, Software Defined Network (SDN)-enabled-Cloud Radio Access Network (C-RAN) has been proposed to take advantage of the RAN splitting to achieve smaller and power-efficient Remote Unit (RU) [206]. This scenario can take advantage of an ADT-based Radio-over-Fibre (RoF) solution, also called Delta-Sigma over Fiber ($\Delta\Sigma$oF) [117]. In this case, RAN is the structure responsible for connecting the user equipment, generally a cellphone, to the core network, or internet.

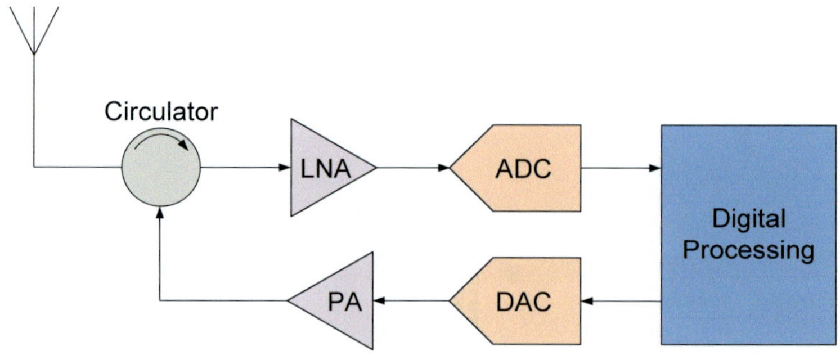


**FIGURE 1.2:** Simplified block diagram illustrating the architecture of an ideal SDR transceiver.

### 1.1.2 The Ideal Software-Defined Radio Paradigm

The rising trend in the digitalization of conventional analog front-ends has motivated the radio research in a concept idealized in the late '90s: the SDR [130]. This idea envisioned a novel architecture for the radio systems, with the waveforms synthesized in the digital domain, achieving greater flexibility, reconfigurability, and efficiency. The opportunities arising from this paradigm are unimaginable. The idea of having a single, completely independent radio transceiver front-end, fully independent of the standard, with a wide range of frequency bands and full agility, together with the capability of being real-time reconfigurable and upgradeable, is so ambitious that it has sparked significant interest among the research community. However, to accomplish this, the RF signal generation and analysis need to move closer to the antenna.

Evolving the software and digital signal generation closer to the antenna, and removing most of the conventional RF hardware from the path, would be a fundamental improvement for both mobile terminals and base stations. The digitalization of radio technology allows for the massive miniaturization of radio hardware, as digital logic blocks continue to shrink with technological advancements. Although Moore's Law has slowed down, the trend towards miniaturization in digital circuits remains stronger than in analog components. While decreasing the area of analog circuits is usually a requirement for mobile terminals, more often than not, shrinking the analog circuit leads to worse component tolerances and deteriorates the performance. This will eventually decrease the analog circuit performance or render it impractical for some radio applications.

Software-Defined Radio is already a key technology in modern communication systems, and this trend towards the full digitalization of radio systems is likely to continue in the future. The complete integration of a flexible radio in a System on Chip (SoC) seems to be a logical step for the development and deployment of SDRs. FPGA-based ADTs possess a set of convenient characteristics such as hardware integration, high performance, reconfigurability, and flexibility that make them a viable approach for a modern and fully integrated SDR. Fig. 1.2 illustrates an ideal SDR implementation where all signal processing is done in the digital domain and only amplification and radiation are performed in the analogue domain.

### 1.1.3 Towards Fully Digital RF Transceivers

The path towards fully digital RF transceivers can be traced back to the SDR concept introduced by Mitola in [130]. This concept was already analyzed in the previous subsection; however, to summarize, it involves synthesizing waveforms completely in the digital domain, which provides greater flexibility and simplicity to the final transmitter. In the pursuit of

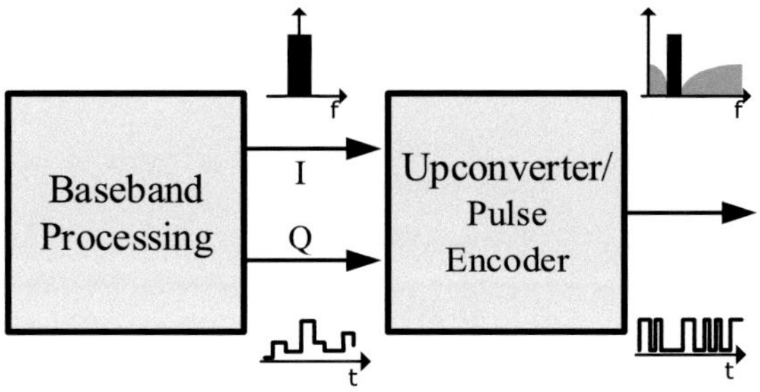


**FIGURE 1.3:** Basic all-digital transmitter architecture.

implementing a fully digital transmitter, whose general architecture is shown in Fig. 1.3, the first step was to implement digital modulators.

The first implementations were made in Application-Specific Integration Circuits (ASICs) and were capable of delivering high-order modulated signals, up to 256-Quadrature Amplitude Modulation (QAM), at modest sample rates, 8.75 MBaud [26]. Some of these attempts were already capable of modulating Global System for Mobile Communications (GSM) and Enhanced Data Rates for GSM Evolution (EDGE) signals [208]; however, all relied on external upconverters to elevate the output Intermediate Frequency (IF) frequency signal into RF frequencies. This need for an external analog upconversion circuit meant that not only was the final transmitter bulky and less efficient, but it also inherited the nonlinear characteristics of traditional analog transmitters. To bypass the need for an external upconversion chain, some authors proposed the idea of using Delta-Sigma ($\Delta\Sigma$) or PWM to reduce the number of quantization bits of the signal without losing signal quality. This, in conjunction with a digital up-converter, would result in a one-bit signal which could be transmitted by a high-speed serializer. After that, the signal only needed power amplification and filtering. This technique resulted in the first ADT capable of delivering an RF signal where all signal manipulation was performed digitally [107]. However, until then, all implementations relied on the design of bespoke digital circuits, ASICs, which, despite their high performance, were less flexible than FPGAs. This meant that the development of such implementations was slow, expensive and required high levels of Very-Large-Scale Very-large-scale Integration (VLSI) knowledge.

To improve development time, some researchers have started using MATLAB, in conjunction with Simulink, to simulate new designs. When the results returned by the simulation were within acceptable margins, other programs would take the Simulink project and convert it into Very High Speed Integrated Circuit Hardware Description Language (VHDL) that could be used to configure an FPGA [182]. This workflow presents many different benefits. Some examples include the already wide use of Matlab by the engineering community, and the use of automatic processes of conversion from conceptual simulation (Matlab/Simulink) to a hardware description (VHDL/Verilog) with a high degree of confidence in the comparison between the simulation results and real-world performance. These advantages, combined with the improvements of FPGA technologies regarding frequencies of operation, increased digital resources, and faster Inputs/Outputs (I/Os), allowed this workflow to be widely used, even today. However, some of the early papers that presented this technique did not implement the full ADT [182], while others, despite implementing a full ADT, did so using an external parallel-to-serial converter. Only in [226] can the first real-time implementation of a PWM-based ADT fully based on an FPGA be found. The reason that other authors chose to use an external parallel-to-serial converter is that,

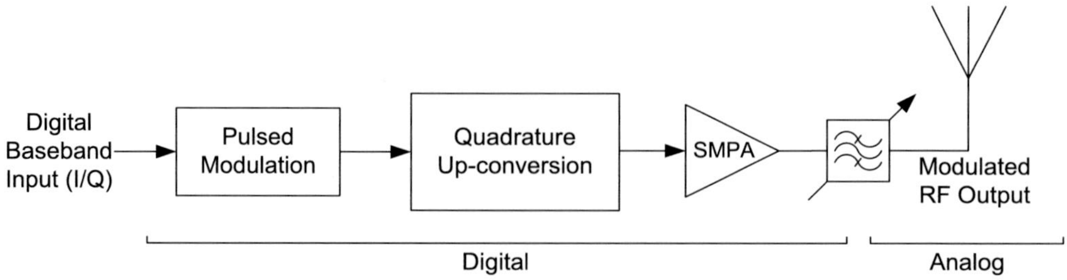


**FIGURE 1.4:** Simplified structure of a conventional all-digital transmitter.

prior to the introduction of Multi-Gigabit Transceivers (MGTs), there were no high-speed Input/Outputs (I/Os) in an FPGA capable of delivering the sample rate required to implement an ADT.

### 1.1.4 FPGAs as Cost- and Time-effective Enabling Technologies for SDR

In an ideal SDR, all the processing is done in the digital domain using powerful Digital Signal Processor (DSP) techniques, while the analog components are reduced to the minimum and are mainly used as signal conditioning and as data bridging between the analog and the digital domain, see Fig. 1.2.

In fact, shifting the radio processing architecture towards the digital domain has huge potential since the major signal processing can now be done by using DSP or FPGA devices, thus making easier configuring and controlling the radio transmitter, and therefore, allowing the radio to adapt to different communication scenarios in an easier way, while also facilitating improvements in terms of spectral management.

Moreover, the continuous technology evolution fosters the development of smaller Integrated Circuit (IC) with improved power efficiency and with massive transistor densities that provide unparalleled integration. As introduced before, while digital ICs exhibit better performance as they become smaller, the same does not apply to analog ICs [136], where it is well known that the analog properties deteriorate as the transistor size decreases. In this sense, it seems clear that transmitter designs having predominantly digital datapaths provide higher integration and cope better with the IC technology evolution.

For all the above-mentioned reasons, it becomes evident that further research leading to the development of innovative transmitters based on flexible digital radios such as the SDR concept is of paramount importance. The latest advances in this field include the development of novel all-digital transmitters [70, 84, 226], where the datapath is digital from the baseband up to the RF stage, as shown in Fig. 1.4. Such a concept has inherent high flexibility and poses an important step towards the development of SDR transmitters. Moreover, this new approach enables the use of FPGA devices for implementing the digital datapath of such RF transmitters, providing additional flexibility as well as field-upgradeability.

As shown in Table 1.1, DSP-enhanced FPGAs provide an interesting compromise between performance, power efficiency, and flexibility, which makes FPGAs an interesting choice as an enabling technology for SDR transmitters. Current FPGAs have an equivalent logic capacity of millions of logic gates in addition to large RAM blocks, embedded processors, DSP modules, and multigigabit I/O standards. If efficiently explored, these valuable resources can be utilized to facilitate the development of agile, all-digital transmitters.

**TABLE 1.1:** Summary comparison of DSP implementation technologies, partially adapted from [197].

| | DSP | ASIC | ASSP | FPGA* |
|---|---|---|---|---|
| Processing Speed | ●●●●○○ | ●●●●●● | ●●●●●○ | ●●●●●○ |
| Power Efficiency | ●●●○○○ | ●●●●●● | ●●●●●○ | ●●●●○○ |
| Reprogrammability | ●●●●●○ | ●○○○○○ | ●○○○○○ | ●●●●○○ |
| Design Flexibility | ●●●●●● | ●●●●●● | ●○○○○○ | ●●●●●○ |
| Area Efficiency | ●●●○○○ | ●●●●●● | ●●●○○○ | ●●●○○○ |
| Development Savings | ●●●●○○ | ●○○○○○ | ●●●●○○ | ●●○○○○ |
| DSP Tools Support | ●●●●●● | ●○○○○○ | ●○○○○○ | ●●●●○○ |

*DSP-enhanced FPGA.

### 1.1.5 Summary

This chapter introduced the key motivations and foundational concepts that support the remainder of this book. The rapid evolution of wireless communication standards, from LTE to 5G and the emerging 6G, demands increasingly flexible, power-efficient, and wideband radio transceivers. Meeting these demands requires a fundamental rethinking of conventional RF front-end architectures, as they are too bulky, costly, and rigid to scale with massive MIMO, multi-band operation, and network densification requirements.

Mitola's SDR paradigm was presented as the guiding vision toward fully digital RF transceivers. Within this framework, ADTs emerges as a practical realization by performing all signal processing in the digital domain and relying on a single-bit DACs to interface with the analog front-end. ADTs drastically reduces the transmitter's footprint and power consumption. Using a SMPAs at the output further reinforces the efficiency advantages of this approach.

FPGAs are identified as a particularly suitable platform for implementing ADTs, offering an attractive trade-off among processing speed, design flexibility, reconfigurability, and development cost. Their continued evolution, characterized by growing logic density, embedded DSP resources, and multigigabit I/O capabilities, makes them a natural enabler for the SDR vision.

Finally, the broad applicability of ADTs was highlighted, spanning cellular base stations, satellite payloads, RoF systems, and future AI-integrated network architectures. Together, these considerations motivate the in-depth exploration of ADT design techniques presented throughout this book.

---

## 1.2 Outline of this Book

This first chapter aims to provide the reader with a comprehensive understanding of the current state of the modern telecommunications landscape. As such, it started by focusing the reader's attention on concepts that will be the main focus of this book, such as SDR and ADTs. The chapter then continued to explain why these concepts are relevant and how they are intertwined with the challenges of tomorrow. This chapter concludes by briefly explaining the concept behind the ADT and the most suitable platforms for its development.

The second chapter provides a more in-depth explanation of some basic concepts that will be used throughout the book, starting with classical transmitter architectures, both for Single Input Single Output (SISO) and MIMO, and then following with a thorough explanation of ADTs. Details from the ADTs general architecture, the principles that they rely on, encoding techniques, and figures of merit used to build and evaluate their performance are some of the topics that will be covered. This chapter concludes with the main research topics on ADTs, and for each research topic, it provides a connection to the chapters that follow.

Starting with the third chapter, the reader enters the core of the book. In this chapter, several techniques are presented that aim to reduce the requirements for analog filtering. The first one focuses on both PWM and ΔΣM and utilizes digital signal manipulation to create a polyphase signal, which, when combined in the analog domain, reduces the noise characteristic of both modulations. The second technique utilizes different PWM waveforms stored in Look-up tables (LUTs) to minimize out-of-band noise again. Then, a third technique utilizes a mixed-domain combination filter, and finally, the fourth technique combines a cascade of modulators with the second technique to significantly reduce the noise introduced by the ΔΣM.

Fully agile modulators are the focus of the fourth chapter. Here, the reader is presented with different ADT's architectures that enable carrier frequency agility. The main goal of this chapter is to demonstrate the ability of ADTs to change the transmitting frequency in real time. First, Baseband (BB)-stage designs are presented, where the upconversion takes place after the modulation. Then, the RF-stage design is presented, where the relative places of the modulator and upconverter are swapped. This triggers a higher degree of agility in achieving real-time frequency flexibility.

The fifth chapter is focused on techniques that enhance the bandwidth of ADTs. Time-interleaved techniques are initially presented as an early attempt at increasing the sampling rate of a modulator without increasing the FPGA's clock frequency. It will also be demonstrated that this technique presents a critical path that increases with the number of parallel phases, which imposes a limit on their number, effectively limiting the overall sampling rate. To address this, the multi-core architecture was introduced. This architecture solved the problem of the critical path being related to the number of phases used, which meant that more parallel phases could be used without degradation of the FPGA's frequency, allowing for a greater increase in sampling rate. One drawback of this architecture was the noise spurs introduced. To minimize them, the state propagation technique was added to the architecture. Finally, moving into an RF-stage architecture paired with the previous technique allowed for a substantial bandwidth increase.

Multi-band transmission is the focus of chapter six. Similarly to the fourth chapter, where it was divided into BB-stage and RF-stage designs, the same division was followed here. This division was made not only for consistency, but also because the techniques presented in this chapter are built upon the concepts and designs introduced in the fourth chapter. The chapter starts by proposing an architecture for a BB-stage ADT that can be real-time reconfigured to transmit up to two bands of modulated data. Then, the architecture is enhanced by including a linear precompensation mechanism that can improve the dual-band figures of merit. The chapter is then concluded with the proposal of an RF-stage architecture designed for up to three bands.

Chapter 7 is the final core chapter and presents several use cases for ADTs. It starts by introducing some RoF designs. These designs take advantage of several key properties of optical fibres, such as the large available bandwidth and the low medium attenuation. Due to the transmission being performed in the optical domain, optical upconversion is also used to produce high millimeter wave carrier frequencies without complex analog designs that would be required in the electrical domain.

The concept of all-digital transceivers is also presented. This expands on the concept of ADTs by also introducing an all-digital reception. Finally, to prove the capability of using an ADT in a MIMO scenario, an antenna array driven by multiple ADTs is presented. This architecture is capable of steering RF-modulated signals with a fair degree of accuracy, all this while being compact and efficient.

# Chapter 2

# RF Transmitter Architectures

This chapter surveys the evolution of RF transmitter architectures, from classical single-band solutions to emerging all-digital approaches. It introduces heterodyne, homodyne, and direct-RF synthesis techniques, emphasizing their limitations when scaling toward multi-antenna systems and advanced beamforming. The discussion then evolves to ADTs, where digital pulse encoding, oversampled converters, and delta-sigma modulation are explored as enabling technologies for flexible, multi-band, and energy-efficient radios. Key challenges such as bandwidth, carrier agility, filtering requirements, and efficiency are addressed, alongside research trends on scalability, integration, and applications in 5G, 6G, antenna arrays, and C-RAN architectures.

## 2.1 Classical RF Single-band Transmitter Architectures

The following section will review the main single-band RF transmitter architectures and their evolution. It starts with classical SISO schemes such as heterodyne and homodyne transmitters, highlighting their reliance on analog upconversion stages and the limitations posed by I/Q imbalance, filtering complexity, and integration costs. With the advent of high-speed DACs, direct-RF synthesis emerged, shifting upconversion into the digital domain and reducing analog front-end requirements. The discussion then extends to antenna arrays, where different beamforming approaches, RF, baseband, and digital path, are compared in terms of flexibility, scalability, and hardware complexity. Finally, the concept of an ideal beamforming architecture is introduced, pointing toward all-digital transmitters as a promising pathway for achieving fully reconfigurable and scalable multi-antenna systems.

### 2.1.1 Single-Input Single-Output Architectures

Traditionally, wireless transmitters are based on a baseband DAC pair followed by an I/Q modulator together with a 2-stage upconversion scheme, referred to as heterodyne architecture (Fig. 2.1a). The I/Q digital signal is converted to an analog one by a DAC pair, filtered, upconverted to an IF carrier, filtered, and upconverted again to the RF or microwave carrier. After this, the signal must be filtered before being amplified by a Power Amplifier (PA). If the amplifier's behavior is not fully linear, another filter must be included before radiation to reduce the out-of-band distortion that may cause interference issues. This approach suffers from several problems, such as the requirement for complex image rejection filters, I/Q imbalance, and due to the large number of external components, it is a costly and complex solution that does not fit well with the integration of SoCs [135], and difficult to scale for Massive MIMO systems.

Similar to the heterodyne architecture, but with only one upconversion and filtering stage, the homodyne architecture depicted in Fig. 2.1b can be introduced. Even though

DOI: 10.1201/9781003225713-2

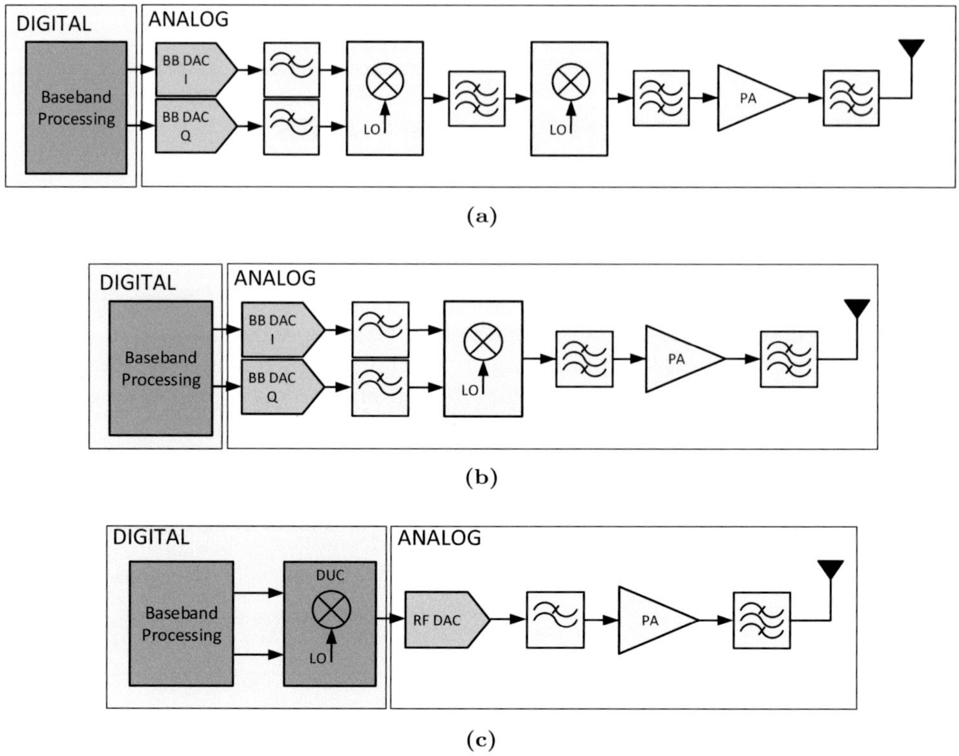


**FIGURE 2.1:** Illustration of the different Transmitter (TX) architectures: (a) heterodyne upconversion; (b) homodyne upconversion; (c) direct-RF upconversion.

there is a significant reduction in the number of external analog circuits, the truth is that the main disadvantages of the heterodyne counterpart are still present, such as the I/Q imbalance and the need for complex image rejection filters. Nonetheless, it is one of the most well-known and dominant architectures in typical RF systems.

As the DACs sampling rates have been increased, the inherent flexibility from the digital domain has been valued and preferred over the traditional analog chains, causing the DACs to be brought closer to the antennas. For some applications where the DACs sampling rate is high enough to generate signals in the RF range, the analog I/Q modulator can be dismissed, and the signal produced from the RF DAC is just filtered and amplified before being radiated. This architecture is depicted in Fig. 2.1c, where it can be seen that the upconversion stage is included in the digital domain (referred to as Digital Up-Conversion (DUC)). It is worth noting that throughout this manuscript, although not explicitly stated in the figure, this DUC may involve digital interpolation to adjust the sampling rates involved as a first step before performing the frequency shifting. In addition to the flexibility of designing the signals in the digital domain, several other advantages can be pointed out, such as no amplitude and phase imbalances in the modulator, elimination of the unwanted image at the output of the upconverting mixer, easy filtering requirements, reduced area footprint, lower costs in the analog front-end, among others.

After reviewing the different transmitter architectures, it is understandable that the DAC is their core. The analog front-end is completely conditioned by the DAC's characteristics, and, in the extreme case, if the DAC is able to reach the RF or microwave range, the analog front-end can be reduced to only filtering, amplification, and radiation.

### 2.1.2 Increasing the Number of Radiating Elements

As antenna arrays are indispensable components in high-data-rate wireless communication systems, not to mention their key role in Radar Systems, it is of utmost importance to highlight how the classical RF transmitter architectures can be adapted to drive such systems.

An antenna array is defined as any system with more than one radiating element. They were first proposed as an innovative way to design antennas with very high gain, without an increase in electrical size. Beamforming, also referred to as Spatial Filtering, is one of the most well-known applications of antenna arrays, and one of the most complex ones, due to the challenging gain/phase/delay matching between multiple outputs. For this reason, this document will just focus on the beamforming architectures.

To accomplish this spatial capability, a Beamforming Network (BN) must be included within the RF front-ends. The BN is required to distribute the power to each element, at the same time that it adjusts the amplitude and phase of each output. These adjustments are typically performed using phase shifters and amplifiers/attenuators. Depending on the placement of this BN, different architectures have emerged in the literature [67, 159, 222].

#### 2.1.2.1 RF-path Beamforming Architecture

The most well-known architecture is referred to as RF-path Beamforming Architecture and is depicted in Fig. 2.2a. As the name states, the BN is directly applied in the RF domain, more precisely, before the amplification to achieve higher efficiencies. One of the main advantages of this architecture is that, as the BN is placed closer to the radiation elements, all the remaining components/systems to the left of it can be shared among all the radiant elements. This architecture is particularly suitable for millimeter- and sub-millimeter-wave systems, where the integration of phase shifters is simpler and cheaper [159]. At lower frequencies, phase shifters are typically bulky, noisy, and introduce nonlinearities.

#### 2.1.2.2 Baseband-path Beamforming Architecture

In the BB-path Beamforming Architecture, the BN is directly applied to the analog BB signal, before any up-conversion stage; see Fig. 2.2b. This leads to the finest phase resolution with the lowest power penalties [159]. However, the different Local Oscillators (LOs) and mixers may cause significant I/Q imbalances, which can jeopardize system performance. This typically implies the use of feedback loops for each TX chain for calibration purposes. Thus, the scalability of such systems in large-scale arrays is typically limited [159].

#### 2.1.2.3 Digital Beamforming Architecture

The Digital Beamforming Architecture is one of the most promising techniques. It is quite similar to the BB-path one, with the difference that the BN is placed in the digital domain, within a DSP (Fig. 2.2c). Thus, this architecture shares the advantages of the latter one, while also enabling a simplification of the RF circuit design. However, two DACs per each array element are required, and besides that, other issues such as I/Q mismatch and different group delays for I- and Q-path can also be pointed out [159].

#### 2.1.2.4 The Ideal Beamforming Architecture

After reviewing the conventional architectures for antenna arrays, it can be noted that the inclusion of BNs leads to a twofold reality: a desired flexibility and reconfigurability can be achieved, albeit at the expense of an extra level of complexity that may limit the system's scalability.

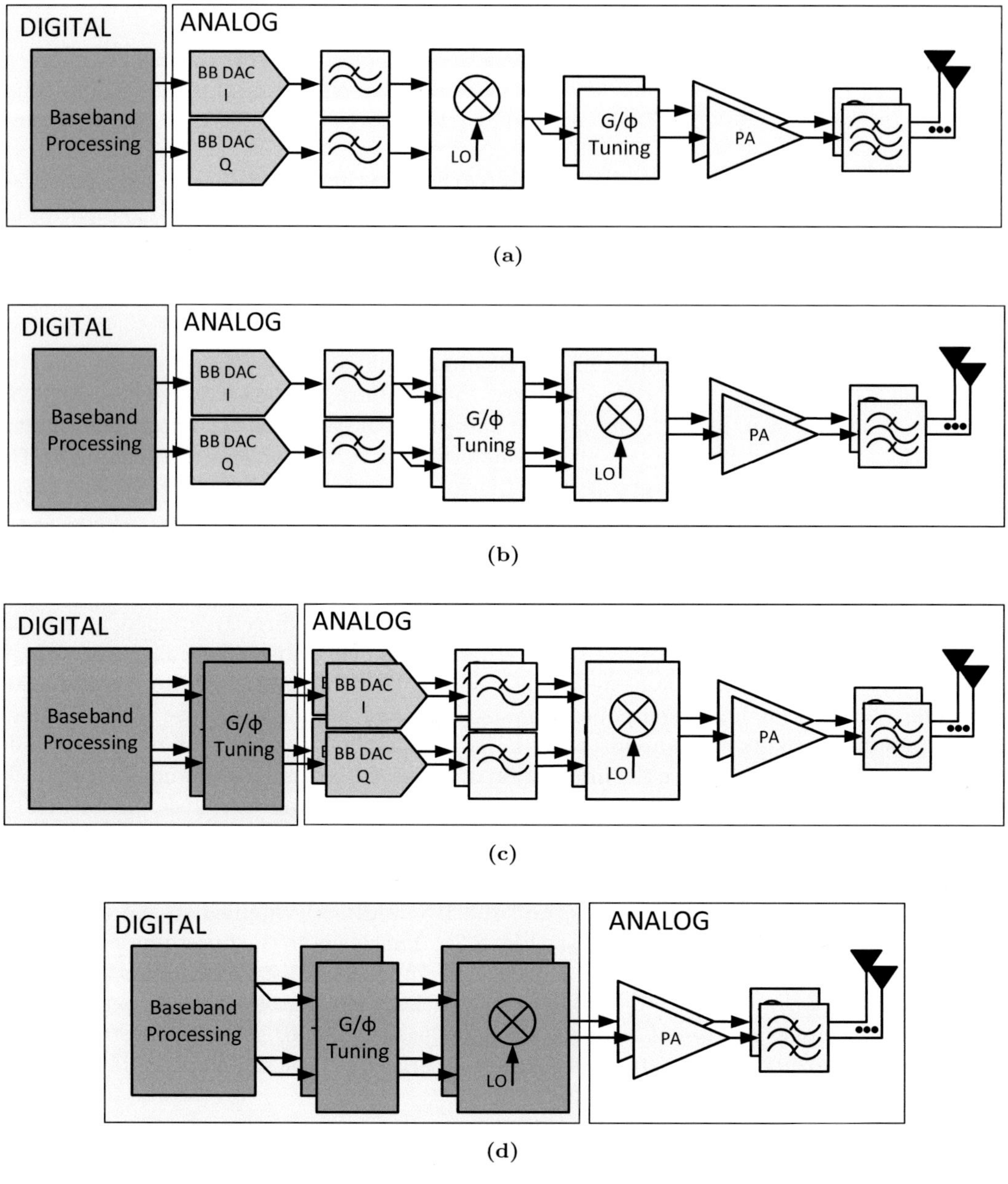


**FIGURE 2.2:** Illustration of the different TX beamforming architectures for antenna arrays: (a) RF-path beamforming; (b) BB-path beamforming; (c) digital beamforming; (d) ideal beamforming.

Upon examining the aforementioned architectures, the importance of digital beamforming techniques has been consistently emphasized throughout the literature [67, 222], and they are identified as the key enabling technology for providing high performance, along with high flexibility and reliability [68]. This recent interest is due to the fact that, by including the BN in the digital domain, the RF circuit/layout design is relaxed, while the software can start to play a leading role in the design and control of antenna arrays. However, one can recognize that the need for two external DACs per radiant element, together with the use of analog upconversion stages, is too burdensome and hampers the scalability and integration capability of such systems from the area footprint's viewpoint. At least, theoretically, it is possible to formulate an ideal architecture that does not jeopardize the system's scalability.

Such architecture is depicted in Fig. 2.2d. It is based on the digital beamforming techniques to enable the long-awaited reconfigurability and flexibility. However, the upconversion stages were moved into the digital domain, at the same time that the external DACs were removed. This leads to a neat layout, and in practice, amplification, filtering, and radiation are all that are required. As will be seen in Section 7.5, the concept of ADTs may be a promising solution for reaching this ideal architecture. Lastly, it must be emphasized that Fig. 2.2d aims to illustrate the conceptual idea; thus, for convenience, the embedded buffer that converts the digital signal into an analog single-bit waveform is omitted.

## 2.2 All-Digital RF Transmitters

As previously introduced, the rising trend in the digitalization of conventional analog front-ends has opened the doors to the concept of SDR. This concept aims to have a completely independent, agile radio transceiver front-end, capable of handling any number of frequency bands, regardless of specific standards. Relevant state-of-the-art research points towards ADTs as the ultimate target for this technology, due to their inherent fully digital behavior from end to end. The research on this class of transmitters focuses on the development of reconfigurable, agile, multi-band, multi-standard, and highly efficient solutions [85,186]. The recent advances in the architectures of these transmitters have undoubtedly contributed to this, largely by enabling the transmission of radio signals using a fully digital datapath. This section elaborates on the main foundations of this concept, as well as the research and development tracks followed by the research community.

### 2.2.1 General Architecture

All-Digital RF Transmitters are a particular subset of the SDR techniques that have attracted renewed interest from both industrial and academic parties. The underlying idea is the encoding of an $n$-bit digital signal into a representation with a lower number of bits, enabling the removal of external DACs and external upconversion stages. Most of the reported transmitters explore the use of reconfigurable and on-chip 1-bit topologies to transmit signals [22, 186, 187]. The use of single-bit transmitters leads to signals with a constant time envelope, which can be amplified by highly efficient and nonlinear amplifiers, such as SMPAs. With such a concept, the analog front-end is reduced to only amplification, filtering, and radiation.

The first attempts to achieve single-bit and efficient power amplification through the utilization of $\Delta\Sigma$ or PWM modulators were applied to low-frequency applications [88]. These implementations have demonstrated that highly efficient and nonlinear amplifiers can be utilized, which has motivated the exploration of the RF scenario, and several academic research lines have begun. The need to achieve the GHz frequency range has driven intensive research in this area. In [139], pulse encoding techniques were applied to the signal magnitude to change the power amplifier bias network. Others have designed a pulse-shaping stage for the in-phase and quadrature components of the signal, followed by analog wideband mixers, through the multiplication of the signal with a binary waveform representation of the carrier frequency [132], or with bulky and high-cost external high-speed multiplexers [74, 85, 210].

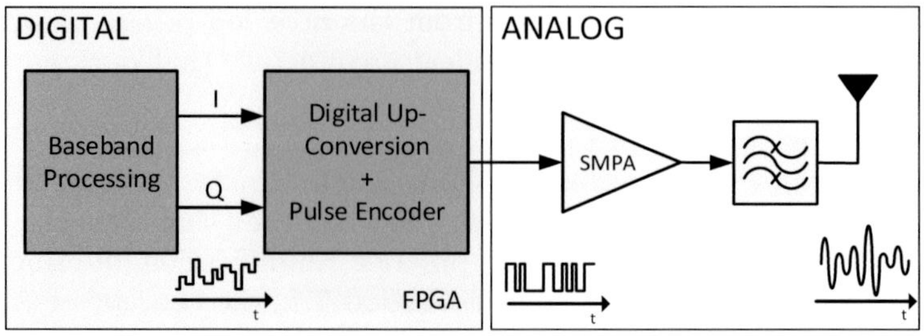


**FIGURE 2.3:** Illustration of the general all-digital transmitter architecture. For convenience, the mixed-domain serializer, which converts the digital signal into an analog pulsed waveform, was omitted.

While some have opted for the implementation of these types of modulators in ICs [23,135], others have emphasized the inherent flexibility and upgradeability of FPGAs. Among the latter, the idea of discarding the external components required for the upconversion stage (either analog wideband mixers or high-speed multiplexers) sparked intense investigation, leading to further exploration and the design of DUC stages within a single FPGA. This idea was achieved through the utilization of MGTs that are present in the medium/high range FPGAs [191,192]. These MGTs are power-efficient transceivers with embedded serializers, highly configurable, and tightly integrated with the FPGA's programmable logic resources. They are typically used to support the implementation of general-purpose and high-speed serial interconnects. As an example, line rates from 500 Mbps up to 32.75 Gbps are available in the GTY Transceivers from Xilinx Ultrascale+ FPGAs [220]. Furthermore, new Pulse Amplitude Modulation (PAM)-4 MGTs from Xilinx reach 58 Gbps [221], while 57.8 Gbps are achieved in identical transceivers from Intel [93]. This small step was of paramount importance to get closer to the goal of having completely flexible and integrated transmitters.

The general block diagram of an FPGA-based ADT is depicted in Fig. 2.3. The baseband complex input signal is divided into the I/Q data components before being converted into a 2-level representation. This conversion is done with an oversampled Pulse Encoder. As will be seen throughout this document, the DUC can be placed either before or after the Pulse Encoder. In short, the output from the ADT is a pulsed waveform that contains the digitally modulated signal at a given RF carrier frequency, together with the out-of-band noise. After amplification, a BandPass Filter (BPF) is required to reconstruct the desired modulated signal before radiation.

### 2.2.2 Oversampled Converters

As the concept of fully digital transmission relies on the use of 1-bit waveforms, without significant degradation of system performance, it is of paramount importance to quantify the impact of quantization.

In the case of Nyquist-Rate Converters, assuming a full-scale sine wave input, the SQNR is given by [100]:

$$SQNR = 6.02N_{bits} + 1.76 \tag{2.1}$$

where SQNR is in dB and $N_{bits}$ is the number of bits from the converter. Eq. (2.1) demonstrates that the hard truncation to 1-bit leads to an inferior system performance due to the low achieved SQNR (= 7.78 dB). Thus, the use of such hard truncation seems to be ill-suited to the design of the desired transmitters. Therefore, in an attempt to achieve better

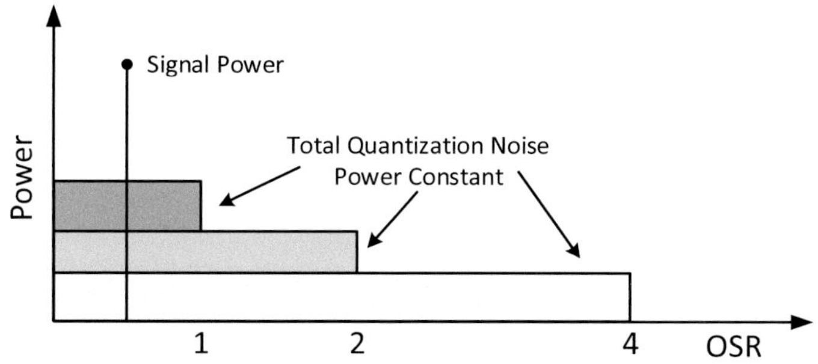


**FIGURE 2.4:** Illustration of the Signal-to-Quantization Noise Ratio (SQNR) gains in oversampled converters. One can see that oversampling affects only the noise spectral density, without changing either the signal power or the total quantization noise power [96].

performance, oversampling has been successfully applied to data converters. Oversampling is the process of acquiring more samples per second than those required by the Nyquist-Shannon criterion. The underlying idea is that, even though the total quantization noise power remains unchanged with the sampling rate, its distribution will change [96]. Thus, this means that the same noise power will be spread in a larger frequency range, reducing the spectral density of the quantization noise (Fig. 2.4). Under the assumption of a full-scale sine wave input [100]:

$$SQNR = 6.02N_{bits} + 1.76 + 10\log(OSR) \tag{2.2}$$

where $OSR$ is OverSampling Ratio (OSR) and is given by $fs/(2BW)$, and $fs$ is the sampling rate and $BW$ is the signal bandwidth. In practice, the spectral density of the quantization noise is reduced by 3 dB/octave, where 1 octave is twice the sampling rate. Equivalently, this means 0.5 bits/octave [100].

In addition to the enhancement in the system performance, oversampled converters have the potential to relax the accuracy requirements from the analog components at the cost of added digital circuitry. As the advancement of digital IC technologies leads to cheaper digital circuitry, this class of converters is increasingly receiving attention from the industry community in various scenarios and applications [179]. However, as shown in Fig. 2.5,

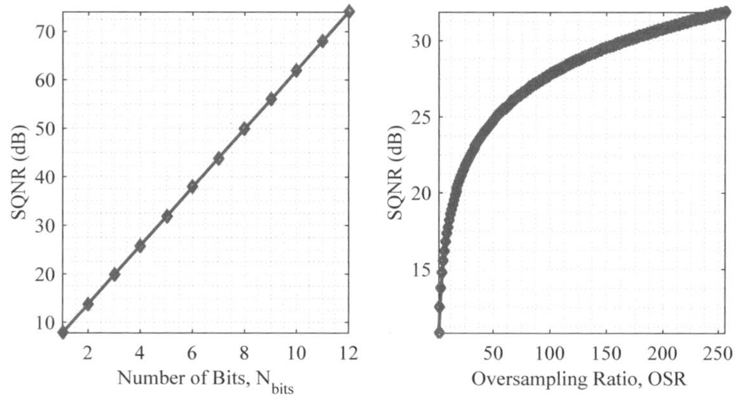


**FIGURE 2.5:** SQNR in a Nyquist-Rate converters (a), and in a 1-bit Oversampled converters. It is understandable that the OSR provides an extra degree of freedom in the enhancement of the SQNR.

challenging OSR can be required when 1-bit quantization is used. While such techniques are suitable for low-frequency applications (such as audio [77], or power electronics [168]), they are not an ideal solution for the case of RF/microwave systems. Even though, as will be seen, this SQNR enhancement due to the OSR will serve as the starting point in designing Pulse Encoders that can provide a reasonable system performance.

### 2.2.3 Pulse Encoding Techniques

In the context of ADTs, Pulse Encoding Techniques refer to a set of architectures that reduce the number of levels while maximizing in-band signal integrity. All of them are based on oversampling converters, and most of them employ ΔΣM, PWM, or some combination thereof.

#### 2.2.3.1 Pulse Width Modulation

One well-known pulse encoder is the PWM architecture that was first introduced in the '40s [112]. The underlying idea is the encoding of amplitude and phase information in the width and timing of pulses. The traditional mechanism to generate a PWM signal is based on the comparison between the modulating signal $x(t)$, and the reference signal $r(t)$, which can be described by:

$$y(t) = \mathrm{sign}(x(t) - r(t)) \tag{2.3}$$

where sign is the sign function.

A thorough mathematical formulation of the PWM is out of the scope of this book. However, in an attempt to shed some insight into this technique, let us briefly demonstrate why it is suitable for use as a Pulse Encoder. A mathematical model that enables the approximation of Eq. (2.3), for every $x(t)$, was proposed in [133] and is reported in [162]:

$$\begin{aligned} y(t) &= x(t) + n_q(t) \\ &= x(t) + \frac{2}{\pi} \sum_{\substack{m=-\infty \\ m \neq 0}}^{\infty} \frac{1}{m} e^{jmw_c t} \sin\left(\frac{m\pi}{2}(1 + x(t))\right) \end{aligned} \tag{2.4}$$

where $w_c$ is the angular frequency from the reference signal, and $n_q(t)$ is the non-linear distortion.

It is understandable from Eq. (2.4) that the output signal $y(t)$ is composed of the original input signal $x(t)$ together with some nonlinear distortion. This non-linear distortion is composed of distorted replicas of the input signal, centered at multiples of the reference frequency, along with the inherent noise from the quantization process [25, 133]. At Direct Current (DC), the equation also demonstrates that there is no noise-distortion, and thus, the in-band signal integrity can be reasonably maintained.

This technique can be transferred to the digital domain. Fig. 2.6 illustrates the block diagram of the digital PWM. In this case, the digital PWM can be seen as a sampled version of the above-described analog PWM. This means that spectral aliasing will occur due to the infinite bandwidth of $y(t)$ [203]. This, depending on the placement of the aliases, may cause performance degradation of the digital PWM as compared to the analog version [83].

An illustration of the typical spectrum is shown in Fig. 2.7. One can see the distorted replicas that corrupt the spectrum. The suitability of this technique for use as a pulse encoder lies in the ideally intact copy of the input signal that is generated between the first harmonic distortion peaks. Several relevant works for this technique have been reported in the literature [34, 187, 189, 229].

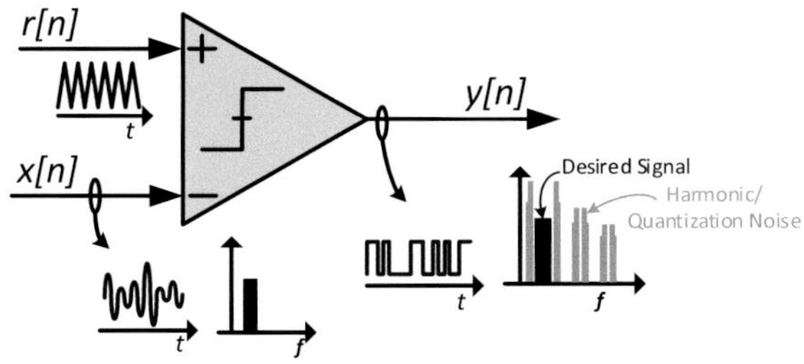


**FIGURE 2.6:** Illustration of the PWM concept: the comparison between the input signal with a reference waveform that produces a 1-bit output signal.

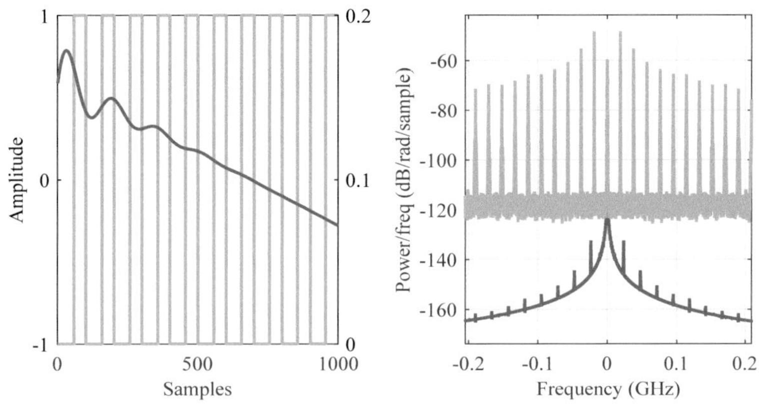


**FIGURE 2.7:** Illustration of the time-domain (left) and frequency spectrum (right) of an PWM-encoded signal. One can see the original interpolated signal (black) and the PWM-encoded signal (gray).

Note that due to the open-loop characteristic inherent to the PWM operation principle, the concept of noise-shaping is not present. Nonetheless, there are some reported works where feedback loops were integrated to provide a noise-shaping behavior [19, 125, 173].

In the previous paragraphs, it was assumed that the input signal was low-pass. However, the same technique can be applied for bandpass signals [62, 80, 142, 149, 161, 202, 211]. Two different approaches are typically followed in the State-of-the-Art (SoA). On the one hand, upconversion can be viewed as part of the PWM process by considering one of the harmonics as the fundamental signal, rather than distorted replicas [142]. In this case, it is required to perform some pre-compensation in the input signal to compensate for the fact that the RF amplitude of the first harmonic is proportional to the sine function and not to the input signal itself [202]. On the other hand, the PWM process can be applied to an upconverted RF signal instead of the BB signal [161]. In this case, depending on the location of the input signal, frequency components from other Nyquist Zones (NZs) may fold back into the band of interest, which can jeopardize the integrity of the in-band signal. As first pointed out in [161], and as further demonstrated in this work, the noise distribution is highly dependent on the reference signal frequency.

#### 2.2.3.2 Delta-Sigma Modulation

Another well-known pulse encoder is the $\Delta\Sigma$ architecture that was first introduced in the '60s [92]. This type of mechanism is widely used to transform a time-varying signal into a

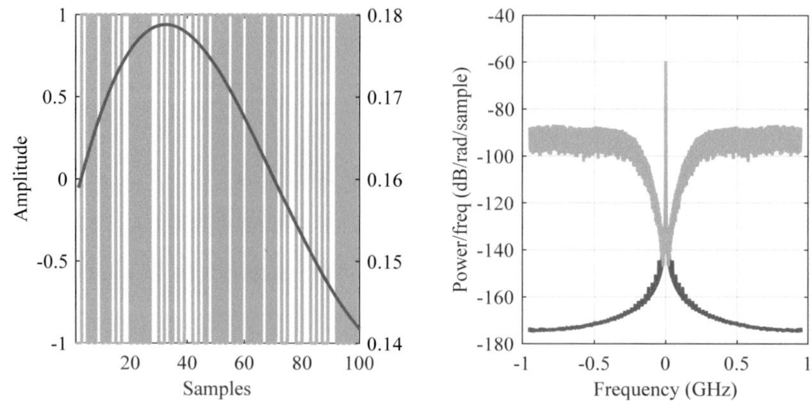


**FIGURE 2.8:** Illustration of the time-domain (left) and frequency spectrum (right) of an ΔΣM-encoded signal. One can see the original interpolated signal (black) and the ΔΣM-encoded signal (gray).

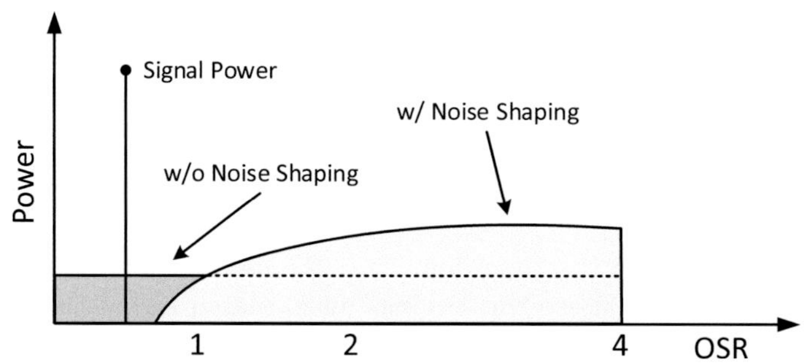


**FIGURE 2.9:** Illustration of the oversampled gain in terms of the SQNR with, and without, noise shaping.

low-resolution equivalent (typically in a bi-level representation, as illustrated in Fig. 2.8). To achieve this, these modulators employ signal oversampling, quantization, and a feedback loop for shaping the quantization noise in frequency (referred to as *Noise Shaping*). As will be seen, this Noise Shaping will break the spectral uniformity of the quantization noise, providing an SQNR enhancement in the in-band signal [57]. Fig. 2.9 is an illustration of the impact of Noise Shaping in terms of increasing the in-band signal performance.

An illustration of the typical spectrum of a modulated signal encoded by a ΔΣM is depicted in Fig. 2.8. One can see that the distortion and quantization noise are pushed away from the input signal.

The non-linear behavior of the quantizer presented in the ΔΣM hinders the mathematical analysis in a deterministic way. The simplest technique to analyze such systems replaces the quantizer by just an additive white noise source, treating the system as a linear model. Even though this model lacks robustness due to its assumption that the quantization noise has certain convenient properties, such as a white spectrum or uniform distribution [178], it is the most well-known model and the one applied by every designer to gain preliminary insights.

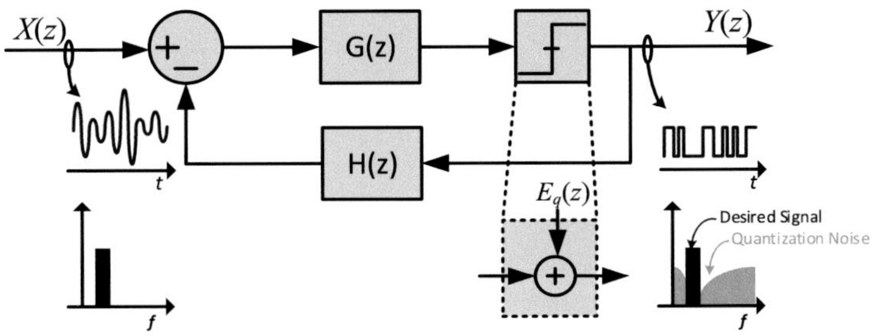


**FIGURE 2.10:** Block diagram illustrating the general model of a ΔΣM.

Fig. 2.10 illustrates the general block diagram of a ΔΣM. Replacing the quantizer by an additive white noise source, the output signal is given by:

$$\begin{aligned} Y(z) &= STF(z)X(z) + NTF(z)E_q(z) \\ &= \left[\frac{G(z)}{1+G(z)H(z)}\right] X(z) + \left[\frac{1}{1+G(z)H(z)}\right] E_q(z) \end{aligned} \tag{2.5}$$

where $STF(z)$ and $NTF(z)$ are referred to *Signal-Transfer Function (STF)* and *Noise-Transfer Function (NTF)*, respectively. The key idea is the application of different transfer functions to both the input signal and the quantization error, with the overall aim of minimizing the spectral overlap between them. In this way, it is possible to reshape the quantization noise so that it is pushed away from the signal band.

A slightly different architecture opts for directly computing the quantization noise and forwarding it to the transfer function from the feedback loop. This architecture is designated as Error-Feedback ΔΣM and is illustrated in Fig. 2.11.

Based on the figure and through the replacement of the quantizer by an additive white noise source, the output signal is given by:

$$\begin{aligned} Y(z) &= STF(z)X(z) - NTF(z)E_q(z) \\ &= G(z)X(z) - (1 + G(z)H(z))\, E_q(z) \end{aligned} \tag{2.6}$$

As will be seen throughout this document, this architecture leads to low-complex modulators, with the shortest critical paths. The critical path in digital design refers to the path throughout the entire design that incurs the maximum delay. Considering that the maximum sampling rate for any architecture is inversely proportional to its critical path, this is of paramount importance.

Regardless of the chosen architecture, the signal's SQNR is proportional to the order of the NTF (simply referred to as modulator order), and to the OSR. As increasing the

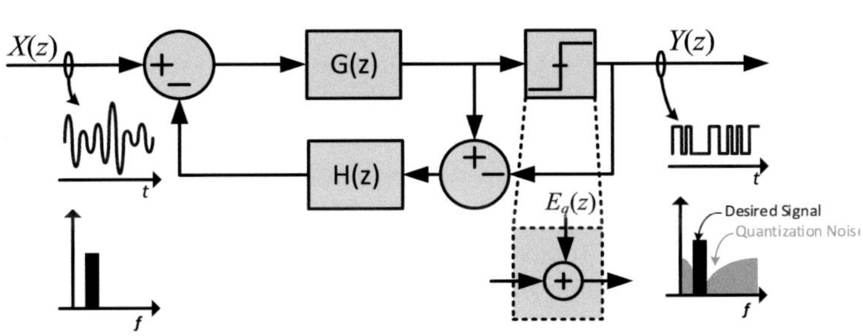


**FIGURE 2.11:** Block diagram illustrating the general model of an Error-Feedback ΔΣM.

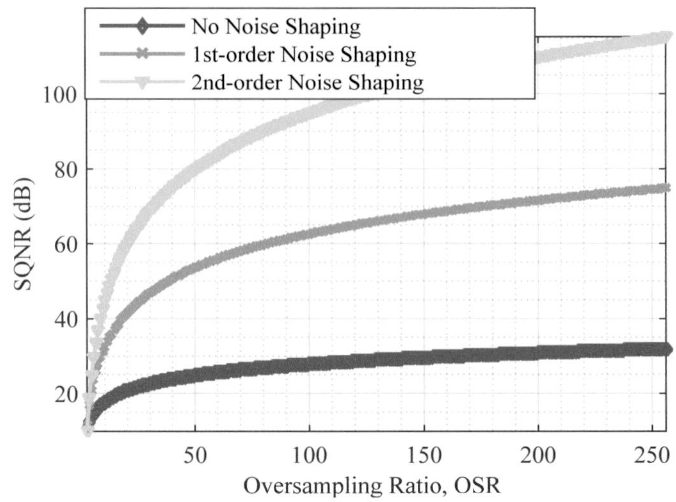


**FIGURE 2.12:** SQNR in 1-bit Oversampled converters for different orders of noise shaping.

sampling rates may not be very practical, in most cases, one solution could be to improve the Signal-to-Noise Ratio (SNR) at the expense of designing higher-order modulators [111].

Under the assumption of a full-scale sine wave input for an oversampled converter with first-order noise shaping [100]:

$$SQNR = 6.02N_{bits} + 1.76 - 5.17 + 30\log(OSR) \tag{2.7}$$

It provides extra 9 dB or, equivalently, a gain of 1.5 bits/octave.

In the case of a second-order noise shaping, and under the same input characteristics [100]:

$$SQNR = 6.02N_{bits} + 1.76 - 12.9 + 50\log(OSR) \tag{2.8}$$

It provides extra 15 dB or, equivalently, a gain of 2.5 bits/octave. This should be compared with the 0.5 bits/octave in the oversampled converters with no noise shaping (Fig. 2.12).

One can demonstrate that an $L^{th}$-order noise shaping modulator improves the SQNR by $6L + 3$ dB/octave, or equivalently, $L + 0.5$ bits/octave [100].

To achieve higher-order modulators, two typical structures are used: the interpolative and the Multistage Noise Shaping Modulator (MASH). On the one hand, the interpolative architecture relies solely on replacing low-order transfer functions with high-order ones. Unfortunately, specifically for input signals with higher PAPR, and for single-bit quantizers, the high NTF's gain typically leads to unstable systems. And once unstable, they may never return to stability, even when the large input signals are no longer present [100]. This implies the insertion/placement of poles to control the NTF's gain, leading to the common architectures such as Cascaded-of-Integrators, Feedback Form (CIFB), Cascaded-of-Integrators, Feedforward Form (CIFF), Cascaded-of-Resonators, Feedback Form (CRFB), and Cascaded-of-Resonators, Feedforward Form (CRFF) [179]. For certain input signals, stability is ensured in these techniques, albeit at the expense of increased interconnection complexity. This ultimately increases the critical path from the modulator, reducing the achievable sampling rate, and, consequently, the achievable OSR. Moreover, the introduction of poles also leads to a significant reduction in the notch bandwidth, resulting in a

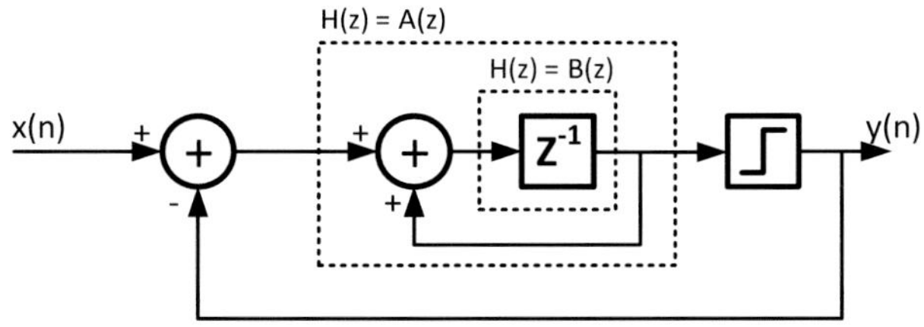


**FIGURE 2.13:** Illustrative first-order DSM with an integrator as loop filter.

decrease in the overall encoded bandwidth [180]. On the other hand, the MASH architectures rely on the use of a cascade-type structure where the overall higher-order modulator is built upon the use of lower-order modulators [27]. Then, all the different outputs are filtered and combined, leading to the desired $L^{th}$-order global transfer function. As low-order (hence, more stable) modulators are the essential blocks of this architecture, it ultimately becomes more robust to stability issues [127]. Unfortunately, the final combination network leads to a higher number of levels, which hinders the applicability of these architectures in the design of single-bit ADTs.

So far, it was assumed that the interest signal was placed at BB, requiring an NTF to work as a high-pass filter to ensure proper noise shaping. This is one of the most common cases, also referred to as baseband ΔΣ architecture, and it can be found in [23, 36, 73, 74, 78, 81, 85, 189]. If the interest signal is in another carrier frequency, the concept of bandpass ΔΣM must be applied. In this case, the NTF is designed as a bandstop filter. Examples of such topology can be found in [74, 101, 103, 107, 169].

### 2.2.4 Time-Interleaved Delta-Sigma

Parallel ΔΣM techniques have been employed to achieve higher working frequencies at the expense of hardware resources and more complex implementations. Using these techniques, it is possible to increase the modulator OSR and consequently improve its SNR and signal bandwidth.

In Section 5.2, a Time-Interleaved Delta Sigma (TIDS) approach for ADT will be analyzed in terms of frequency performance. In this approach, two modulators, a first- and second-order ΔΣM, are decomposed into time-interleaved equivalents. This is done with a polyphase decomposition of the internal filter of the ΔΣM, particularly the integrator, A(z), as shown in the example in Fig. 2.13. Low-order modulators were used since they provide better coding efficiency, as shown in Chapter 2 from [38], and typically have simpler hardware architectures, which favor higher sampling frequencies.

The ΔΣM internal filter loop $H(z)$ can be decomposed into its M×M polyphase equivalent. $\overline{H}(z)$ is a set of different transfer functions $\overline{P}_{ij}$ such that each parallel output, $O_i$, depends on the input $I_j$ as shown in equation Eq. (2.10), see also Fig. 2.14. The transfer functions $\overline{P}(z)_{ij}$ are given through the polyphase decomposition of the original $H(z)$ [205]. Re-arranging the new transfer functions into the matrix form, $\overline{H}(z)$ has the form as shown in Eq. (2.11).

Given any transfer function $H(z)$ from a Finite Impulse Response (FIR) or Infinite Impulse Response (IIR) filter, it is possible to decompose it into an equivalent $M \times M$ transfer function $\overline{H}(z)$ [205]. $\overline{H}(z)$ is an equivalent of the transfer function $H(z)$ with $M$ different inputs and outputs (Fig. 2.14). $\overline{H}(z)$ shows all the relations between the polyphase system $M$ inputs and $M$ outputs. Each output phase $O_i(z)$ is now given by a sum of the

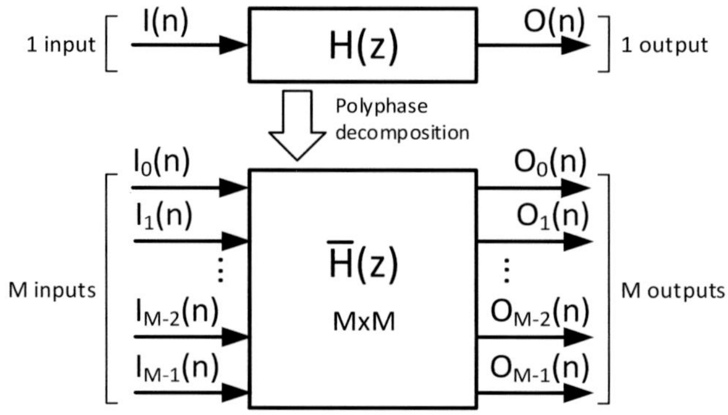


**FIGURE 2.14:** Polyphase decomposition.

input phases $I_j(z)$ as in Eq. (2.10). In Eq. (2.11), $P_k(z)$ are the polyphase components of from $H(z)$ re-written as the sum in equation Eq. (2.9).

$$H(z) = \sum_{k=0}^{M-1} z^{-k} P_k(z^M) \tag{2.9}$$

$$O_i(z) = \sum_{j=1}^{M} \overline{H}_{ij}(z) I_j(z) \tag{2.10}$$

$$\overline{H}(z) = \begin{bmatrix} \overline{P}_{1,1}(z) & \overline{P}_{1,2}(z) & \cdots & \overline{P}_{1,M}(z) \\ \overline{P}_{2,1}(z) & \overline{P}_{2,2}(z) & \cdots & \overline{P}_{2,M}(z) \\ \vdots & \vdots & \ddots & \vdots \\ \overline{P}_{M,1}(z) & \overline{P}_{M,2}(z) & \cdots & \overline{P}_{M,M}(z) \end{bmatrix} \tag{2.11}$$

However, the polyphase decomposition changes depending on the function $H(z)$, and there are multiple approaches when using a ΔΣM loop filter. Furthermore, one can easily understand that $\overline{H}(z)$ can be much more complex than $H(z)$ depending on the original transfer function and the $M$ factor. In fact, most of the time this decomposition is done to the loop integrator, which might not be the optimum solution. In Section 5.2 two different approaches will be implemented, decomposing the integrator or just the delay in the integrator.

As explained prior, the polyphase decomposition matrices are different depending on the chosen $H(z)$, which results in different hardware implementations as shown in Eqs. (2.12) and (2.13).

$$A(z) = \begin{bmatrix} \frac{z^{-1}}{1-z^{-1}} & \frac{1}{1-z^{-1}} & \frac{1}{1-z^{-1}} & \frac{1}{1-z^{-1}} \\ \frac{z^{-1}}{1-z^{-1}} & \frac{z^{-1}}{1-z^{-1}} & \frac{1}{1-z^{-1}} & \frac{1}{1-z^{-1}} \\ \frac{z^{-1}}{1-z^{-1}} & \frac{z^{-1}}{1-z^{-1}} & \frac{z^{-1}}{1-z^{-1}} & \frac{1}{1-z^{-1}} \\ \frac{z^{-1}}{1-z^{-1}} & \frac{z^{-1}}{1-z^{-1}} & \frac{z^{-1}}{1-z^{-1}} & \frac{z^{-1}}{1-z^{-1}} \end{bmatrix} \tag{2.12}$$

$$B(z) = \begin{bmatrix} 0 & 1 & 0 & 0 \\ 0 & 0 & 1 & 0 \\ 0 & 0 & 0 & 1 \\ z^{-1} & 0 & 0 & 0 \end{bmatrix} \tag{2.13}$$

Matrices Eqs. (2.12) and (2.13) give the relationship between the input and the output phases such that an output phase is equal to the sum of all the terms of a column, each one multiplied by the respective input in each row [205]. Comparing the two approaches, some differences are observable from the matrices. In the integrator ($A(z)$ in Fig. 2.13), each output depends on multiple inputs multiplied by a different $\overline{P}_{ij}$. In the delay ($B(z)$ in Fig. 2.13) approach, most of these partial transfer functions $\overline{P}_{ij}$ are zero; in this case, each output depends on only one of the inputs. This way, the hardware complexity remains reduced since no additional adders are needed to sum the multiple contributions of each input.

The different hardware implementations resulting from these approaches create significantly different critical path delays, which have a substantial impact on the achievable sampling frequency. In [36], the two decomposition solutions were implemented in an FPGA circuit, and an analysis of the maximum clock frequency was made for both the first and second-order modulators. For each architecture, the maximum achievable frequency was determined for different numbers of parallel paths, ranging from 2 to 16. The equivalent sampling frequency, $f_s$ is defined as the circuit clock frequency $f_{clk}$ times the number of parallel paths $M$. For each different polyphase ΔΣM implementation mapping, place-and-route routines were used to maximize the hardware clock frequency in the used FPGA device.

### 2.2.5 BB- vs RF-stage Modulators

All-digital RF transmitters can be split into two main classes, BB- or RF-stage, depending on the placement of the Pulse Encoder (relatively to the DUC module), and on the clock rate at which the quantization is done.

#### 2.2.5.1 RF-stage Modulators

As the name suggests, RF-stage modulators perform the pulse encoding directly at RF rates, after the input signal is upconverted to the RF carrier. The block diagram of such modulators is illustrated in Fig. 2.15a). This leads to a required sampling rate of at least twice the desired RF carrier frequency in both the pulse encoder and in the DUC modules (hereinafter referred to as $fs_{RF}$). This requires the use of high-sampling-rate pulse encoders capable of transforming the $m$-bit upconverted data into a 1-bit equivalent. The design of such encoders results in layout constraints and timing closure difficulties, leading to the costly and time-consuming development of custom integrated circuits [66,126,144,145,175]. In addition to this, as all the logic needs to be clocked at these sampling rates, the power consumption may typically be too high.

#### 2.2.5.2 BB-stage Modulators

To avoid the complex design process, BB-stage modulators were proposed in the literature [85]. This type of modulator applies the pulse encoder at low sampling rates (hereinafter referred to as $fs_{BB}$), digitally interpolates the signal to $fs_{RF}$, and, subsequently, upconverts it to the RF carrier frequency (Fig. 2.15b). As a conventional DUC would lead to a non-constant envelope waveform, a "special" DUC must be used to perform the signal upconversion while maintaining the signal with two levels. The only technique reported in the literature to design this specific DUC is based on the replication and combination of 1-bit BB samples ($x_I[n]$ and $x_Q[n]$) together with their inverted equivalents ($-x_I[n]$ and $-x_Q[n]$), simply referred to as *Select&Combine* [36,171–173].

In an attempt to shed some insight into this technique, it is of paramount importance to demonstrate why this technique is suitable for use as a DUC. Formally, this operation can be described with three main blocks: an $L$-factor up-sampler followed by a Zero-Order

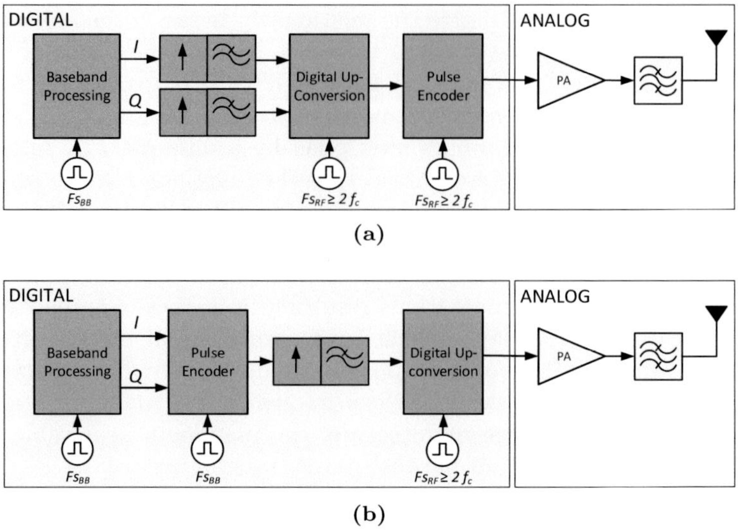


**FIGURE 2.15:** Illustration of the general block diagram of a BB-stage ADT (a) and an RF-stage ADT (b). In these specific figures, and for the sake of shedding insight into the practical implementations of such architectures, it was decided to explicitly include the digital interpolators before the DUC. However, it is essential to note that throughout this book, digital interpolations may be included within the DUCs.

Hold (ZOH) filter, $h(n)$, together with a real upconverter module (depicted in Fig. 2.16). The data is first up-sampled from the pulse encoder's sampling frequency ($fs_{BB}$) to the serializer's sampling frequency ($fs_{RF}$), resulting in an up-sampler factor of $L = fs_{RF}/fs_{BB}$. In practice, this operation is done by inserting $L$-1 zero-valued samples between the original samples with the aim of increasing the sampling rate. However, a major disadvantage emerges: undesired spectral images are created and centered on multiples of the original sampling rate. After that, an ideal low-pass filter should be included to smooth out the discontinuities, replacing the zeros with medium values and minimizing the undesired spectral images. However, its application is not possible owing to the need to maintain the signal with only two levels, which means that a discrete rectangular filter with the following impulse response, must be applied [171]:

$$h[n] = \begin{cases} 1, & 0 \leq n \leq (L-1) \\ 0, & n \geq L \end{cases} \tag{2.14}$$

After this interpolation, an upconversion is required to translate the carriers from BB/IF to RF. In a more formal way, the complex data is multiplied by an RF carrier, and just the real part is considered:

$$\begin{aligned} \widehat{x}_{RF}[n] &= Re\left(x_{INT}[n]e^{jw_c n}\right) = \\ & x_{INT_I}[n]\cos(w_c n) - x_{INT_Q}[n]\sin(w_c n) \end{aligned} \tag{2.15}$$

By considering only the real part, the complex multiplier reduces to just two multiplications. Moreover, considering the RF center frequency as one-fourth of the sampling frequency, the design of the upconversion chain can be even more simplified. In this frequency, $e^{\frac{j\pi n}{2}} = \cos\left(\frac{\pi n}{2}\right) + j\sin\left(\frac{\pi n}{2}\right)$, which results in the tristate pulsed quadrature signals $\cos(\frac{\pi n}{2}) = [1, 0, -1, 0]$ and $\sin(\frac{\pi n}{2}) = [0, 1, 0, -1]$. Replacing these signals in Eq. (2.15) leads

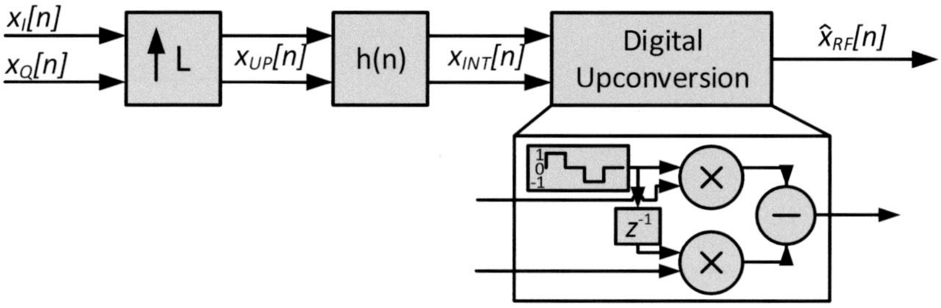


**FIGURE 2.16:** Block diagram of the specific digital upconverter module used in BB-stage modulators.

to a signal in the form of $[x_I[n], -x_Q[n], -x_I[n], x_Q[n]]$ with the inherent advantage that the multipliers can be replaced by logical inverters. Thus, one can conclude that this strategy can effectively perform the digital upconversion from BB to one-fourth of the serializer's sampling rate.

Despite the neatness and elegance of the BB-stage modulators, these architectures possess inherent drawbacks. First, the change of the carrier frequency means that the serializer's sampling rate must be changed. This leads to an unintentional variation of the modulator's OSR, which will considerably modify the system Figures of Merit (FoM). Second, as the pulse encoder operates at very low sampling rates, the OSR is inherently quite limited, which ultimately bounds the modulator's performance. Finally, the agility and multi-band capability in these techniques are quite limited, which is also a consequence of the low sampling rate from the pulse encoder, resulting in a limited Relates to the first Nyquist Zone NZ.

### 2.2.6 System-level Figures of Merit

As in every system design, there are some important FoM that must be valued and taken into consideration to assess the system performance. The design of an ADT is not an exception, and thus, the most important FoMs for the research line will be introduced.

It is adequate to split the FoM into two main categories—in-band performance and out-of-band performance—depending on the part of the spectrum that must be assessed. The key performance indicators in terms of the in-band signal are:

- Modulation Error Ratio (MER) is a measure of the SNR in digitally modulated signals. It is computed over $N$ symbols and is defined as [58]:

$$MER = \frac{\sum_{j=1}^{N} |S(j)|^2}{\sum_{j=1}^{N} |Z(j) - S(j)|^2)} \tag{2.16}$$

  where $Z$ is the actual received symbol, and $S$ is the ideal symbol.

- Error-Vector Magnitude (EVM) is another measure of modulation quality and error performance in digitally modulated radios [21]. It is proportional to the difference between the measured and the ideal symbols, and inversely proportional to the MER [75]:

$$EVM = \sqrt{\frac{\sum_{j=1}^{N} |Z(j) - S(j)|^2)}{\sum_{j=1}^{N} |S(j)|^2}} \times 100 \tag{2.17}$$

In terms of out-of-band performance FoM, one can state:

- Coding Efficiency ($C_{eff}$) is typically used to quantify the quantization noise on the transmitter efficiency [57], and is computed in the frequency domain, by integrating the in-band power, $P_{fund}$, over the total output power without filtering $P_{tot}$.

$$C_{eff} = \left(\frac{P_{fund}}{P_{tot}}\right) \times 100 \tag{2.18}$$

  $C_{eff}$ can provide an initial insight into the spectral purity of the transmitter. Moreover, some authors consider that it is also relevant for the prediction of the average efficiency of a given RF transmitter, with a bi-level output [74]:

$$\eta_{Average} = C_{eff}\,\eta^{BW}_{PA_{peak}}\,\eta_{duty-cycle} \tag{2.19}$$

  where $\eta^{BW}_{PA_{peak}}$ is the peak efficiency of the PA averaged over the signal bandwidth, and $\eta_{duty-cycle}$ is the duty-cycle efficiency effect. However, the impact of the $C_{eff}$ is still quite dependent on the PA stage, because, depending on matching bandwidths, not every out-of-band power will be dissipated. Taking into account this issue, some authors introduced the Limited Coding Efficiency ($C_{eff1-5}$),, which is the coding efficiency within the range of one-tenth of the carrier frequency up to 5 times the carrier frequency [126].

- Adjacent-Channel Power Ratio (ACPR) provides the ratio between the total output power in the fundamental zone, $P_{fund}$, and the total power integrated in a lower, $P_{LA}$, and upper, $P_{UA}$, adjacent-channel bands [21]. Can be mathematically described as:

$$ACPR_T = \frac{P_{fund}}{P_{LA}+P_{UA}} = \frac{\int_{\omega_{L2}}^{\omega_{U1}} S_o(\omega)d\omega}{\int_{\omega_{L1}}^{\omega_{L2}} S_o(\omega)d\omega + \int_{\omega_{U1}}^{\omega_{U2}} S_o(\omega)d\omega} \tag{2.20}$$

  where $S_o(\omega)$ is the power spectral density.

In the context of this book, all the reported results are quantified either in terms of EVM-MER pairs when the in-band performance is the dominant indicator that needs to be highlighted, or in ACPR and/or $C_{eff}$ when the out-of-band noise is the key factor to be focused on. $C_{eff}$ is not always presented as a performance indicator because, again, its impact is quite dependent on the amplification stage (a stage that is not focused in the context of this book).

## 2.3 Main Research Topics on All-Digital RF Transmitters

This section tries to provide an overview on the relevant works from the SoA in the research field of ADTs. The main research topics followed by the research community will be reported, along with the specific and relevant works that have been proposed. It was decided to organize the relevant reported works from the SoA into a set of main research topics that have been addressed by the research community.

### 2.3.1 Transmission Bandwidth

To meet the high-data-rate requirements of the new wireless communication standards in an efficient manner, the first step involves increasing the available bandwidth. The achievable bandwidth in ADTs is proportional to the Pulse Encoder's sampling rate and is also dependent on the notch bandwidth from the $\Delta\Sigma$M (or spectral gap in the PWM case). Thus, it is understandable that one of the most relevant research topics attracting considerable attention is related to novel techniques for achieving higher pulse encoder sampling rates.

As the increase of sampling rates through the use of single-rate approaches requires the custom design of ASICs, the use of multi-rate methodologies started to be valued, with some deriving novel modulators based on the node equations [110], but it has been demonstrated that, for a high number of phases, the sampling rate gains are limited [124].

A single-bit semi-parallel processing structure for high-frequency processing is also proposed in [81]. The proposed structure achieves a higher OSR by processing $N$ constant samples in parallel by combining $N$ closed-loop processing blocks of a regular $\Delta\Sigma$M. However, the last processing element of the proposed architecture leads to an increased critical path, resulting in reduced sampling rates. Thus, although promising results are obtained for low bandwidths (around 0.4 MHz and 1.6 MHz), the proposed architecture is not scalable with increasing bandwidth.

Another pseudo-parallel $\Delta\Sigma$ was proposed in [99]. The architecture is based on the previous one [81], but the block-to-block dependence that limits the clock speed was removed. To accomplish this, the authors proposed using speculation to alleviate the connection between the modulator's critical path and the clock frequency. Thus, there is no need to wait for the results from the previous iterations, at the expense of replicating logic for each possible set of output bits. However, for a large set of output bits, the area and power overheads can be too heavy.

Then, an optimized second-order Time-Interleaved $\Delta\Sigma$M with 488 MHz of Bandwidth (BW) was reported in [200]. In addition to the use of speculation, bit separation was also proposed to enable higher sampling rates. Bit separation enables the use of smaller accumulators to accommodate both the higher- and lower-order bits, leading to a reduction in the overall critical path. As a downside, achieving such wide bandwidths entails the use of challenging serializer sampling rates of around 20 Gbps. In addition to the increased complexity and cost, designing analog front-ends for such high sampling rates becomes quite challenging. The need for high serializer's sampling rates is quite common in the majority of the SoA, due to the application of the DUC after the pulse encoder (the concept of the BB-stage modulator). In particular, 28 Gbps was reported in [201], 20 Gbps in [200] and 60 Gbps in [126].

Chapter 5 will demonstrate that the design of RF-stage modulators allied with fully parallel $\Delta\Sigma$M can show promising results in terms of bandwidth with reasonable serializer's sampling rates. In particular, a time-interleaved architecture will be presented, which is capable of increasing the OSR without raising the FPGA's clock. Additionally, a multicore-based architecture that achieves 125 MHz of bandwidth with a single bit will be presented. Finally, through the optimization of the previous architectures and the transition to an RF-stage modulator, a bandwidth of up to 1.25 GHz will be demonstrated.

### 2.3.2 Carrier Agility

To meet the RAN expectations in a compact and efficient manner, there is a strong need for the development of flexible, agile, and reconfigurable radio transceivers with native support for multiple standards [61]. Real-time frequency agility is one of the main research topics in the field of ADTs, and it can be split into two main sub-areas, depending on the type of modulator: BB- or RF-stage.

On the one hand, the carrier agility in the BB-stage modulators is quite limited by several reasons. As introduced earlier, in this class of modulators, the DUC is based on the combination and replication of the I/Q samples, along with their inverted versions.

This leads to the carrier frequency being one-fourth of the serializer's sampling rate. The most straightforward way to transmit at other carrier frequencies is based on the real-time adaptation of the number of I/Q samples [191]. However, the agility continues to be limited to a set of possible carrier frequencies ($fs_{RF}/4$, $fs_{RF}/8$, $fs_{RF}/16$, and so on). Equivalently, the serializer's sampling rate can also be changed in real-time [185]. Again, even though the serializer sampling rate can be, in theory, updated to achieve the desired carrier frequency, there is a drastic consequence: the OSR from the Pulse Encoder will be changed as well. Thus, any change in the serializer's sampling rate leads to an unintentional variation of the modulator's sampling frequency, resulting in a change of the OSR, which will considerably modify the system's FoM. A novel technique that allows continuous adjustment of $f_c$ in the digital domain was proposed in [173]. The technique comprises a quadrature bandpass noise shaping stage in addition to a closed-loop PWM that enables the shaping of quantization noise around the desired IF. A similar architecture composed of a two-stage upconversion chain together with an IF image rejection was also proposed in [186]. Nonetheless, the tunability of the carrier placement is still limited to a small first NZ due to the low pulse encoder's sampling rate.

On the other hand, the agility of the RF-stage modulators can be quite wide, because the DUC is placed before the Pulse Encoder. Thus, the pulse encoder is clocked at a higher sampling rate, typically equal to the serializer's sampling rate, which means it has a higher first NZ. Thus, the carrier agility would only be limited by the frequency resolution from the DUC, typically imposed by the number of bits from the phase accumulators from the Digital Direct Synthesis (DDS). However, as far as FPGA-based modulators are concerned, there is no architecture suitable for the design of such modulators because it is not possible to clock the programmable logic subsystems within the FPGAs, at least two times the desired RF carrier frequency, when focusing the typical range of 500 MHz up to 6 GHz. Thus, wideband carrier agility is only found in specific custom ICs [66, 144, 145, 175].

Chapter 4 will demonstrate the first architecture that enables the synthesis of an RF-stage modulator within an FPGA. The underlying idea involves exploring highly parallelized structures to achieve high global sampling rates. In particular, real-time agility in the range from 100 MHz up to 6 GHz with a maximum frequency resolution of 1.5625 MHz will be demonstrated.

### 2.3.3 Multi Band Capability

To meet the data rate requirements efficiently, the first step involved increasing the available bandwidth in Third Generation (3G) systems [94]. Then, in an attempt to achieve scalable wider bandwidths without spectrum allocation constraints, the concept of Carrier-Aggregation (CA) was introduced in Fourth Generation (4G) systems, such as Long-Term Evolution (LTE)-Advanced [94]. By standardizing both the contiguous and non-contiguous CA capabilities, the combination of multiple frequency bands enabled high-speed data transmission. Due to the commercial success of LTE-Advanced features, it is expected that they will continue to evolve as a part of 5G technologies [114]. Nevertheless, despite the apparent ideal and native support for the multi-band capability, design challenges associated with the non-contiguous CA transmission have hampered the proposal of multi-band solutions.

The reported literature on non-contiguous CA ADTs is quite scarce. Generally speaking, multi-band transmission can only be achieved with integer multiples of the modulators' sampling frequencies [185], or with reduced sampling rate topologies [109, 121]. Others employ bulky and inefficient power combiners to join different bands before transmission [18, 109]. All these difficulties in designing multi-band transmission arise from the placement of the DUC after the pulse encoding (recall Fig. 2.15b). Following this approach, as the bandwidth of encoded signals is technically infinite, the upconversion to the different bands typically leads to degraded system performance.

In addition to that, as far as the FPGA-based architectures are concerned, the limited sampling rate from the programmable logic subsystems, typically less than 200 MHz, limits the first NZ. This considerably limits the maximum distance between different bands to twice this value, assuming complex Single-Side Band (SSB) upconversion schemes. Thus, the only methodology to achieve higher spans between bands is based on the utilization of replicas from different NZs [185]. However, the inherent decrease in terms of SNR associated with the need to maintain an integer multiplicity in all the involved sampling rates/frequencies leads to reduced performance.

Chapter 6 presents a range of architectures suitable for concurrent dual-band and triple-band transmission. Moreover, a pre-compensation is also proposed to minimize the amount of out-of-band noise that may be folded back to the interest bands in BB-stage modulators. As will be seen, in the dual-band case, real-time agility in the placement of up to two bands in the range of 300 MHz around the carrier frequency is demonstrated. In the triple-band case, up to three bands can be located in the range from 100 MHz up to 2.5 GHz with a frequency resolution of 4.88 MHz.

### 2.3.4 Analog Filtering Requirements

Recalling Fig. 2.15, the BPF is of utmost importance to reconstruct the analog signal before radiation and to minimize the out-of-band noise that is radiated. As the design of tunable filters with a high Q-factor is quite challenging, there has been interest in proposing novel strategies to relax the analog filtering requirements, either through the integration of embedded reconstruction filtering techniques or by easing the Q-factor requirements.

Concerning the development of architectures with an embedded filtering capability, one can refer to the use of power combiners to combine different outputs, and somehow, to generate a mixed-domain combination filter where the delay and the weight from each output were carefully chosen to implement a FIR filtering scheme [69, 199]. Nonetheless, these architectures are not quite appealing in terms of efficiency, due to the need for using multiple SMPAs and a power combiner, or a power combiner and a linear PA.

As far as the relaxing of the Q-factor is concerned, a method for canceling the bandpass quantization noise in envelope pulsed transmitter architectures, also called *carrier bursting*, was proposed in [78]. The architecture explores the overlap of a controlled amplitude component in the pulses. Nonetheless, this technique leads to fluctuations in the signal envelope, hindering the use of SMPA. A similar architecture was proposed in [29, 31, 55] for quadrature transmitters. However, the architecture requires the use of one extra PA, and one power combiner placed after the PAs, which, from the viewpoint of the efficiency, can be problematic. A method for reducing the Q-factor requirements was proposed in [185] for architectures based on LUTs, such as PWM. The idea is based on the random circular shifting of the pulsed waveforms in order to maintain the same energy, but with different words. This approach enables the utilization of low-stringent bandpass filters for the output RF reconstruction filter, which allows for an increase in signal bandwidth. The increase of the modulator's order to push the quantization noise away from the carrier frequency in $\Delta\Sigma$ architecture was also proposed in [170]. However, this tends to increase the modulator's critical path which means that lower sampling frequencies are achieved. Furthermore, instability problems may also appear.

Chapter 3 will demonstrate several architectures that allow the reduction of out-of-band noise in ADTs. The first one uses a 7-bit PWM-based transmitter with pattern randomization to increase coding efficiency and lower EVM. The second one uses improved PWM waveforms, while the third uses mixed-domain filtering. Both allow for an increase in coding efficiency through the reduction of out-of-band noise. Finally, an architecture that maintains the single-bit output while achieving a drastic reduction of the out-of-band noise around the carrier frequency. The proposed architecture relies on the use of a 2-stage $\Delta\Sigma$M with different output levels and different sampling rates.

### 2.3.5 System Efficiency

All-digital transmitters rely on the exploration of the Class-S transmitter architecture, where a constant-envelope waveform is driving SMPAs [30]. Ideally, in SMPAs, the transistor acts as a closed switch with a zero resistance during the ON-state, and as an open switch with an infinite resistance during the OFF-state. By doing this, no overlap between the drain current and voltages occurs, and ultimately, the power dissipation is zero, resulting in an efficiency of 100%.

The typical SMPAs falls in the classes D, E, and F of amplification [181]. In short, Class-D amplification employs two transistors as switches, and two sub-classes can be designed—Voltage-Mode Class-D (VMCD) or Current-Mode Class-D (CMCD)—according to the type of the waveform of the voltage and current (square and sinusoidal, or vice versa). Class-E amplification employs a single transistor as a switch. The drain-voltage waveform results from the sum of the DC and RF currents charging the drain-shunt capacitance. Class-F enables higher efficiency by employing harmonic resonators in the output network that shape the drain waveforms [165].

Although Class-S operation can theoretically achieve 100% power efficiency, independent of power back-off, due to the non-overlap of the voltage and current waveforms [215], several characteristics cause significant degradation. First, one should point out the presence of design impairments (such as finite-length transmission line connections and skew in the switches' input signal [157]) as well as switch impairments (such as non-zero internal resistance [102]). Second, the inclusion of a high-Q BPF with a narrower bandwidth than the PA behaves like a complex, time-varying load impedance to the PA [214]. This may cause an overlap between the voltage and current waveforms, reducing the system efficiency [134]. To minimize this effect, the Pulsed Load Modulation (PLM) amplifier architecture was proposed in [97] by replacing the SMPA with 2 class-B PA controlled by the same envelope modulator[1]. However, implementations for such modulators can be found in [104, 105]. At last, one should also emphasize that the duty cycle from the pulsed waveform also leads to a modulation of the load seen by the PA. One promising way to minimize this effect is proposed in [149], in which the imaginary part of the load impedance is varied along with the input duty-cycle, in order to maintain the high-efficiency independent from the input signal.

Also related to the interconnection between pulsed modulators and SMPAs is the minimum width that an input pulse must have to be correctly amplified. In this sense, some RF-PWM variants were investigated [125]: Limited Pulse-Width Modulation (LPWM) and Quantized Pulse-Width Modulation (QPWM). While in the first, pulses shorter than the minimum width are skipped, in the second, only $N$ discrete and non-zero pulse-widths are allowed in the SMPA input.

The dependence between $C_{eff}$ and the overall system efficiency was investigated in [102] and [74] (recall Eq. (2.19)). As an example, one ADT with a highly-efficient SMPA of around 80% and an overall system efficiency of 5% is reported in [74]. Even though the research community also points out that the impact of the $C_{eff}$ in the system efficiency cannot be so straightforwardly generalized, because depending on matching bandwidths, not every

[1]However, it must be pointed out that this class of amplification falls out of the scope of the Class-S operation, which is the reason why this research line will not be further elaborated.

out-of-band power will be dissipated [126], one can understand that the $C_{eff}$ cannot be neglected. Thus, in this sense, it is quite important to propose new techniques to reduce the out-of-band noise, or in other words, increase the $C_{eff}$. While some authors presented the development of architectures with an embedded filtering capability based on analog power combiners as promising solutions for the enhancement of the $C_{eff}$, and ultimately, the increase of the system efficiency [35,95,189], the fact is that these claims are not entirely true, due to the inclusion of the analog power combiner after the SMPAs which may jeopardize the entire system efficiency. However, such architecture can be interesting if the power combination is performed efficiently, such as with extended H-bridges SMPAs [229, 230]. Nonetheless, to date, the complexity inherent in the development of efficient multi-level Class-S PAs has limited the number of levels to less than three [30]. Similarly, techniques to reduce the quantization noise around the carrier frequency were proposed in [55]. The work demonstrates that by removing a specific part of the quantization noise while maintaining a quasi-constant signal envelope, it is possible to improve the $C_{eff}$ without introducing nonlinearity and distortion into the system.

### 2.3.6 Scalability, Integration and Applications

The potential of ADTs modulators has also been demonstrated in various scenarios, including centralized RAN architectures, antenna arrays, and radars. The current trend in RAN architectures is to move baseband processing to the cloud, enabling centralized signal processing and management. C-RAN architecture is an example of these new cellular network architectures designed for the future mobile network infrastructures, which focuses on the separation of the digital Base Band Units (BBUs) from the Remote Radio Heads (RRHs). The interface between the BBUs and the RRHs is typically based on optical links and is referred to as *fronthaul*. ADTs have been proposed as promising solutions to reduce the complexity of the RRH. By sending optical pulses with the encoded RF information, the RRH is just responsible for filtering, amplification, and radiation. Relevant works were reported in both the downlink [33] and in the uplink [122, 162]. Similarly, Pulse Encoders are also being highlighted as key enablers for the throughput explosion in digital mobile fronthauls [207,213]. The overall idea is to encode the BB I/Q data (with no upconversion to RF) into pulsed waveforms before being sent to the optical link. In this way, the throughput in the fronthaul can be maximized because fewer bits are transmitted per user without a significant loss of performance.

The exploration of all-digital RF transmission techniques in antenna arrays has also been targeted by the research community, for both non-communication and communication purposes. On the one hand, the design of radars based on Pulse Encoders was first introduced in [216]. The key idea is to reduce the number of external DACs to boost the development of scalable radars, without challenging requirements in the analog front-end design. Similar relevant implementations have been proposed in [59, 137, 138]. On the other hand, the integration of ADTs in antenna arrays for communication purposes was first introduced by our group in [42]. The primary objective is to relax the stringent requirements in designing multiple front-ends for antenna arrays. Afterwards, [126] also proposed the same idea as the main foundation for the design of Massive-MIMO front-ends. Section 7.5 will further develop on this research topic by demonstrating the first real and complete system, experimentally validated in terms of beamforming and beamsteering capabilities, where antenna arrays are directly driven by multiple ADTs fully integrated into the same FPGA. The remainder of Chapter 7 will focus on the analysis of RoF solutions based on ADTs, as well as some new all-digital techniques for all-digital signal reception, which might enable true all-digital transceivers.

# Chapter 3

## Relaxing the Analog Filtering Requirements

### 3.1 Introduction

The unprecedented attention given to wireless communications over the last few decades has led to the proliferation of diverse wireless standards operating at different frequencies, using different coding and modulation schemes, and targeted for various purposes. Such vast proliferation pushes an additional research effort toward the development of new smart radios [129], capable of adapting to different communication scenarios. In this new wireless communication paradigm, the well-known SDR concept [130] holds the potential for implementing the supreme universal radio, that is, a fully reconfigurable radio that can adapt its own communication parameters to meet user demands as well as channel and network conditions.

However, a key challenge for achieving this new communications paradigm involves the development of a highly flexible physical layer to support the transmission of multi-band, multi-rate, and multi-standard signals, which in practice is very difficult to implement using conventional approaches.

A new all-digital approach, where its datapath is digital from the baseband up to the RF stage, has been proposed as a solution for such flexible transmitters [30]. Such a concept has inherent high flexibility and poses an important step toward the development of SDR transmitters. Moreover, this new approach enables the use of FPGA devices for implementing the RF transmitter, providing additional flexibility as well as field-upgradeability.

In fact, modern FPGAs provide an equivalent logic capacity of millions of logic gates, complemented by large Random-Access Memory (RAM) blocks, embedded processors, DSP modules, and multigigabit I/O standards. These resources have already enabled the implementation of LUT-based all-digital transmitters, demonstrating the feasibility of agile and fully digital RF generation within reconfigurable platforms [156].

Despite the advantages mentioned in the previous paragraph, current state-of-the-art digital transmitter architectures are still very restrictive for developing a truly SDR-based system. In fact, most current ADTs require very high-quality filters to remove out-of-band noise [64, 200]. The introduction of an analog filter reduces the flexibility of the ADT by binding the transmitter output to the filter bandwidth and central frequency, essentially negating the gains in flexibility acquired by using an ADT.

This chapter is supported by the conference papers [34, 52, 152] and by the journal articles [35, 156, 183], and in it, several solutions to reduce the out-of-band noise of ADTs will be presented. These solutions are based on polyphase noise canceling, waveform optimization, LUT size reduction, mixed-domain combination filters, PAM-4, and cascading modulators. Some have the objective of further spreading the noise to frequencies away from the signal, such as waveform optimization and cascading modulators, or filtering out the noise without the use of analog filters, such as polyphase noise canceling and mixed-domain combination

DOI: 10.1201/9781003225713-3

filters. Others pretend to reduce the quantization noise directly, with the implementation of PAM-4, and finally others intend to reduce the resource usage in terms of RAM blocks within the FPGA. Ultimately, these techniques enable the use of lower-quality-factor analog filters with larger bandpass bandwidths, not only reducing the cost of the final transmitter but also allowing for greater carrier flexibility.

## 3.2 Delta-Sigma and Pulse-Width Modulators

In this section, we address two key limitations of current approaches: the poor coding efficiency and the combined high quality factor and tunability requirements of the RF output filter, which significantly limit the practical usability and dissemination of these delta-sigma-based flexible RF transmitters. The work presented here is built upon previous work [184, 193] by presenting and evaluating a novel FPGA-embedded agile transmitter based on pulse modulators. This architecture significantly improves important figures of merit, such as coding efficiency and EVM, while maintaining support for multichannel, multi-standard data transmission. The RF output carriers operating in the microwave frequency range are still generated within a single FPGA device. Moreover, the high flexibility of this reconfigurable transmitter enables easy modification of key parameters, such as carrier frequencies, usable bandwidth, and dynamic range.

### 3.2.1 Delta Sigma Modulator Design

This subsection outlines the architecture of a second-order low-pass $\Delta\Sigma$M that will be utilized in the construction of a flexible digital radio transmitter. The $\Delta\Sigma$M used is a CIFB type [141] with a 3-level quantizer, as illustrated in Fig. 3.1 by its general representation in the $z$-domain. The CIFB structure was chosen due to its good stability when used in a low-pass configuration and also because of its short critical path, which allows a higher sampling rate. Regarding the chosen 3-level quantizer, it was shown in [193] that using one extra level in addition to the conventional bi-level quantizer is possible to increase the SQNR by approximately 5 dB.

The $\Delta\Sigma$ coefficients were precomputed using the $\Delta\Sigma$-toolbox [177], for a sampling rate of $f_s = 250$ MHz and a bandpass of 20 MHz, resulting in an effective OSR of 12.5. These coefficients were then recomputed by a convergence process where all coefficients are rounded to powers of two, or if not suitable, to a two's-complement binary representation

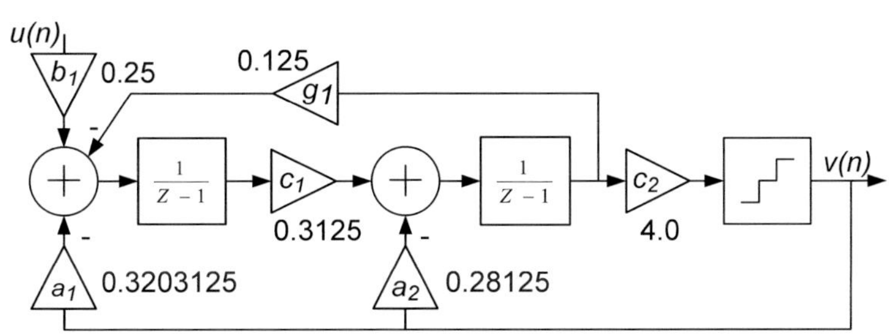


**FIGURE 3.1:** z-domain representation of the implemented CIFB second-order $\Delta\Sigma$M using a 3-level quantizer.

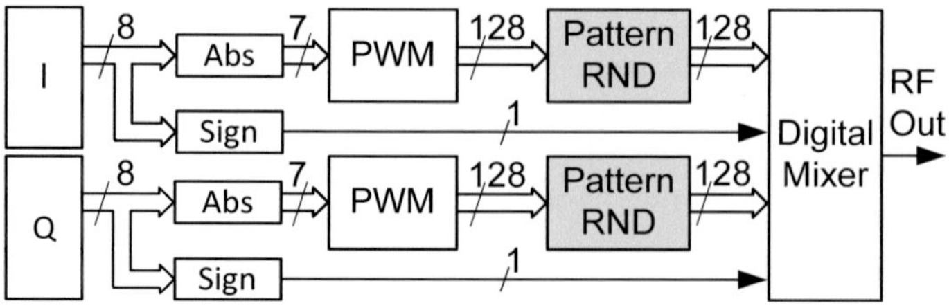


**FIGURE 3.2:** Block diagram illustrating the architecture of the PWM-based transmitter with pattern randomization.

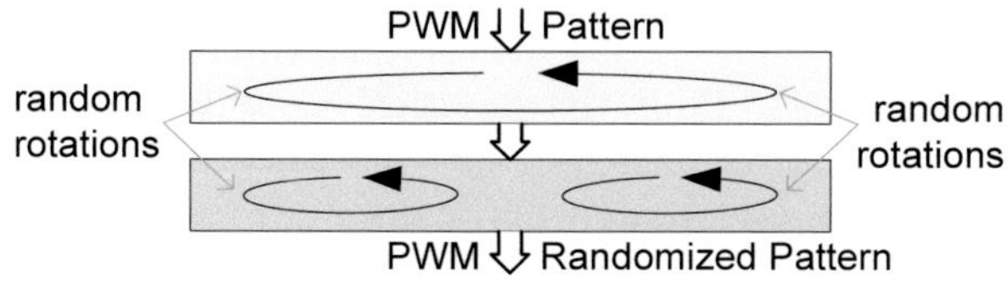


**FIGURE 3.3:** Block diagram illustrating the PWM pattern randomization process.

using up to eight bits. This approach contributes to reducing the critical path, as powers of two coefficients can be implemented using logical shifts, thereby avoiding the need for additional hardware resources. The resulting type-2 Chebyshev highpass NTF can be given by:

$$\mathrm{NTF}(z) = \frac{z^2 - 2z + 1.039}{z^2 - 0.875z + 0.3145}. \tag{3.1}$$

### 3.2.2 Pulse Width Modulator Design

This subsection details the design of the PWM-based transmitter. In this architecture, each In-phase (I) and Quadrature (Q) components of the baseband signal are quantized using 8 bits. The 7-bit magnitude data is encoded into a 128-bit PWM word, corresponding to 129 different levels of energy (from 128 bits all zeros up to all ones). This PWM word combined with the signal information will then allow performing the RF up-conversion, see Fig. 3.2. However, if a conventional PWM approach is used, such as when using thermometer coding, there is a specific waveform pattern corresponding to each 7-bit input magnitude, which in turn corresponds to the desired duty cycle. In this scenario, the RF transmitter will generate harmonics spaced at the PWM sampling frequency. On the other hand, if the desired duty cycle is kept but the waveform changes arbitrarily, then it is possible to spread the energy of these harmonics over the RF spectrum, which in turn reduces the RF output filtering requirements. This way, the base architecture of the transmitter was extended by adding a pattern randomization block after the PWM waveform generation, as shown in Fig. 3.2. This block performs several rotations for each sample of the 128-bit PWM word in a pseudo-random manner; see Fig. 3.3.

Regarding the chosen PWM word length, experimental results show that by increasing from 6 to 8 the number of quantization bits of the baseband signal, that is, by using a 128-bit PWM word instead of a 32-bit word, it is possible to increase the dynamic range of the carrier in approximately 3 dB.

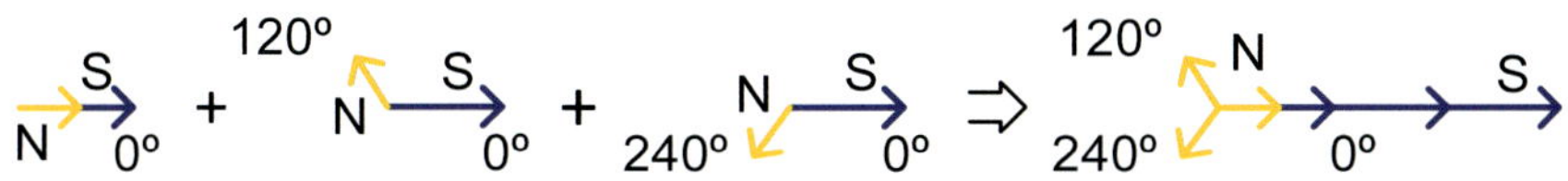


**FIGURE 3.4:** Illustration of the noise canceling idea when using polyphase noise.

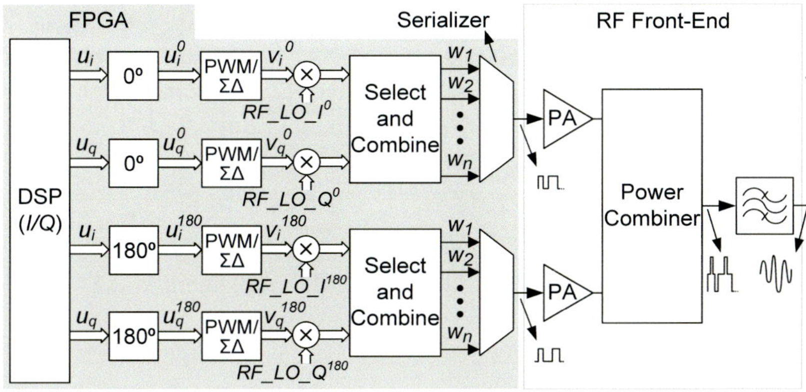


**FIGURE 3.5:** Single channel illustration of the proposed polyphase multipath agile transmitter architecture.

### 3.2.3 Proposed All-Digital Transmitter

In this subsection, we present a new multichannel transmitter architecture that significantly improves coding efficiency and reduces RF filtering requirements while maintaining the flexibility of conventional all-digital transmitters.

A key idea for improving the coding efficiency of this transmitter without sacrificing flexibility is to reduce the overall noise by using polyphase multipath circuits, as illustrated in [128] and [16].

The base idea of noise reduction through polyphase circuits relies on having multiple paths where the phase of the desired signal is kept constant but where the unwanted noise has different canceling phases; see Fig. 3.4. This way, when the output signals of the multipath circuits are combined, the energy of the desired signal will grow proportionally to the number of multipath circuits, while, ideally, the noise will cancel. This will directly improve the overall coding efficiency of the transmitter. The high-level block diagram of the proposed polyphase multipath agile transmitter is shown in Fig. 3.5, where the multichannel operation and a higher number of phases are omitted for the sake of simplicity.

The proposed transmitter shown in Fig. 3.5 is constituted by two independent paths, one containing noise at 0° and another one where the noise was shifted to 180°. Each independent path applies a specific phase shift to the baseband I and Q components ($u_i$, $u_q$). These phase-shifted signals are then shaped using PWM or ΔΣM in order to make them suitable for the RF up-conversion process. The first step to generate $w$ requires each PWM/ΔΣM to be multiplied by the corresponding *RF_LO* signal. However, each *RF_LO* vector has a specific phase for each independent path. This way, the mixing process also performs a phase shift to the entire signal (both the desired information and the unwanted noise). For the 0° datapath, the $RF\_LO\_I^0$ and $RF\_LO\_Q^0$ vectors should produce a -0° phase shift. In the proposed architecture, these vectors are equal to (+1,0,-1,0,+1,...) and (0,+1,0,-1,0,...), respectively. The *Select and Combine* block will simply add the *RF_I* and the *RF_Q* components, resulting in a vector containing the four components of the desired signal ($+v_i$, $+v_q$, $-v_i$, $-v_q$, ...), with the dimension of $w$. The serialization of this vector at a

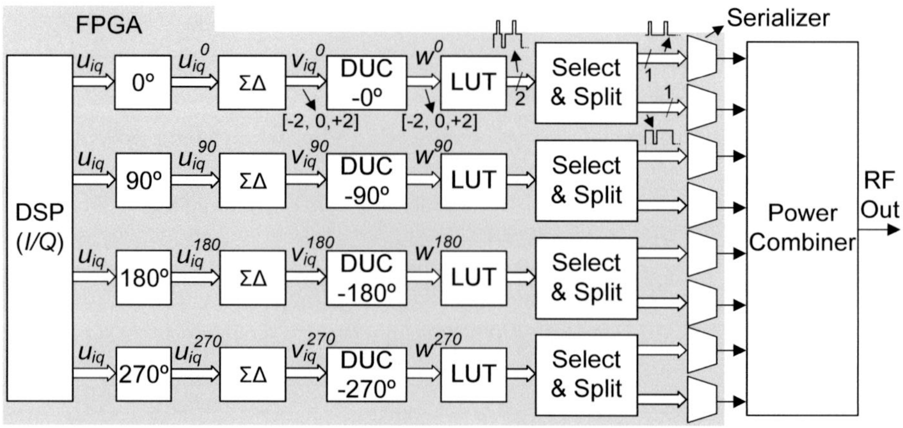


**FIGURE 3.6:** High-level block diagram illustrating the implemented ΔΣ-based polyphase multipath agile transmitter.

bitrate of $4 \times f_c$ will generate a single RF carrier centered at $f_c$. The allowable output carrier frequency is given by:

$$fc = N \times fs \ , \ N \in \mathbb{N}. \tag{3.2}$$

For the 180° datapath, the $RF\_LO\_I^{180}$ and $RF\_LO\_Q^{180}$ vectors should produce a—180° phase shift. In the proposed architecture, these vectors are equal to (-1,0,+1,0,-1,...) and (0,-1,0,+1,0,...), respectively. Similarly to the 0° datapath, the serialization of the resulting vector at a bitrate of $4 \times f_c$ will generate a single RF carrier centered at $f_c$.

After the RF up-conversion stage, the desired signal of all independent paths is now phase aligned, while the noise has different canceling phases. This way, it is now possible to combine the energy of all RF paths using a power combiner, thereby enhancing the signal's dynamic range.

#### 3.2.3.1 Delta-Sigma-based RF Transmitter

Here, the design stage of the ΔΣ-based agile and coding-efficient all-digital transmitter will be detailed. This transmitter extends the proposed architecture shown in Fig. 3.5 by utilizing 90° and 270° phases in addition to the 0° and 180° phases; see Fig. 3.6. This way it is possible to further improve the SNR of the desired signal.

The designed transmitter uses the same baseband information (I and Q) for each specific path. As a good design trade-off between SNR and the ΔΣ operating speed, each baseband component is quantified using 10 bits. This information is then shifted to the proper phase and then shaped to a 3-level representation using ΔΣM. Then, each specific path will perform the DUC using a proper phase so that all paths have the desired signal shifted to the same phase. The *LUT* block will convert the 3-level two's complement signal into a 2-bit word where the integers *-2*, *0*, and *2* are converted into *00*, *01*, and *11*, respectively.

The *Select and Split* block will then send to one serializer a vector containing only the Most Significant bit (MSb) while another vector containing only the Least Significant bit (LSb) will be sent to another serializer, as shown in Fig. 3.6.

Again, after the serialization process, the desired signal of all independent paths is now phase-aligned and centered at $fc$, while the noise has different canceling phases. This way, when all paths are combined, the desired signal will have an improved dynamic range.

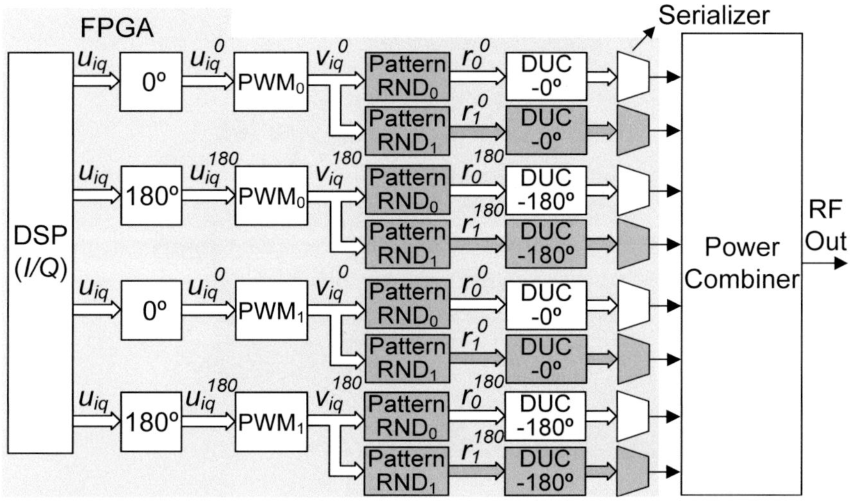


**FIGURE 3.7:** High-level block diagram illustrating the implemented PWM-based polyphase multipath transmitter. The optional blocks are shown in dark gray.

#### 3.2.3.2 PWM-based RF Transmitter

The design stage details regarding the PWM-based transmitter will be provided in this subsection, as shown in Fig. 3.7. This transmitter requires a baseband signal, where each component is quantized using 8 bits: 1 bit for the signal and 7 bits for the magnitude. For each individual path, the 7-bit magnitude information is directly translated into a PWM word accordingly to a specific PWM matrix. For the first two independent paths, the 128-by-128 PWM matrix used can be described as a triangular lower filled with ones, as shown below:

$$PWM_0 = \begin{bmatrix} 1 & & 0 \\ \vdots & \ddots & \\ 1 & \cdots & 1 \end{bmatrix} \tag{3.3}$$

For the last two independent paths, the 128-by-128 PWM matrix used is a horizontal mirror of the first one, as shown below:

$$PWM_1 = \begin{bmatrix} 0 & & 1 \\ & \iddots & \vdots \\ 1 & \cdots & 1 \end{bmatrix} \tag{3.4}$$

Since each row of $PWM_1$ has the same number of ones as in $PWM_0$, the resulting signal after the PWM modulation will have the same energy, regardless of the used PWM matrix. On the other hand, since the ones are placed at different positions for the chosen PWM matrices, the PWM noise will appear with different phases. This way, when combining the outputs, the energy of the desired signal will add in a proportion that is higher than the unwanted noise.

Additionally, when the optional *Pattern Randomization* is used, the PWM noise is spread along the RF spectrum, which can be interesting for reducing the RF filter requirements.

The designed transmitter applies in *Pattern* $RND_0$ a set of pseudo-random rotations for each PWM word (as previously shown in Fig. 3.3). For the used sampling frequency, the same random rotations are applied to several PWM samples, as can be seen in Fig. 3.7. On the

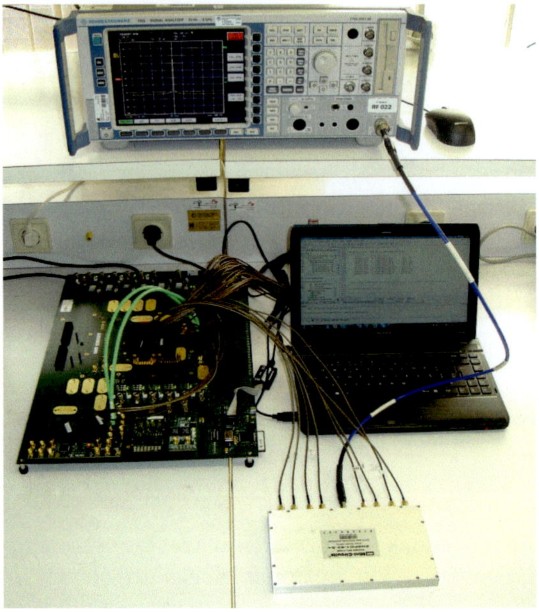

**FIGURE 3.8:** Setup of the evaluated FPGA-based all-digital transmitter.

other hand, since shifting the position of the ones in each PWM word will change the PWM noise phase, we expanded the transmitter to include an additional *Pattern* $RND_1$. This block will implement a different set of pseudo-random rotations, allowing the duplication of the number of output paths with different phase noise, and therefore improving the dynamic range of the desired signal.

### 3.2.4 Experimental Results

This subsection presents and discusses the main experimental results for three different variants of the proposed agile and coding-efficient all-digital transmitter. The first variant will regard $\Delta\Sigma$-based data transmission, while the other variants will present PWM-based data transmission with and without pattern randomization.

The proposed RF transmitter was prototyped on an *ML628* development board containing a *Virtex6 HX380T* FPGA. For the purpose of this proof-of-concept, we implemented the base datapath shown in Fig. 3.5 and connected its digital RF outputs to a *Mini-Circuits* 8:1 RF power combiner model *ZN8PD1-53+*. Finally, the power combiner was directly connected to a Vector Signal Analyzer (VSA), as can be seen in Fig. 3.8.

The baseband test signals used in this setup are a 64-QAM signal with a symbol rate of 1.25 MSPS and a PAPR of approx. 7 dB and a WiMAX signal (OFDM-based) with a 1.5 MHz bandwidth and a PAPR of 8 dB. Moreover, all baseband signals were stored inside the FPGA's internal memory.

#### 3.2.4.1 $\Sigma\Delta$-based Data Transmission

As a first experiment, the implemented RF transmitter architecture, illustrated in Fig. 3.6, was configured to transmit one carrier centered at 1.25 GHz, containing a 64-QAM signal. For this implementation example, all $\Delta\Sigma$Ms were set to operate at a sampling frequency of

156.25 MHz. This sampling frequency, $fs$ is given by:

$$fs = \frac{4 \times fc}{\text{width}(w)}, \tag{3.5}$$

where $fc$ is the chosen carrier frequency, and width($w$) is the width of the word $w$ that will be fed to the serializer.

For the first independent path, the *RF_LO* vectors and the *Select and Combine* block of the *DUC -0°* were configured to generate the following 32-bit output word:

$$w^{0} = [\text{-}v_i,\ \text{-}v_q,\ +v_i,\ +v_q,\ \text{-}v_i,\ \text{-}v_q,\ +v_i,\ +v_q,\ \text{-}v_i,\ \text{-}v_q,\ +v_i,\ +v_q,\ \text{-}v_i,\ \text{-}v_q,\ +v_i,\ +v_q,$$
$$\text{-}v_i,\ \text{-}v_q,\ +v_i,\ +v_q,\ \text{-}v_i,\ \text{-}v_q,\ +v_i,\ +v_q,\ \text{-}v_i,\ \text{-}v_q,\ +v_i,\ +v_q,\ \text{-}v_i,\ \text{-}v_q,\ +v_i,\ +v_{q,}].$$

For the second path, the *DUC -90°* was configured to generate the following 32-bit output word:

$$w^{90} = [+v_q,\ \text{-}v_i,\ \text{-}v_q,\ +v_i,\ +v_q,\ \text{-}v_i,\ \text{-}v_q,\ +v_i,\ +v_q,\ \text{-}v_i,\ \text{-}v_q,\ +v_i,\ +v_q,\ \text{-}v_i,\ \text{-}v_q,\ +v_i,$$
$$+v_q,\ \text{-}v_i,\ \text{-}v_q,\ +v_i,\ +v_q,\ \text{-}v_i,\ \text{-}v_q,\ +v_i,\ +v_q,\ \text{-}v_i,\ \text{-}v_q,\ +v_i,\ +v_q,\ \text{-}v_i,\ \text{-}v_q,\ +v_{i,}].$$

The *DUC -180°* was configured to generate the following 32-bit output word:

$$w^{180} = [+v_i,\ +v_q,\ \text{-}v_i,\ \text{-}v_q,\ +v_i,\ +v_q,\ \text{-}v_i,\ \text{-}v_q,\ +v_i,\ +v_q,\ \text{-}v_i,\ \text{-}v_q,\ +v_i,\ +v_q,\ \text{-}v_i,\ \text{-}v_q,$$
$$+v_i,\ +v_q,\ \text{-}v_i,\ \text{-}v_q,\ +v_i,\ +v_q,\ \text{-}v_i,\ \text{-}v_q,\ +v_i,\ +v_q,\ \text{-}v_i,\ \text{-}v_q,\ +v_i,\ +v_q,\ \text{-}v_i,\ \text{-}v_{q,}].$$

For the last independent path, the *DUC -270°* was configured to generate the following 32-bit output word:

$$w^{270} = [\text{-}v_q,\ +v_i,\ +v_q,\ \text{-}v_i,\ \text{-}v_q,\ +v_i,\ +v_q,\ \text{-}v_i,\ \text{-}v_q,\ +v_i,\ +v_q,\ \text{-}v_i,\ \text{-}v_q,\ +v_i,\ +v_q,\ \text{-}v_i,$$
$$\text{-}v_q,\ +v_i,\ +v_q,\ \text{-}v_i,\ \text{-}v_q,\ +v_i,\ +v_q,\ \text{-}v_i,\ \text{-}v_q,\ +v_i,\ +v_q,\ \text{-}v_i,\ \text{-}v_q,\ +v_i,\ +v_q,\ \text{-}v_{i,}].$$

These output words were then converted into a 2-bit representation and split so that each bit is fed to one MGT serializer. The serializer must be configured to operate at a line rate that is 4 times the desired carrier frequency. This way, for the 1.25 GHz carrier, the serializer was set to operate at 5 Gbps.

After combining the outputs of the 8 serializers, it is possible to obtain the spectrum shown in Fig. 3.9. For comparison purposes, the obtained spectrum resulting from combining eight paths is shown overlapped with that of a single path, and the gain for the single path has been normalized to the results of the eight-path implementation. As can be seen, for the same energy of the desired signal, the 8-path implementation has clearly less out-of-band noise. This way, the proposed transmitter clearly allows for improving the coding efficiency while maintaining the flexibility inherent to the all-digital transmitter concept.

#### 3.2.4.2 PWM-based Data Transmission

Here, the experimental results for two different PWM-based approaches for realizing an all-digital transmitter are presented. One uses a pure PWM-based architecture, while the other combines PWM with a pattern randomizer as illustrated in Fig. 3.7.

In both cases, the transmitters were configured to implement a 1.25 GHz carrier containing a 64-QAM signal so that it is possible to provide a better comparison with the previously implemented $\Delta\Sigma$-based transmitter. Accordingly to Eq. (3.5), all 7-bit PWM modulators were set to operate at the same sampling frequency of 156.25 MHz. Additionally, all independent paths implement the DUC in a manner similar to the previously implemented $\Delta\Sigma$-based transmitter.

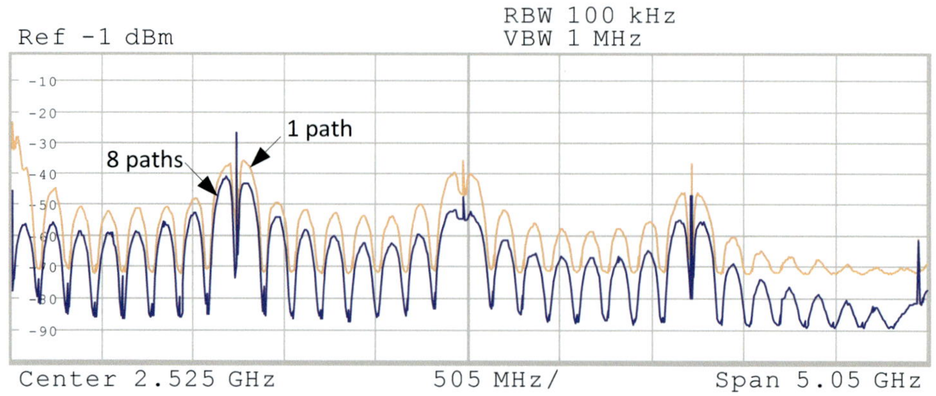


**FIGURE 3.9:** Overlapped measurement results for the ΔΣ-based agile transmitter when using 8 paths (Black) versus a single path (Gray) approach with a gain normalized to the 8 paths.

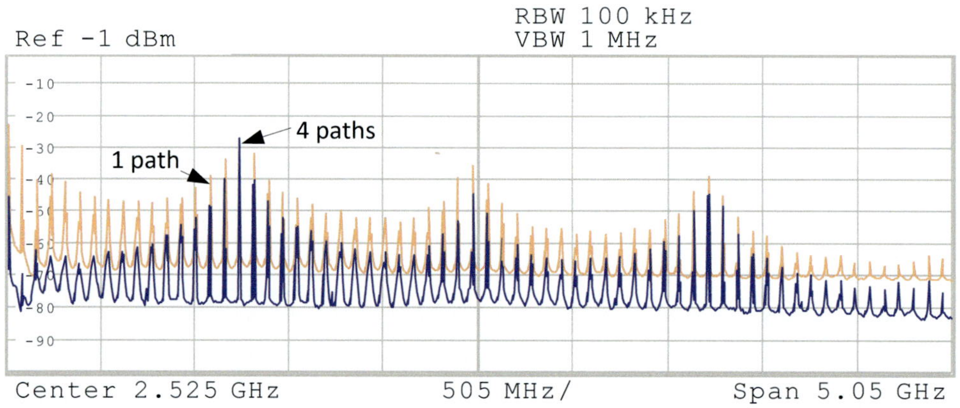


**FIGURE 3.10:** Overlapped measurement results for the PWM-based transmitter without pattern randomization when configured with 4 paths (Black) vs. a single path (Gray) with normalized gain.

After combining the outputs of the 4 serializers for the PWM-based transmitter without pattern randomization, it is possible to obtain the spectrum shown in Fig. 3.10. For comparison purposes, the obtained spectrum resulting from combining four paths is shown overlapped with that of a single path. Again, the single-path spectrum results were normalized to the 4-path implementation results for an easier comparison. As can be seen, for the same energy on the carrier, the single-path spectrum shows considerably higher noise. Moreover, the 8-path implementation has a significant PWM noise reduction for the lower frequency harmonics. This way, it is clear that the proposed PWM-based transmitter allows significant improvements in terms of coding efficiency.

Then, the FPGA hardware was reconfigured to implement the PWM-based transmitter with pattern randomization. In Fig. 3.11, the obtained spectrum of the implemented transmitter is illustrated, overlapping the results for one implementation where 8 paths are combined with a single-path approach, with the gain normalized to the 8-path implementation. The obtained results show clear improvements in coding efficiency. Moreover, it can be seen that using pattern randomization allows for reducing the peak spuriousness,

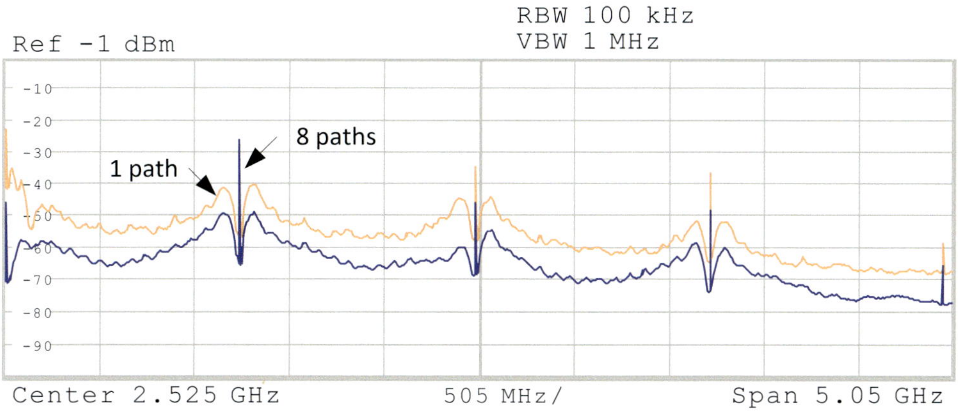


**FIGURE 3.11:** Overlapped measurement results for the PWM-based transmitter with pattern randomization when configured with 8 paths (Black) versus a single path (Gray) with a gain normalized to the 8 paths.

which in turn can be beneficial for alleviating the quality factor requirements of the RF reconstruction filter.

### 3.2.4.3 Multichannel Data Transmission

Here, the implementation results for a ΔΣ-based architecture when configured for multichannel data transmission will be presented. The architecture of the implemented RF transmitter is identical to that previously illustrated in Fig. 3.6. However, instead of having 4 independent paths, each one having a different phase noise for the same baseband signal, the multichannel transmitter uses only two different phases (0° and 180°) and has two baseband signals, a 64-QAM connected to the first two independent paths and a WiMAX signal connected to the last two independent paths.

The multichannel transmitter was configured to simultaneously transmit the 64-QAM signal, centered at 1.25 GHz, and the WiMAX signal, centered at 781.25 MHz. For this implementation example, all ΔΣMs were set to operate at a sampling frequency of 156.25 MHz. A base requirement of this approach is that the sampling frequencies of all ΔΣMs must have an integer relation. This way, it is possible to overlap the noise transfer function nulls of the multiple ΔΣMs, and consequently allow the insertion of the desired signals in those nulls.

Accordingly to Eq. (3.5), transmitting the 64-QAM signal centered at 1.25 GHz requires the serialization of the parallel word $w$ with a width of 32-bits. On the other hand, transmitting the WiMAX signal centered at 781.25 MHz requires $w$ to have a width of 20-bits.

For the first independent path containing the 64-QAM signal, the *DUC -0°* was configured to generate the following 32-bit output word:

$$w^0 = [-v_i,\ -v_q,\ +v_i,\ +v_q,\ -v_i,\ -v_q,\ +v_i,\ +v_q,\ -v_i,\ -v_q,\ +v_i,\ +v_q,\ -v_i,\ -v_q,\ +v_i,\ +v_q,\\ -v_i,\ -v_q,\ +v_i,\ +v_q,\ -v_i,\ -v_q,\ +v_i,\ +v_q,\ -v_i,\ -v_q,\ +v_i,\ +v_q,\ -v_i,\ -v_q,\ +v_i,\ +v_{q,}].$$

For the second path, the *DUC -180°* was configured to generate the following 32-bit output word:

$$w^{180} = [+v_i,\ +v_q,\ -v_i,\ -v_q,\ +v_i,\ +v_q,\ -v_i,\ -v_q,\ +v_i,\ +v_q,\ -v_i,\ -v_q,\ +v_i,\ +v_q,\ -v_i,\ -v_q,\\ +v_i,\ +v_q,\ -v_i,\ -v_q,\ +v_i,\ +v_q,\ -v_i,\ -v_q,\ +v_i,\ +v_q,\ -v_i,\ -v_q,\ +v_i,\ +v_q,\ -v_i,\ -v_{q,}].$$

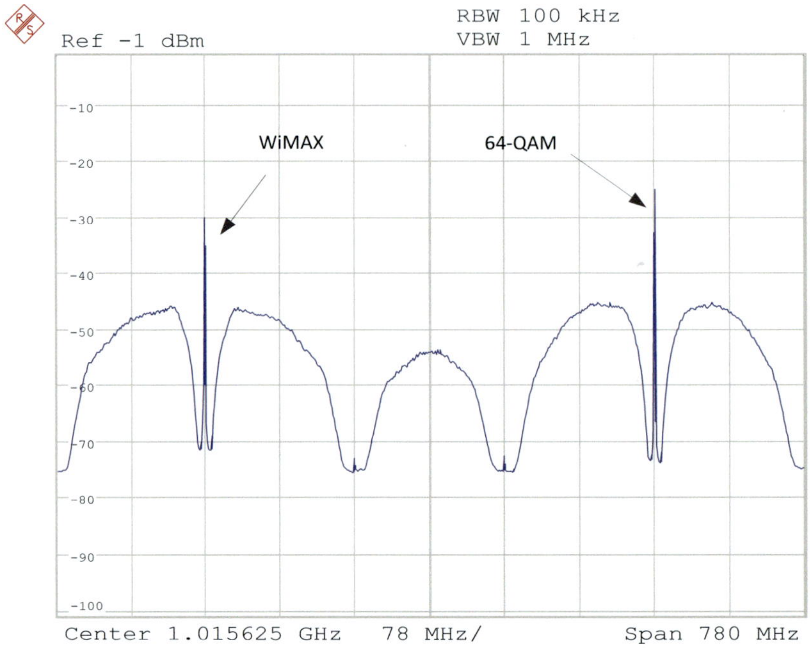


**FIGURE 3.12:** Measured multichannel spectrum of the ΔΣ-based transmitter.

The *DUC -0°* for the third independent path containing the WiMAX signal was configured to generate the following 20-bit output word:

$$w^{0} = [\text{-}v_i,\ \text{-}v_q,\ +v_i,\ +v_q,\ \text{-}v_i,\ \text{-}v_q,\ +v_i,\ +v_q,\ \text{-}v_i,\ \text{-}v_q,\\ +v_i,\ +v_q,\ \text{-}v_i,\ \text{-}v_q,\ +v_i,\ +v_q,\ \text{-}v_i,\ \text{-}v_q,\ +v_i,\ +v_{q,}].$$

For the last path, the DUC $-180°$ was configured to generate the following 20-bit output word:

$$w^{180} = [+v_i,\ +v_q,\ \text{-}v_i,\ \text{-}v_q,\ +v_i,\ +v_q,\ \text{-}v_i,\ \text{-}v_q,\ +v_i,\ +v_q,\\ \text{-}v_i,\ \text{-}v_q,\ +v_i,\ +v_q,\ \text{-}v_i,\ \text{-}v_q,\ +v_i,\ +v_q,\ \text{-}v_i,\ \text{-}v_{q,}].$$

These output words were converted into a 2-bit representation and then split so that each bit is fed to a separate serializer. At last, since the line rate of each serializer must be four times the carrier frequency, the first four serializers were set to operate at 5 Gbps, while the last four serializers were set to operate at 3.125 Gbps. After combining the outputs of the 8 serializers, it is possible to obtain the spectrum shown in Fig. 3.12.

#### 3.2.4.4 Comparative Analysis

In Fig. 3.13, the spectrum results for the PWM-based approach with and without pattern randomization, as well as the ΔΣ-based transmitter, are overlapped to facilitate a better comparative analysis between the implemented transmitters.

As can be seen in Fig. 3.13, the worst implementation in terms of coding efficiency is the ΔΣ-based, where the quantization noise is clearly higher than in the other approaches. The implementation with lower out-of-band noise within the 145 MHz span is the 7-bit PWM without pattern randomizer. However, this approach generates PWM harmonics with considerably high levels of energy spaced at the PWM sampling frequency. By spreading the

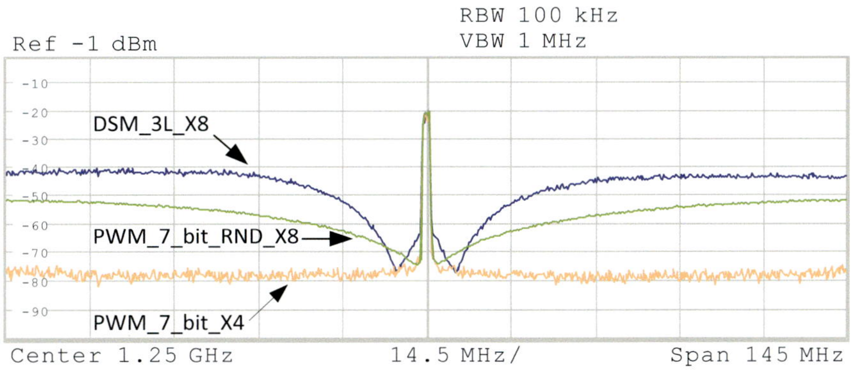


**FIGURE 3.13:** Overlapped measurement results comparing the PWM-based approach with and without pattern randomization ant the ΔΣ-based transmitter.

PWM noise across the frequency spectrum, the 7-bit PWM implementation with pattern randomization successfully eliminates the undesirable PWM harmonics.

The evolution of coding efficiency and modulation accuracy, according to the number of transmission paths used for the three implemented transmitters, is shown in Figs. 3.14 and 3.15, respectively. As anticipated by the previous results shown in Fig. 3.13, the ΔΣ-based transmitter approach has the lower coding efficiency and lower EVM. Moreover, the power combining of multiple independent paths with different phase noises also has a less significant impact in this approach, especially when combining more than four paths. A possible explanation may rely on the fact that in ΔΣ modulation, the variation of the output signal is higher than when using a PWM-based approach. This way, these higher variations at the output of the transmitter, combined with non-ideal output waveforms of the serializers, and also with small timing mismatches between the independent paths, are a possible explanation for this worse performance when combining a higher number of independent paths.

The PWM-based approach with pattern randomization achieves the best results in terms of modulation accuracy while maintaining high coding efficiency. In fact, discarding the typical 1.5 dB insertion losses of the power combiner, the total energy resulting from combining the 8 independent paths is only 1.4 dB below when compared with the ideal addition. This means that almost all the energy of the desired signal is added constructively in this proposed all-digital transmitter. At last, the PWM-based transmitter without pattern randomization has also proven to be an interesting approach when combining multiple independent paths, as evidenced by the high coding efficiency and modulation accuracy achieved.

A comparison between several state-of-the-art all-digital transmitters is shown in Table 3.1. As can be seen, all the proposed transmitters offer very high coding efficiency

**TABLE 3.1:** Comparison with other all-digital transmitter approaches

| | Johnson 2006 [101] | Hung 2007 [91] | Ghannouchi 2010 [72] | Schmidt 2011 [174] | PWM_7_bit | PWM_7_bit_RND | DSM_3L |
|---|---|---|---|---|---|---|---|
| Signal | W-CDMA | CDMA | WiMAX | N/D | 64-QAM | 64-QAM | 64-QAM |
| Cod. Eff. | 3.4% | 32.0% | 6.6% | 35.0% | 78.7% | 79.3% | 46.3% |
| SNR | 74.5dB | 43dB | 45dB | N/D | 50dB | 50dB | 45dB |
| Modulation | 2-level ΣΔ | 3-level ΣΔ | 2-level ΣΔ | 3-level ΣΔ | 7-bit PWM | 7-bit PWM | 3-level ΣΔ |
| Validation | Simulation | Simulation | Hardware | Simulation | Hardware | Hardware | Hardware |

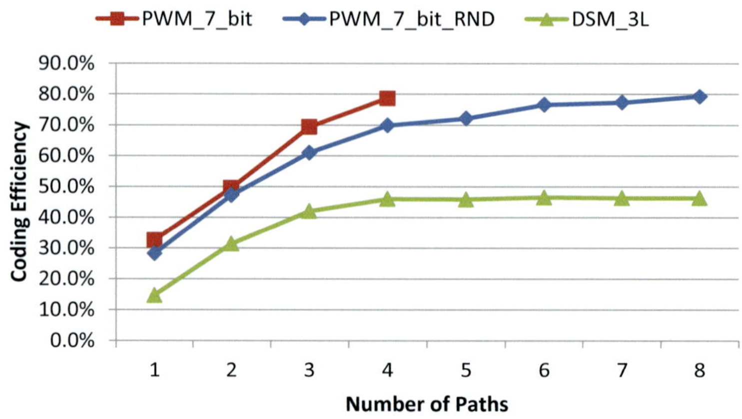


**FIGURE 3.14:** Coding efficiency evolution accordingly to the number of used transmission paths for the three implemented transmitters.

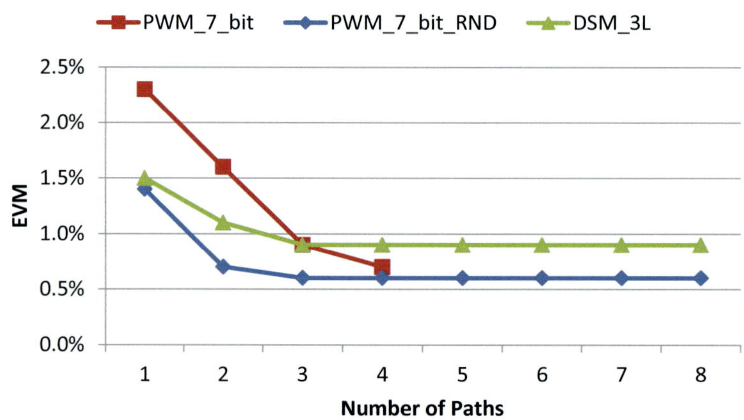


**FIGURE 3.15:** Error vector magnitude evolution accordingly to the number of used transmission paths for the three implemented transmitters.

and a reasonably high SNR, while maintaining the high flexibility inherent to all-digital RF transmitters.

The main occupied FPGA resources for the three implemented transmitters are shown in Table 3.2. The 7-bit PWM-based approach with pattern randomization is clearly the larger design, which is due to the additional hardware complexity for implementing the pattern randomization blocks. However, even for the larger design, the entire datapath takes only about 5% of the FPGA fabric, providing a large area available for other system functionalities and a vast degree of integration, including, for instance, the processing of baseband protocols.

### 3.2.5 Key Remarks

In this section, an all-digital transmitter architecture utilizing various pulse shaping techniques in combination with polyphase multipaths is presented. The proposed transmitter combines multichannel data transmission with high coding efficiency and SNR while maintaining the flexibility.

**TABLE 3.2:** Main occupied resources of the implemented FPGA-based all-digital transmitters.

| Logic resources | DSM_3L | PWM_7_bit | PWM_7_bit_RND |
|---|---|---|---|
| LUTs | 965 (1%) | 3513 (1%) | 11290 (5%) |
| Flip-Flops | 616 (1%) | 2803 (1%) | 9540 (2%) |
| RAM36E1 | 12 (2%) | 28 (2%) | 28 (2%) |
| GTHE1_QUADs | 2 (33%) | 1 (17%) | 2 (33%) |

From the three proposed variants, the 7-bit PWM-based transmitter with pattern randomization has the highest complexity. However, it provides the higher coding efficiency and the lower EVM. Moreover, the PWM noise spreading over the spectrum enables relaxing the design constraints for the RF reconstruction filter.

The obtained results also show the feasibility and potential of using FPGA-based polyphase multipath architectures for implementing digital RF transmitters.

## 3.3 LUT-based Modulators

In this section, we propose a change to the PWM technique used in ADTs that reduces the quantization noise power near the carrier frequency. This new technique enables the reduction of the quality factor for the transmitter output filter without altering the hardware complexity when compared to conventional PWM-based ADTs.

### 3.3.1 All-digital transmitter architecture

This subsection details the FPGA-based architecture used to implement an all-digital transmitter with PWM similar to the one presented in [183]. This architecture uses PWM modulation to generate a two-level RF output. The RF transmitter prototype block diagram is shown in Fig. 3.16.

The baseband complex input signal is divided into In-phase (I) and Quadrature (Q) data components, respectively, $s_i$ and $s_q$. Because the PWM has a fixed number of levels, in this particular case 64 levels, the signal must be quantized to a signal with the same

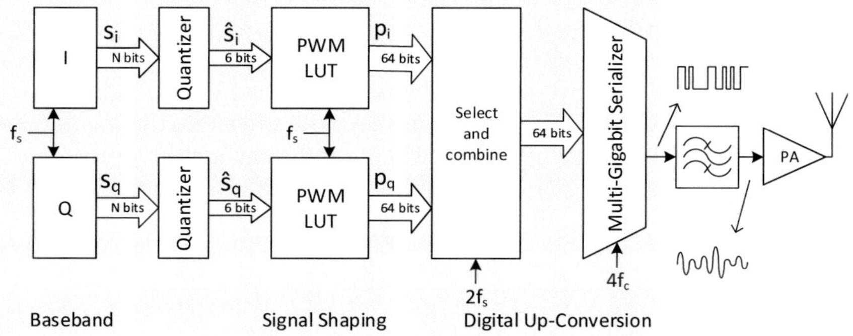


**FIGURE 3.16:** FPGA-based all-digital transmitter architecture with PWM modulation.

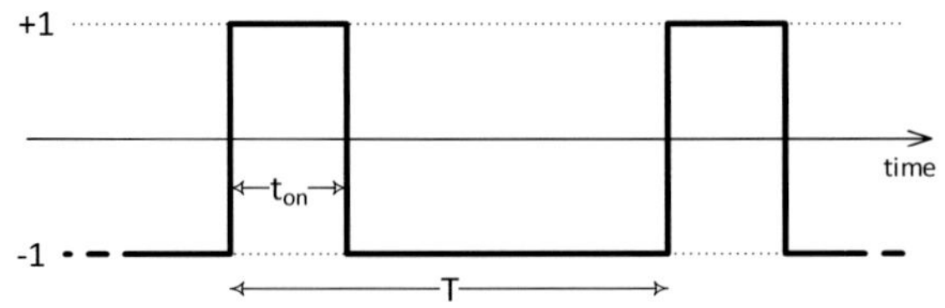


**FIGURE 3.17:** Rectangular PWM wave with period $T$ and duty cycle $t_{on}/T$.

number of levels. Hence, $s_i$ and $s_q$ are quantized into $\hat{s}_i$ and $\hat{s}_q$ signals with $log_2(64) = 6$ bits.

The PWM is done using a LUT with the associated PWM level matrix stored. Each memory line corresponds to a signal level, and its output has a rectangular waveform with the respective duty-cycle width. The digital PWM implementation forces the duty-cycle width to be discretized into possible 64 rectangular waves. The 64-bit outputs, $p_i$ and $p_q$, have a duty cycle defined by the number of '1's in the sequence.

In this architecture, the up-conversion is implemented using the select and combine block and Multi-Gigabit Serializer (MGS), also known as MGT. The select and combine generates a parallel vector that is the combination of $p_i$, $p_q$, and their inverted versions $\overline{p}_i$ and $\overline{p}_q$. The vector is defined as a repetition of the ($p_i$ $p_q$ $\overline{p}_i$ $\overline{p}_q$) pattern sampled at a $4 \times f_c$ frequency. The serialization of this vector with the multi-gigabit serializer allows a high sampling frequency at the system output while maintaining a lower clock frequency, $f_s$, for the baseband hardware blocks. Once serialized at a frequency $4 \times f_c$, this pattern is equivalent to the multiplication of the I and Q signals with two square waves at $f_c$, with 90-degree phase shift between them [226]. The ADT-generated output is the baseband PWM signal up-converted to an RF carrier at $f_c$.

### 3.3.2 Pulse width modulator design

This subsection explains the design of the regular PWM for the used ADT architecture and the proposed improvement to the modulator. The first part focuses on analyzing the regular PWM frequency response and the problems it poses for signal filtering. The second part presents the proposed improvement to the PWM method, aiming to reduce the harmonic distortion power closer to the carrier frequency.

#### 3.3.2.1 Conventional PWM

The uniform-sampling PWM is a nonlinear process that modulates a signal into a pulse time span. In time, a PWM signal is a rectangular wave, as in Fig. 3.17, with a variable duty cycle defined as $t_{on}/T$.

Considering the PWM high oversampling ratio, we can assume that for time spans closer to the period T, the input signal $s(t)$ is constant and the waveform can be approximated by a periodic rectangular wave. This approximation is sufficient for the purposes of this work. Hence, for a normalized signal $s(t) \in [-1, +1]$, it is possible to encode it as a PWM wave with the duty cycle given by Eq. (3.6).

$$\frac{t_{on}}{T} = \frac{s(t) + 1}{2}, \quad \frac{t_{on}}{T} \in [0, 1] \tag{3.6}$$

PWM quantizes a continuous amplitude signal into a two-level signal, but, as a strongly nonlinear process, it introduces distortion to the signal. Assuming a high oversampling ratio, an approximation to this time wave can be described by its Fourier series approximation

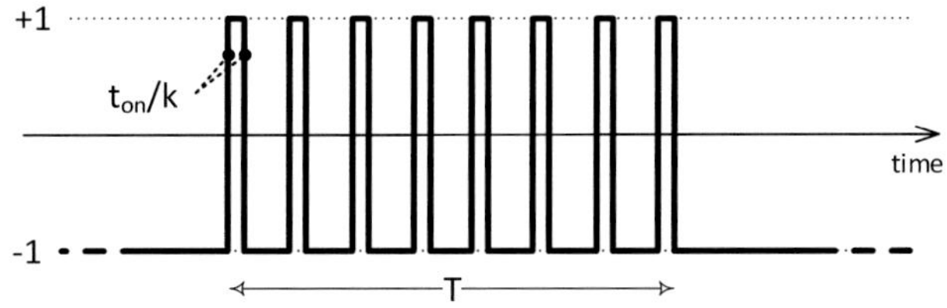


**FIGURE 3.18:** Improved PWM wave with period $T$ and duty cycle $t_{on}/T$, but divided into various rectangular waves to maximize the number of transitions.

$p(t)$, as in Eq. (3.7), with the Fourier coefficients $a_0$, $a_n$ and $b_n$ given by Eqs. (3.8) and (3.9), respectively.

$$p(t) = \frac{a_0}{2} + \sum_{n=1}^{\infty} a_n \cos\left(2n\pi\frac{t}{T}\right) + b_n \sin\left(2n\pi\frac{t}{T}\right) \tag{3.7}$$

$$a_0 = 4\frac{t_{on}}{T} - 2; \quad a_n = \frac{2}{n\pi}\sin\left(2n\pi\frac{t_{on}}{T}\right) \tag{3.8}$$

$$b_n = \frac{2}{n\pi}\left(1 - \cos\left(2n\pi\frac{t_{on}}{T}\right)\right) \tag{3.9}$$

The solution for $p(t)$ is expressed in Eq. (3.10). The resulting signal is a sum of the original modulating signal $s(t)$ and amplitude- and phase-modulated signals centered at integer multiple frequencies of $1/T$, here defined as $n_q(t)$.

$$p(t) = \underbrace{\frac{t_{on}}{T} - 1}_{s(t)} + \underbrace{\sum_{n=1}^{\infty} \frac{4}{n\pi}\sin\left(n\pi\frac{t_{on}}{T}\right)\cos\left(2n\pi\left(\frac{t}{T} - \frac{t_{on}}{2T}\right)\right)}_{n_q(t)} \tag{3.10}$$

The sum of the modulated signals $n_q(t)$ can be interpreted as the PWM quantization noise. The quantization noise magnitude in the frequency domain is given by its Fourier transform as in Eq. (3.11).

$$N_q(f) = 4\sum_{n=1}^{\infty} \frac{4}{n\pi}\left|\sin\left(n\pi\frac{t_{on}}{T}\right)\right|\left(\delta(f - \frac{n}{T}) + \delta(f + \frac{n}{T})\right) \tag{3.11}$$

From Eq. (3.11), it is clear that the PWM quantization noise is centered around integer multiples of $1/T$, with amplitude decreasing with the harmonic index $n$. Therefore, the higher magnitude noise peak is located near the signal, in the first harmonic $1/T$, which makes it the most crucial and also the hardest harmonic peak to filter.

#### 3.3.2.2 Improved PWM

Since the PWM noise power is higher for frequencies at multiples of $1/T$, it is desirable to diminish the value of $T$ to push the noise away from the signal. This can be achieved by increasing the sampling frequency of the PWM; however, the maximum frequency is typically limited by hardware clock constraints. Instead of increasing the sampling frequency, the rectangular wave period $T$ can be decreased by modifying the PWM waveform used to encode the signal.

The pulse width $t_{on}$ over the period $T$ can be divided into smaller time steps, while maintaining the relationship $t_{on}/T$ over the period $T$. Fig. 3.18 shows an example of the

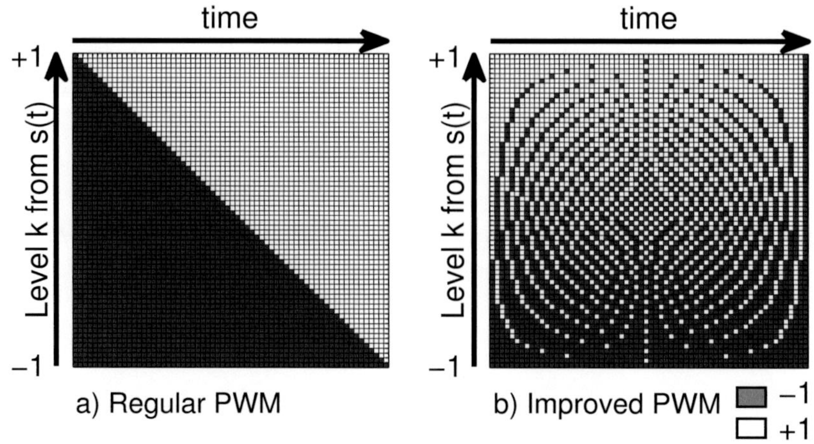


**FIGURE 3.19:** PWM LUT representation. (a) At the left it is presented the LUT used in the conventional uniform-sampled PWM. (b) In the right, the improved LUT to reduce the harmonic distortion near the carrier frequency.

conventional PWM wave, as presented in Fig. 3.17, modified according to the proposed approach.

The number of rectangular pulses over the period $T$ depends on the PWM level to be transmitted and the PWM time resolution. A PWM LUT with $N$ levels has a minimum pulse time duration of $T/N$. The $t_{on}$ value is given for the level $k$ (where $k$ is an $N$ level quantized form of the signal $s(t)$ accordingly to Eq. (3.6)) as $kT/N$, assuming $k$ is an integer from 1 to the PWM LUT size.

The improved PWM LUT in Fig. 3.19(b) was constructed by maximizing the number of transitions between -1 and +1 during the period $T$ while maintaining the relation $t_{on}/T$. The index values that have the value +1 in the improved PWM LUT can be found using the equation Eq. (3.12). The index values with amplitude -1 are the remaining $N - k$ time slots.

$$LUT^{i}_{+1} = round(\frac{N}{k} \times [1...k]) \tag{3.12}$$

Fig. 3.19 shows the differences between the regular and the proposed PWM LUT bitmaps. The gray squares represent the equivalent time slot with the value -1 and the white squares the ones with the value +1.

In Fig. 3.19 b), the time $t_{on}$ is evenly distributed over the PWM word period. This transformation changes the rectangular wave period but keeps the PWM word energy and preserves the relation $t_{on}/T$ over the word period. According to Eq. (3.10) and Fig. (3.11), reducing $T$ shapes the quantization noise to higher frequencies, away from the pulse-width modulated baseband signal.

### 3.3.3 Experimental results

A proof-of-concept prototype was designed using the architecture presented in Fig. 3.16, featuring both the regular uniformly sampled PWM and the proposed improved PWM. The FPGA-based ADT was implemented in a KC705 board from Xilinx with a Kintex-7 XC7K325T FPGA. The RF ADT two-level output signal was directly connected to a VSA from Rohde & Schwarz (model FSQ 8) for signal and spectral analysis. Several baseband

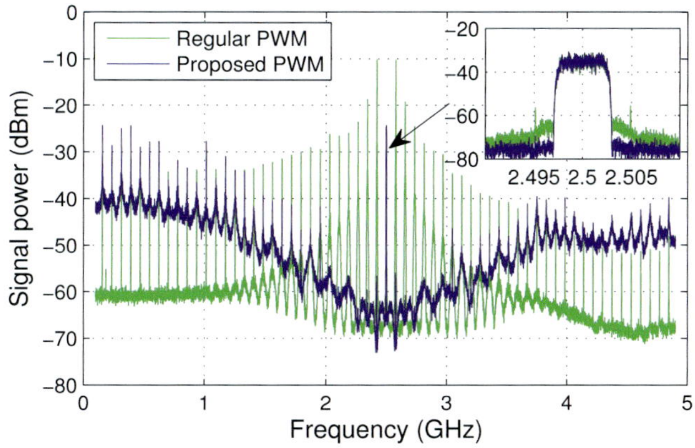


**FIGURE 3.20:** Measured ADT spectrum with regular pulse-width modulation and the proposed PWM. The carrier, centered at 2.5 GHz, can be viewed in more detail in the small figure in the top right corner. The resolution bandwidth is 300 WkHz, and the video bandwidth is 3 kHz.

signals were generated using MATLAB and loaded into the FPGAs internal I and Q memories. The ADT up-conversion was set to shift the baseband signal to a 2.5 GHz carrier frequency. The maximum PWM sampling frequency achievable for a 64-bit PWM word, as in Fig. 3.19, is 78.125 MHz.

The measured spectrum results at the FPGA output for the regular PWM-based ADT signal and the proposed PWM are depicted in Fig. 3.20. The harmonic noise for the proposed method is shaped away from the carrier frequency as intended, with most of the quantization noise power concentrated in the farthest harmonics.

A detailed view of the harmonic noise reduction for the twelve PWM harmonics closer to the carrier frequency is shown in Fig. 3.21 over a 1 GHz span. It is easily observable that the difference between the regular PWM and the proposed PWM spectral noise. In fact, for the presented signal, the peak noise power for the regular PWM was 18 dB above the carrier located at the first harmonic. On the other hand, with the proposed method, the

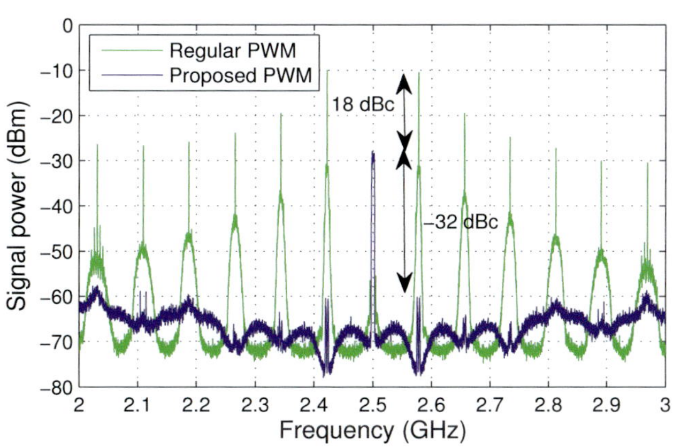


**FIGURE 3.21:** Detailed view of the peak power noise from the regular PWM and the proposed PWM. Resolution bandwidth is 100 kHz, and video bandwidth is 10 kHz.

**TABLE 3.3:** PWM noise power peak reduction

| | | Carrier Offset | Test Signal Bandwidth 3MHz | 6MHz | 12MHz |
|---|---|---|---|---|---|
| Lower harmonics | 6th | -468.7 MHz | 28.2 dB | 31.2 dB | 32.5 dB |
| | 5th | -390.6 MHz | 32.6 dB | 32.2 dB | 36.2 dB |
| | 4th | -312.5 MHz | 33.0 dB | 33.7 dB | 34.8 dB |
| | 3rd | -234.3 MHz | 39.3 dB | 42.3 dB | 39.9 dB |
| | 2nd | -156.2 MHz | 46.5 dB | 44.0 dB | 42.8 dB |
| | 1st | -78.1 MHz | **51.8 dB** | **50.5 dB** | **45.5 dB** |
| Higher harmonics | 1st | 78.1 MHz | **52.1 dB** | **49.7 dB** | **45.3 dB** |
| | 2nd | 156.2 MHz | 44.7 dB | 44.6 dB | 41.8 dB |
| | 3rd | 234.3 MHz | 38.5 dB | 40.7 dB | 39.5 dB |
| | 4th | 312.5 MHz | 30.5 dB | 33.1 dB | 33.8 dB |
| | 5th | 390.6 MHz | 31.5 dB | 31.2 dB | 34.9 dB |
| | 6th | 468.7 MHz | 25.5 dB | 28.2 dB | 30.3 dB |

peak noise power is now more than 30 dB below the carrier, and it's located at the sixth harmonic.

Table 3.3 shows the peak noise power reduction between regular PWM and the proposed PWM for three different QAM signals with 3 MHz, 6 MHz, and 12 MHz of bandwidth at 2.5 GHz. The measured values suggest that the peak quantization noise reduction reduces as the signal bandwidth increases. However, it should be noted that the noise reduction in the first harmonic was more than 45 dB for all tested signals.

### 3.3.4 Key Remarks

In this section, a new pulse-width modulation method for ADTs was presented. Its complexity is the same as regular PWM, but it allows for drastically reducing the harmonic quantization noise near the carrier frequency. Results show not only the feasibility of the proposed PWM-based ADT approach, but also the reduction of the quality factor requirements for the output RF reconstruction filter.

## 3.4 Resource Usage Reduction Through the Use of LUTs

As stated in the previous section, to achieve wider bandwidths and a broader range of carrier frequency agility, the sampling frequency must be increased. To achieve that last one, more parallel cores must be implemented, leading to a substantial increase in resource usage, which is supported only by the largest FPGAs. To overcome this problem, a solution is proposed in [223] to generate binary sequences that approximate the up-conversion and $\Delta\Sigma$M. These binary sequences are then loaded into a LUTindexLUT-based modulators. The baseband signal is used as the table address to recover the appropriate output binary sequence for that input. This technique allows not only for complex $\Delta\Sigma$M modulations to be implemented in lower-grade FPGAs but also to reduce the power consumption due to the lower usage of digital signal processing blocks. In [224], this technique was improved

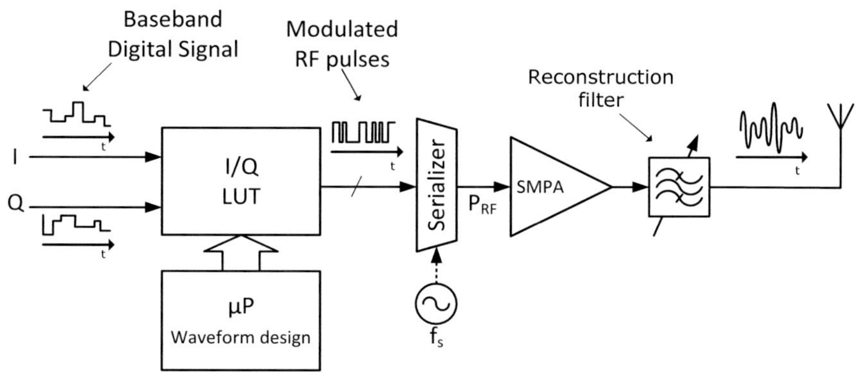


**FIGURE 3.22:** Block diagram of the proposed all-digital transmitter.

to support up to two carriers. However, the presented implementation still used 510 Block Random-Access Memorys (BRAMs) and 88 DSPs. The amount of resource usage presented is acceptable for a single antenna implementation. However, 6G is expected to rely heavily on massive MIMO, making this implementation unsuitable for the next generation of mobile telecommunications [115].

Resource usage needs to be brought further down to make ADTs suitable for 6G massive-MIMO applications. For that, an improved technique to implement a LUT-based ADT will be presented in this section. This technique utilizes software to generate the LUT offline efficiently and flexibly, while leveraging hardware to achieve high performance at the RF carrier frequency and bandwidth. This software and hardware combination enables an adaptable transmitter that can switch between carrier frequencies, bandwidths, and/or coding efficiencies. Ultimately, this transmitter offers a simpler approach to implementing an RF transmitter based on SDR.

### 3.4.1 RF Waveform LUT Transmitter Architecture

In this work, a bandpass $\Delta\Sigma$M is used to modulate a baseband input signal using a behavioral model. This behavioral model enables the storage of a LUT in the hardware that represents the behavior of the transmitter, without directly implementing its complex hardware. This LUT also encodes the digital upconversion and upsampling to further reduce the real-time computational hardware needed to implement an ADT in FPGA. Each address in the LUT stores the RF digital pulses representing the output that the real-time transmitter would produce for a given input signal.

In Fig. 3.22, a block diagram of the proposed system is presented, where the input signal acts as the address of a LUT, and where the RF pulses that correspond to every I/Q combination possible are stored. Then, those RF pulses pass through the serializer, which acts as a DAC.

The values of the RF pulses are calculated using a pre-determined algorithm that runs offline in a separate processor. After the algorithm finishes calculating all values, the latter are stored in the RAM on the FPGA, which will act as the LUT. When the transmission is active, the baseband signals act as the LUT address, where the I and Q from the baseband signal are concatenated to form the address signal. This address signal then retrieves the corresponding RF wave from the LUT. Since only one sample addresses the LUT, the number of bits retrieved from it represents the OSR present in the final signal.

The LUT waveforms synthesis will be described below.

#### 3.4.1.1 Look-up table generation algorithm

The LUT is generated using an algorithm that performs a narrow time window simulation of the ΔΣM. This is done offline, and the output of the ΔΣM is calculated for all possible I/Q pairs of the input signal. This calculation must be done each time a new carrier frequency, $f_c$ is needed. And so, each new carrier frequency results in a new LUT that is stored in the FPGA. Since the algorithm runs offline in software, there is no need to reprogram the FPGA. Moreover, since the LUT generation is performed separately from signal transmission, switching between different carriers can be done without significant downtime. Another advantage of this LUT approach is that the transmission delay does not depend on the modulator's complexity. This allows for the use of more complex modulators that offer better performance without the added hardware complexity associated with such modulators. The low complexity, modulation, and up-conversion times make this LUT-based ADT ideal for latency-sensitive applications. Another advantage is that only the ΔΣM algorithm in the simulation needs to be changed to modify the modulator topology. This algorithm change can be implemented in the same manner as the change in carrier frequency, making it fast and requiring no hardware modification.

The algorithm used to generate the RF pulses is the following:

- Quantize the signals I and Q with $N$ levels. This results in $I_q$ and $Q_q$, the quantized versions of $I$ and $Q$, respectively.
- Compute the complex carrier wave, $CW$, for the corresponding carrier frequency $f_c$ with $M$ samples.
- For each new carrier frequency $f_c$, calculate the complex carrier wave $CW$ with $M$ samples

$$CW_i(n) = \cos\left(2\pi\frac{fc}{fs}n\right); \quad CW_q(n) = \sin\left(2\pi\frac{fc}{fs}n\right); \tag{3.13}$$

- With $I_q$ and $Q_q$ constant, calculate the RF wave for $M$ samples.

$$S_{RF}(n) = I_q \times CW_i(n) + Q_q \times CW_q(n); \tag{3.14}$$

- Using delta-sigma modulation, convert the RF signal into a pulsed one.

$$P_{RF}(n) = DSM(S_{RF}(n)); \tag{3.15}$$

- Store RF wave in the LUT.

$$LUT(I_q, Q_q, n) = P_{RF}(n); \quad n \in \{0; 1; ...; M-1\} \tag{3.16}$$

To complete the LUT, the algorithm needs to run for all $[I_k; Q_k]$ pairs. Due to the transitory response of the filter present in the ΔΣM loop, the input signal $S_{RF}$ is repeated three times and processed by the ΔΣM. Only the last $M$ samples of the result from the ΔΣM are used and passed to the $P_{RF}$ to be saved in the LUT. After that, the LUT upconverts the signal into the desired carrier frequency. Fig. 3.23 is visualization of a LUT for $N = 4$ quantization levels, where each line represents the output of an $[I_k; Q_k]$ pair.

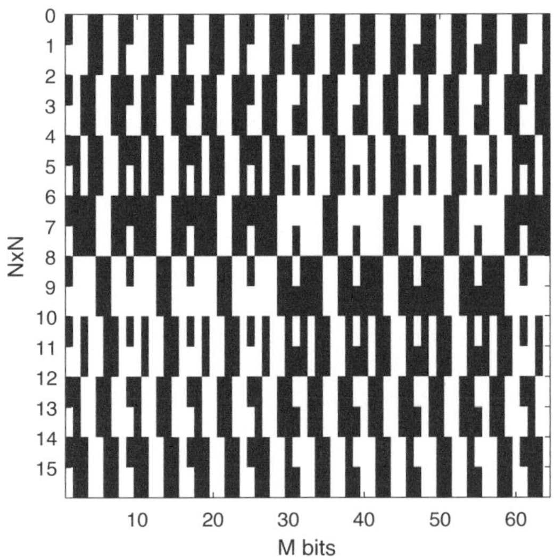


**FIGURE 3.23:** Depiction of a LUT generated for a 2-bit input signal ($N = 4$).

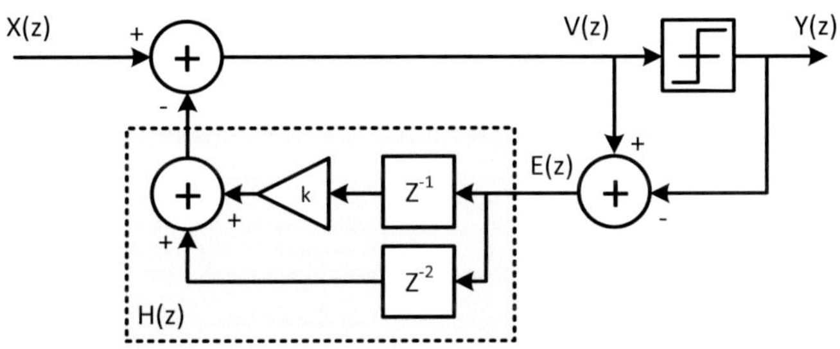


**FIGURE 3.24:** Block diagram of a Delta-Sigma modulator feedback loop architecture.

#### 3.4.1.2 Modulator Topology

Despite being a well-known technique already used as a digital DAC, ΔΣM has some limitations when used for RF transmission. One limitation is the relatively low sampling frequency that can be achieved due to the architecture of most modulators. The loop in the modulator design results in a high critical path delay, which limits both the transmitter's maximum carrier frequency and bandwidth. Leading many authors to implement parallel architectures to increase the overall sampling frequency [36].

A depiction of the architecture of the ΔΣM used in this work can be seen in Fig. 3.24. In the figure, $X(z)$ represents the input signal, $Y(z)$ represents the output of the modulator, which is the quantized version of $X(z)$, $E(z)$ represents the feedback error which, prior to be subtracted back from the input signal, must be filtered by $H(z)$. This subtraction of the already filtered error signal from the input signal means that both will appear on the final quantized output signal. This will be cleared by analyzing Eqs. (3.17) and (3.18).

$$Y(z) = X(z) - \underbrace{(1 + H(z))}_{NTF(z)} E(z) \tag{3.17}$$

$$H(z) = NTF(z) - 1 \tag{3.18}$$

where $NTF(z)$ represents the noise transfer function of the modulator. This $NTF(z)$ is responsible for the noise shaping in the final quantized signal.

The time it takes to process the loop filter, $H(z)$, will dictate the maximum frequency at which the ΔΣM can operate. Increasing the order of the filter $H(z)$ is desirable due to improved in-band noise rejection. However, higher-order filters will require more digital logic to be implemented, which increases the critical path and reduces the maximum frequency of operation. As mentioned above, parallel polyphase implementations of ΔΣM were developed to address the sampling frequency issue. These implementations enabled a low clock rate (operational frequency) in the FPGA while providing a higher equivalent sampling ratio. However, these implementations are resource-intensive and do not scale properly [36]. To address the scalability problem, [37] and [46] presented a ΔΣM core-based polyphase approach where the number of cores was limited by the size of the FPGA and its ability to drive those cores at a reasonably high clock rate. These types of implementations allowed for the reach of GHz-range carrier frequencies. However, they also exacerbated the resource usage problem.

The main advantage of the proposed architecture is its independence from the modulator's complexity in terms of maximum frequency of operation. This is because the modulation is performed offline, not in real-time. Considering this, this architecture allows for a higher-order filter to be implemented even in a low-end FPGA.

The block diagram of a ΔΣM with a simple filter can be seen in Fig. 3.24. This filter uses two delays and a gain to create a bandpass NTF. The filter can be defined as:

$$H(z) = kz^{-1} + z^{-2} \tag{3.19}$$

From both Eqs. (3.18) and (3.19), the final NTF response of the ΔΣM is given by:

$$NTF(z) = 1 + kz^{-1} + z^{-2} \tag{3.20}$$

It is possible to look at the $NTF(z)$ as a filter. In this case, this filter would have two complex conjugate zeros. Solving for its Fourier transform, its frequency response is given by:

$$NTF(f) = NTF(e^{j2\pi}) = 1 + ke^{-j2\pi} + e^{-j2\pi\times 2} \tag{3.21}$$

The zeros on the previous equation, Eq. (3.21), depend on $k$ according to:

$$k = -2\cos(2\pi f_0); f_0 \in [0; 0.5] \tag{3.22}$$

where $f_0$ is the normalized frequency location for the NTF response zero. This means that by changing the value of $k$, the frequency response of the NTF can be changed, essentially making this transmitter tunable in frequency. The final simulated spectrum from this transmitter can be seen in Fig. 3.25.

The input signal must be upconverted to the NTF zero's frequency. A fully tunable ΔΣM-based ADT is achieved by changing both the frequency of upconversion and the NTF zero. As a result, a pulsed RF signal with carrier frequency from DC to $f_s/2$ can be transmitted.

#### 3.4.1.3 Look-up table model

The equivalent system implemented using a LUT can be seen in Fig. 3.26. Note that the architecture represented in the figure is not directly implemented in the FPGA. Still, it is an equivalent system that does approximately the same as the final LUT implementation. The behavior is not strictly the same because a ΔΣM modulator has an infinite pulse length response, while our LUT implementations have a finite one.

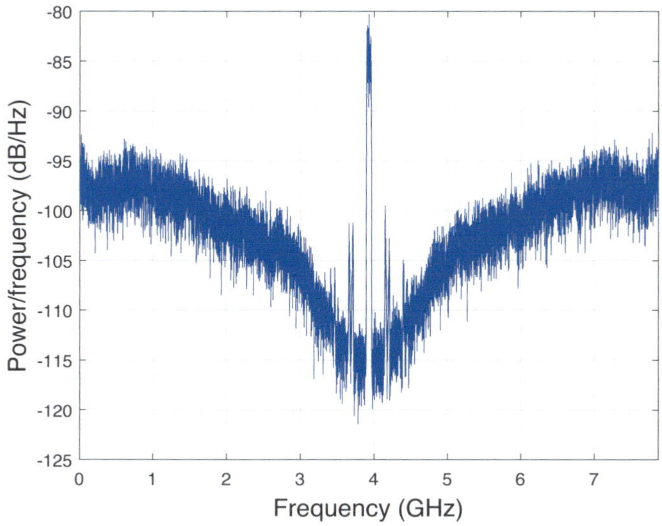


**FIGURE 3.25:** Simulation of the output spectrum of the proposed LUT-based transmitter using the complex conjugated zeros NTF presented in Eq. (3.21), using a 100 MHz bandwidth signal and a final DSM sampling of 15.72864 GHz.

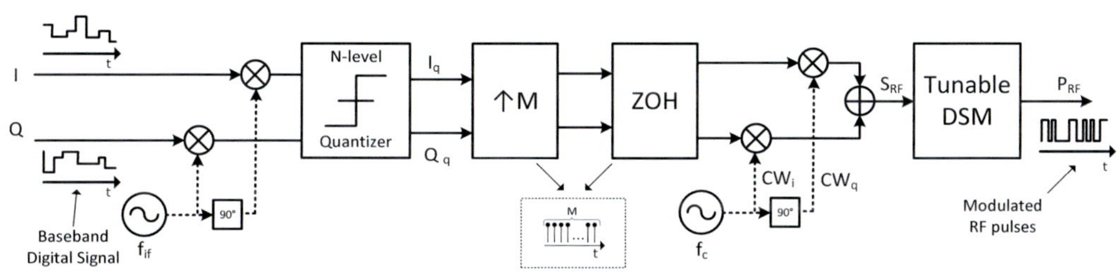


**FIGURE 3.26:** Equivalent signal processing system for the proposed hardware architecture in Fig. 3.22 using the proposed LUT generation algorithm.

The following is a detailed explanation of all blocks from the system presented in Fig. 3.26, which represents an equivalent model of the transmitter implemented in the LUT.

*Pre-upconverter*

The first block's function is the upconversion to an intermediate frequency, which allows the final transmitter to reach frequencies that are not an integer multiple of the original signal's sample frequency.

*Quantizer*

Then follows the signal's quantization to the specified number of levels $N$. The size of the LUT will depend on $N$, with the number of entries in the LUT being equal to $N \times N$, since for every I sample, there are $N$ possible Q samples, and, as was already shown, both I and Q have $N$ possible values. This value will also limit the maximum achievable SNR, with the SNR of a signal with $N$ quantization levels and no noise shaping being given, according to [100], by:

$$SNR_{dB} = \underbrace{20\log(N)}_{quantization} + \underbrace{1.76 + 10\log\left(\frac{f_{bb}}{2BW}\right)}_{process\ gain} \tag{3.23}$$

However, our transmitter will implement a first-order ΔΣM which shapes the noise, and so, Eq. (3.23) can be further improved [100]:

$$SNR_{dB} = \underbrace{20\log(N)}_{quantization} + \underbrace{1.76 - 5.17 + 30\log\left(\frac{f_{bb}}{2BW}\right)}_{first\text{-}order\ process\ gain} \tag{3.24}$$

From Eq. (3.24), it is possible to see that the final SNR value depends on two factors, the number of quantization levels, $N$, and also on the first-order process gain. This first-order process gain increases with $30 \times log(OSR)$, where $OSR = f_{bb}/2BW$. However, in Eq. (3.23), this gain only increases with $10 \times log(OSR)$. This means that the noise shaping has a 3 times gain improvement in terms of SNR compared to a similar signal without noise shaping. Higher-order noise shaping techniques will present an even larger SNR increase for the same OSR, hence the great appeal of higher-order modulators.

*Interpolation*

For the signal to be digitally up-converted to RF, it must be interpolated. In this case, the final sampling rate needs to be $M$ times the baseband sampling rate; this is $f_s = M \times f_{bb}$. A zero-order hold is used to implement this interpolation, which means that the input signal is kept constant for $M$ samples. This $M$ value is a very important parameter in this type of LUT-based ADT because it dictates the size of the RF wave output from the LUT. Larger values of $M$ will allow for a better approximation of the ΔΣM. However, larger values of $M$ also limit the maximum achievable signal bandwidth, assuming the same final sampling rate.

*RF Upconversion*

With the signal already interpolated, the digital upconversion can be done. A quadrature digital up-conversion scheme was used following the operation presented in Eq. (3.14). After the upconversion, the signal is already sampled at a high rate and is in its carrier frequency. However, a reduction in the number of quantization bits is needed before it can be sent to the analog domain using a one-bit DAC.

*Delta-Sigma Modulation*

A ΔΣM is used to reduce the number of quantization levels of the RF signal. After the ΔΣM, the signal is a binary wave, and it can be stored in the LUT to be addressed by the input baseband signal, with the final LUT size being $N \times N \times M$ bits.

Due to the limited size of FPGAs, a trade-off must be made between the values of $M$, $N$, and the total size of the LUT. As seen in Fig. 3.27, the higher both $M$ and $N$ are, the higher the final SNR. However, as seen previously, the LUT size depends on $N \times N \times M$, which means that increasing either $M$ or $N$ will increase the LUT size.

Fig. 3.27 presents the transmitter simulation, at the system level, for multiple values of both $N$ and $M$, for a 5 MHz signal sampled at 100 MHz, using a band-pass second-order ΔΣM. In this simulation, the hardware is considered to be ideal. From the figure, Fig. 3.27, it is possible to see that the SNR varies greatly with the change in both $N$ and $M$. This

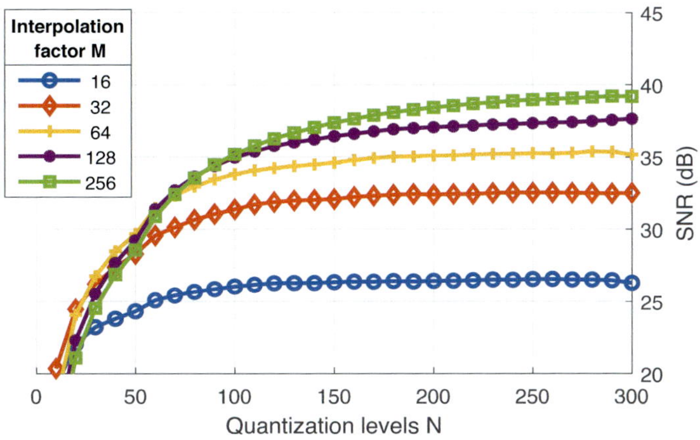


**FIGURE 3.27:** Variation of the proposed transmitter SNR depending on the LUT size in terms of the factors $N$ and $M$ (system simulation results), for a 5 MHz signal sampled at 100 MHz.

means that it is crucial to select the right values. Furthermore, the image also shows that for values of $M$ up to 64, there is not much to gain by increasing the quantization levels past $N = 128$ (7-bits), and so, for those interpolation values, increasing the quantization levels past that value would only increase the LUT size without any gains in terms of signal quality. This will also be demonstrated experimentally and further analyzed in the results section.

### 3.4.2 Reducing the LUT Size

To achieve the lowest possible resource consumption, a new technique to reduce the LUT size will be presented. This technique, which we named "Half LUT Implementation", arises from the inverse symmetry present in the LUT (see Fig. 3.23). In the image, the black rectangles represent a '0', and the white ones represent a '1'. As can be seen, inverting the colors (bits) of the lower half of the image allows for the creation of a fully symmetrical pattern. This color inversion represents a bit inversion in the LUT, which enables the dynamic generation of the lower half of the LUT from the top half, eliminating the need to store the lower half in memory. Another way to understand this symmetry is to analyze (2), where we can see that the input signal to the ΔΣM is comprised of two components, the in-phase and the quadrature component. If we make $I_q = -I_q$ what we are doing is imposing a 180° phase rotation to this signal. However, the magnitude will be the same. This is also what we do when we change the most significant bit in the address signal from '0' to '1' or vice versa. This means that $S_{RF}$ will be inverted by 90°, and to rotate $S_{RF}$ by 180°, we also need to rotate $Q_q$ by 180°. This rotation is not actually performed, and it naturally appears in the LUT due to the way it is constructed. Specifically, in the LUT construction, $I_q$ is kept constant, and all values of $Q_q$ are swept before proceeding to the next $I_q$. This can also be seen in the address signal, which is comprised of the concatenation of the bits of the signals $I_q$ and $Q_q$, with the bits from $I_q$ taking place in the most significant bits of the LUT's address. This means that the LUT's first line will be constructed from the signals $I_q$ and $Q_q$, whose phases are rotated 180° when compared to the last line. A representation of the LUT without the inversion of the bits of the lower half can be seen in Fig. 3.28 (a).

The input signal used as the LUT address is still the same, so it still has $2 \times log_2(N)$ bits. However, the LUT now has only $log_2(N \times (N-1))$ entries. This means that the most

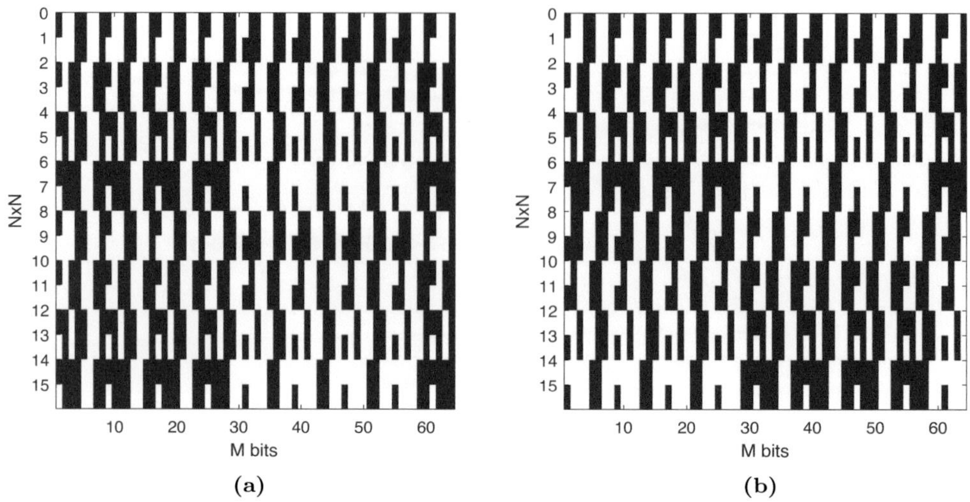


**FIGURE 3.28:** LUT lower half without bit inversion (a), with bit inversion (b).

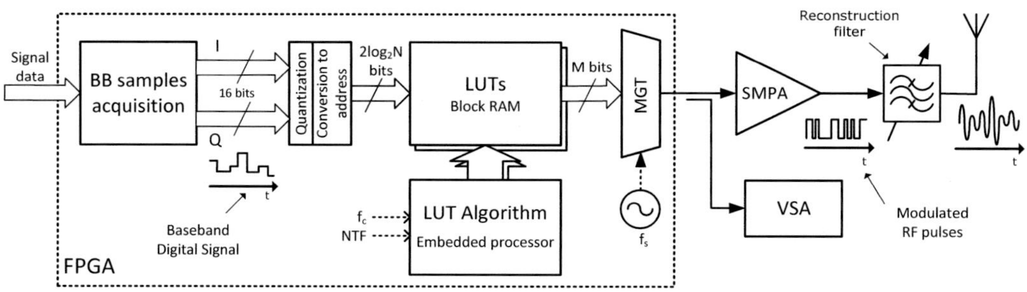


**FIGURE 3.29:** FPGA-based prototype hardware architecture of the proposed transmitter.

significant bit of the input signal can be used to determine if we are in the top or lower half of the LUT. After that, assuming we are in the top half, the address that goes to the LUT is the remaining bits of the input signal. On the other hand, if we are in the lower part of the LUT, we need to invert the input signal bits before using them as the address. This inversion of the input signal's bits is necessary to invert the half LUT. Without this final address inversion, the LUT would be as depicted in Fig. 3.28 (b). In the end, this technique saves half of the LUT size, using only two inverts (not operations).

### 3.4.3 Hardware implementation

The proposed transmitter architecture was implemented in an FPGA as depicted in Fig. 3.29. The baseband signal is received in the form of I/Q samples. The frequency of the MGT dictates the final sampling frequency of the $\Delta\Sigma$M, $f_s$, which determines the baseband frequency by the following relation: $f_{bb} = f_s/M$. After a specific value of the frequency of the MGT is set, the $f_s$ and $M$ values are fixed. In this work, the MGT sampling frequency is always kept constant, which means that $f_s$ and $M$ are also constant.

Due to the signal being received with 16 bits, a truncation prior to the LUT is required. This truncation is a source of error, which degrades the signal. However, because the original

**FIGURE 3.30:** Laboratory setup, with the computer, clock generator, FPGA (red) and VSA.

resolution is 16-bits and the final quantization bits will be around 7, since the number of bits being equal to $log_2(N)$ and $N$ was already mentioned to be around 128, this source of error was deemed acceptable. After this truncation, both I and Q signals are concatenated and used as the index to the LUT. This LUT is implemented using block RAM memory in order to be able to fit a larger one into the FPGA. In this LUT a binary signal containing the quantized RF wave is stored as 0's and 1's. Since this information is stored as words of length $M$, it is the responsibility of the MGT to serialize this word in order to recover the final RF signal.

### 3.4.4 Results

Fig. 3.30 shows a laboratory prototype that was implemented as a proof of concept. Encircled in red in Fig. 3.30 is the board where the ADT was implemented. This board is a ZCU102 from Xilinx with the XCZU9EG FPGA. In green is the clock generator responsible for the input clocks that drive the MGT, in yellow is the VSA from Rohde & Schwarz (model FSW 8), and in blue is the personal computer responsible for the FPGA's programming and for the data acquisition from the VSA.

The output sampling frequency for this implementation is determined by the MGT, which is set to operate at 15.72864 Gsps, resulting in a maximum achievable carrier frequency of approximately 7.8 GHz. The MGT was set to receive 64 bits in parallel. This number corresponds to the wave size $M$, which implies that the baseband sampling frequency $f_{bb}$ is 245.76 MHz. With the specified value of $M$, Fig. 3.27 was used in order to select the value of $N$ such that the SNR was not greatly compromised on one hand, and on the other hand, the LUT was not too large. It is possible to see from Fig. 3.27 that for $M = 64$ the SNR does not increase significantly when $N$ goes above 150. Therefore, three values of $N$ were chosen: 64, 128, and 256, which are powers of 2 and allow for the full utilization of the input bits while simplifying the quantization process through simple truncation. This value choice resulted in three LUT implementations that will be referred to as 6, 7, and 8 bits for $N = 64$, 128, and 256, respectively.

Several signals were generated using Matlab and loaded into the board's RAM. These signals were 16-QAM and 256-QAM modulated at data rates of 8.192, 24.576, and 122.88 Msps, corresponding to bandwidths of 9.8, 29.5, and 147.5 MHz, respectively. The RF ADT's binary output signal was connected to the VSA for signal and spectral analysis, as depicted in Fig. 3.29.

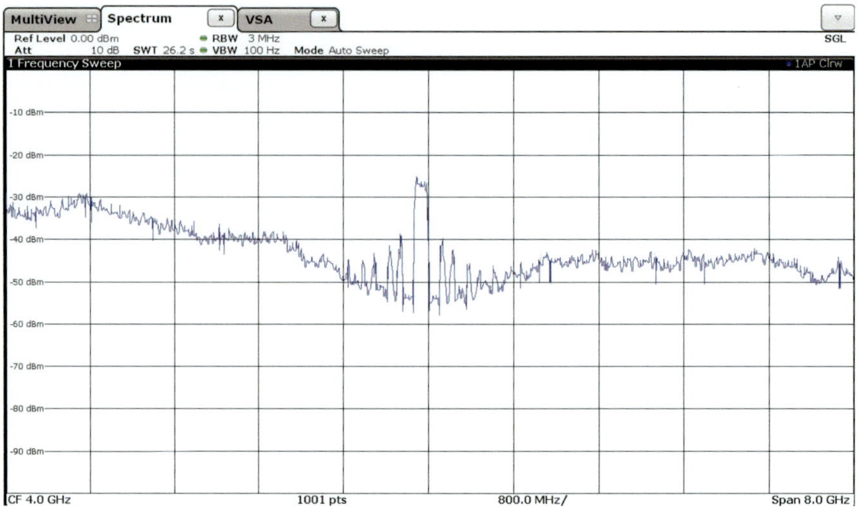


**FIGURE 3.31:** Spectrum of the proposed all-digital transmitter. The image shows a 147.5 MHz signal bandwidth at 3.93216 GHz.

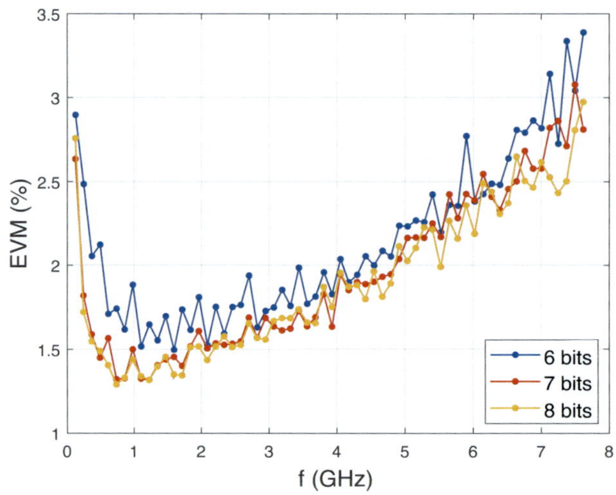


**FIGURE 3.32:** Comparison of the performance of the LUT with multiple word lengths for the input signal. The signal used is a 16-QAM at 24.576 Msps.

### 3.4.4.1 Optimization of the Number of Quantization Levels

Fig. 3.31 shows an example of the output spectrum after the MGT. The figure shows the full spectrum from DC to 8 GHz. This spectrum, if directly radiated, would violate the transmission masks. However, an analog filter can be used to ensure the signal complies with spectral masks.

As mentioned above, the expected best value for the number of quantization levels was between $N = 64$ and 256, with $N = 128$ being the prime candidate. To test this theory and, by extension, the three separate LUT implementations, a 24.576 Msps 16-QAM signal was created and used for testing. For all three implementations, the carrier frequency was varied from 0.1 GHz up to 7.62 GHz. The results of these tests can be seen in Fig. 3.32.

**TABLE 3.4:** Resource Usage of the different implementation of LUTs .

| | BRAMs | LUTs | DSPs |
|---|---|---|---|
| $N$=64 (6 bits) | 8 | 29 | 0 |
| $N$=128 (7 bits) | 29 | 100 | 0 |
| $N$=256 (8 bits) | 114 | 395 | 0 |

The results of the tests presented in Fig. 3.32 show that, as predicted by the simulation, the optimum number of quantization bits is 7 ($N = 128$). As can be seen in the image, the EVM values for both 7 and 8 bits are very close, indicating that the additional number of quantization levels did not significantly improve the final results. On the other hand, the EVM results for 6-bits of quantization are higher than for 7 or 8 bits, which means that there is a penalty in the performance when the quantization levels are lowered from 128 to 64. However, the performance loss is not so significant as to render the output signal useless; therefore, an evaluation of the reduction in resource usage compared to the performance loss is needed.

In terms of performance loss, the EVM presented by the 6-bit implementation ($N$=64) is around 10% worse than when compared to the one presented by the 7-bit ($N$=128) implementation; however, in terms of resource usage, the LUT of the 6-bit implementation uses around a quarter of the BRAMs and 29% of the LUTs used by the 7-bit implementation. This represents a significant reduction in resource usage, which translates into a trade-off that can be highly beneficial, especially for low-end FPGAs. With this trade-off, 90% of the 7-bit transmitter performance can be retained while using only around 25% of the resources. In Table 3.4, it is possible to see the resource usage for all three implementations. In the table, an almost linear increase in resource usage by a factor of approximately four is observed with the increase in the number of quantization bits used. Taking all previous results into consideration, all following tests are done using $N = 128$ levels (7-bits). This implementation was chosen because of its negligible performance loss when compared to the 8-bit implementation; moreover, the 6-bit one was not chosen due to the exact opposite, that is, it presents a performance loss when compared to the 7-bit implementation.

#### 3.4.4.2 Simulation vs Measurements

To better understand the behavior presented by our implementation, a simulation was made, the results of which are presented in Fig. 3.33. As shown in the figure, an increase in EVM is present in the final measurements as the frequency rises; however, it is not present in the simulation. This fact suggests that the increase in EVM is caused by RF constraints as the frequency rises, rather than being inherent to the LUT implementation. Moreover, the results at lower frequencies are very close to the simulated ones, proving that our implementation is working as expected.

#### 3.4.4.3 Half LUT Implementation Results

With our results validated both in simulation and through measurements, it was time to test the "Half LUT Implementation". An EVM sweep from 0.1 GHz to 7.62 GHz, using the same 24.576 Msps signal used in the prior experiments, is presented in Fig. 3.34. The overlap in the presented EVM values for both implementations clearly shows that the "Half LUT Implementation" provides a no-compromise solution in terms of performance, while utilizing only half of the resources. For reference, the implementation using the entire LUT uses 29 BRAMs, while half LUT uses only 15, the same happens in the LUTs used in the FPGA, where we go from 100 to 46. Moreover, the implementation still uses zero DSPs.

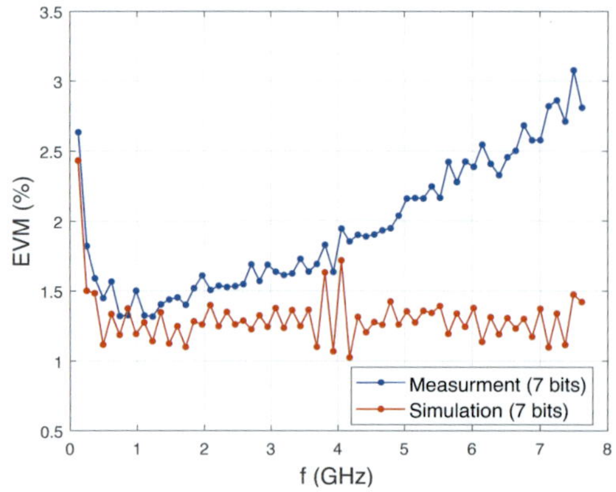


**FIGURE 3.33:** Comparison between the simulation and measurement results for the implementation with $N = 128$ (7 bits). The signal used is a 16-QAM at 24.576 Msps.

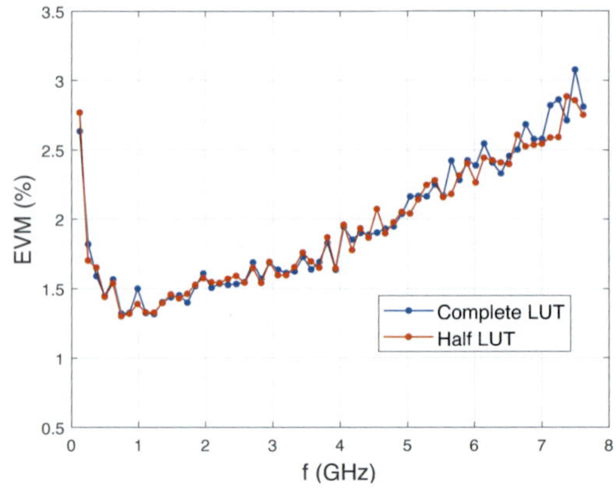


**FIGURE 3.34:** Comparison of the EVM between the full and half LUT implementations, using $N = 128$. The signal used is a 16-QAM at 24.576 Msps.

With this validation, the final two EVM sweeps, presented in Figs. 3.35 and 3.36, are done using the "Half LUT Implementation".

#### 3.4.4.4 High modulation Order and Bandwidth Capabilities

The two final tests that we made are presented in Figs. 3.35 and 3.36. These tests demonstrate that our design can handle not only high-order modulation schemes (Fig. 3.35) but also wide bandwidths (Fig. 3.36). From Fig. 3.35, we can see that the EVM rises linearly with the increase in frequency, which is expected and most likely caused by RF attenuation. Furthermore, we see good EVM behavior, both near DC and also near $Fs/2 = 7.86432$ GHz, meaning that the EVM does not rise as the frequency approaches both ends, proving that our implementation can transmit a 10 MHz 256-QAM signal from 0.1 GHz up to 7.62 GHz. With the 150 MHz bandwidth signal presented in Fig. 3.36, the same good behavior with increasing frequency is observed; however, both near DC and at $Fs/2$, our design exhibits an increase in EVM. Despite this increase, the EVM never passes 12.5%, which is the required EVM by third Generation Partnership Project (3GPP) for a 16-QAM signal [2]. Moreover, if we limit the lower frequency to 0.2 GHz, the EVM never surpasses 8%, which is the 3GPP limit for a 64-QAM signal. In summary, both these results comply with 3GPP specifications for the required EVM in the presented frequency range.

#### 3.4.4.5 Comparison with State of Art

Comparing our work with the state of the art, we observe that we utilize fewer resources and have a lower MGT operation frequency compared to the implementation in [224], while achieving a per-band bandwidth approximately 7.5 times greater and a wider carrier frequency range. Moreover, for frequencies that are multiples of $f_{bb}$, our implementation does not use DSPs, as they are only required to achieve all other frequencies in between. In [223], a different LUT implementation is also presented; however, our implementation achieves a carrier frequency approximately 1 GHz higher while providing a signal with a bandwidth up to around 7 times larger. Moreover, as stated in [224], this type of LUT-base ADT uses much fewer resources when compared to other implementations, such as the ones presented in [46] or [47]. Table 3.5 presents a comparison of this work with the current

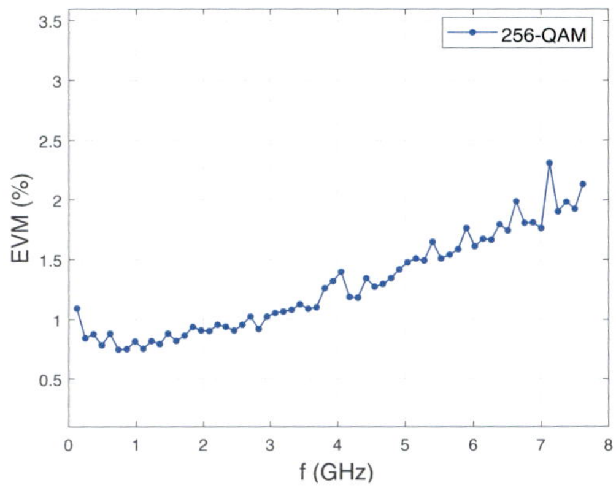


**FIGURE 3.35:** EVM measurements using a 256-QAM, 10 MHz bandwidth signal, with a sample rate of 8.192 Msps, for the 7 bits implementation ($N$=128).

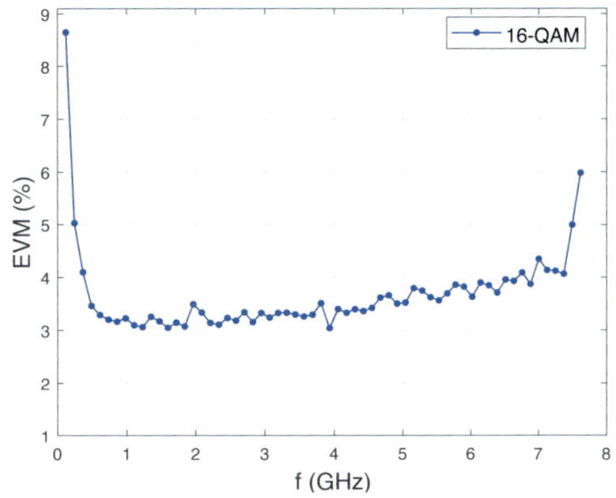


**FIGURE 3.36:** EVM measurements using a 16-QAM, 150 MHz bandwidth signal, with a sample rate of 122.88 Msps, for the 7 bits implementation ($N$=128).

**TABLE 3.5:** Comparison with the state-of-the-art

| | Resource Usage (Qualitative) | Max BW (MHz) | Frequency Range (GHz) |
|---|---|---|---|
| This work | Low (BRAMs, optional DSP48) | 122.88 | 0.1-7.62 |
| [224] | High (BRAMs and DSP48) | 40 (2 × 20) | 1-6.150 |
| [223] | Not Specified | 24 | 0.65-6.85 |
| [36] | Medium (BRAMs and DPS48) | 15.25 | - |

In the table "-" stands for without carrier agility.

state-of-the-art for the most relevant metrics: resource usage, maximum bandwidth, and carrier frequency range.

### 3.4.5 Key Remarks

This work presents an agile LUT-based implementation of an ADT. It utilizes delta-sigma modulation, with offline precomputed output waveforms for all possible input values. This reduces the logic required to be implemented in FPGA, making the final transmitter more energy- and space-efficient.

Ultimately, our implementation achieved carrier agility from 0.1 GHz to 7.62 GHz, while delivering bandwidths of up to 150 MHz, all with EVMs that comply with 3GPP specifications. Furthermore, a new technique, "Half LUT Implementation", was presented and tested. This technique allows for the halving of the LUT size while maintaining the same performance as the complete LUT. This technique relies on the fact that the LUT has symmetry, so the second half of the LUT can be generated using the first half and some NOT logic gates to invert both the LUT content and the address.

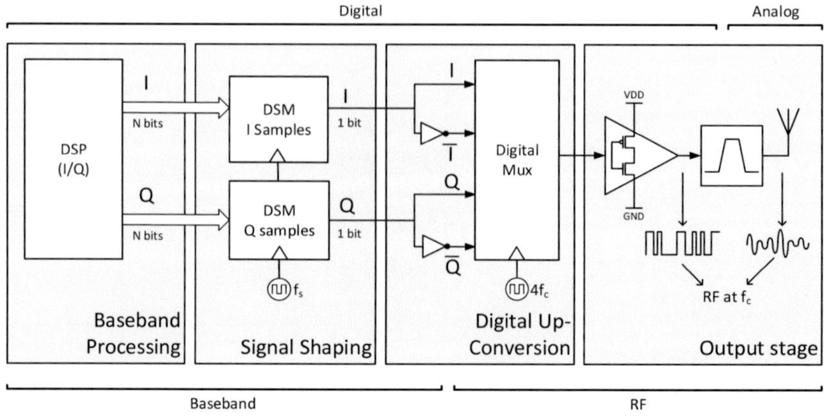


**FIGURE 3.37:** ΔΣM-based ADT architecture. Baseband delta-sigma conversion occurs at $f_s$ 7hile the digital up-conversion works at a frequency of $4 \times f_c$, being $f_c$ the carrier frequency.

## 3.5 Mixed-Domain Combination Filters

In this section, a mixed-domain method is proposed to filter the ADT quantization noise without the use of a high-Q analog filter. This technique uses signal interference to cancel the ADT's quantization noise and improve its $C_{eff}$.

### 3.5.1 All-Digital transmitters and their limitations

This subsection summarizes the ADT architecture used in FPGA-based transmitters and highlights some of the limitations that digital transmitters still have. ΔΣM-based ADTs use delta-sigma pulse modulators to convert high-resolution signals into lower-amplitude resolution, while maintaining an adequate SNR within the signal's bandwidth. The lower bit resolution, specifically two-level signals, enables the generation of RF waves using an entirely digital datapath.

Fig. 3.37 shows a ΔΣM-based digital transmitter architecture. The baseband processing block may vary depending on the protocol and standards used. This block does not differ from the digital baseband processing used for common radio architectures. The signal shaping stage uses ΔΣM to convert the in-phase and quadrature $N$-bit samples to single-bit representation signals. The up-conversion block up-converts the 1-bit baseband signal to RF using a high-speed multiplexing of the I and Q 1-bit signals. The multiplexing is equivalent to multiplying the I and Q samples by two square waves with a 90° phase relation between them.

The resulting frequency spectrum has the RF signal centered at $f_c$ with added quantization noise from ΔΣM. While reducing the added quantization noise within signal's bandwidth is possible using higher-order ΔΣM, removing the noise in the sidebands is impossible without filtering the output signal.

There is a direct relationship between the quantization noise and the overall transmitter efficiency. This relation, called $C_{eff}$, is dependent on the modulator-added noise. The average $C_{eff}$ is defined as the power on the signal's channel $P_s$, over the total transmitted

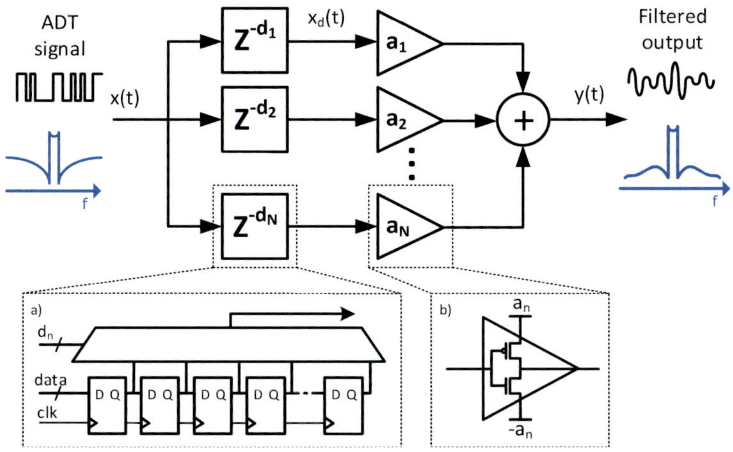


**FIGURE 3.38:** Simplified structure of a combination filter. Each branch has a specific time delay $d_n$ and signal gain $a_n$.

power $P_t$ [72].

$$C_{eff} = \frac{P_s}{P_t} \tag{3.25}$$

The $C_{eff}$ in ΔΣM depends on the input signal characteristics, particularly on its PAPR [55], and is typically low for common wireless communications signals, which frequently use QAM or OFDM modulations with high PAPR. In [55] it is shown that for signals with 3 dB PAPR or more, the transmitter efficiency is below 30 %, regardless of the ΔΣM order, and it will decrease with increasing PAPR. Low $C_{eff}$ means that only a small fraction of the transmitter's power is, in fact, signal power, the remaining being quantization noise, which needs to be filtered. The output analog filter, necessary for these transmitters, must have a high Q to attenuate the quantization noise close to the signal's bandwidth, which imposes difficult filter requirements. The output of the transmitter thus requires a fixed, narrowband filter. For a multi-band and flexible carrier transmitter, this imposes a significant limitation on its flexibility.

### 3.5.2 Proposed filtering method

This subsection describes the proposed method to create a digitally controllable output filtering of the transmitter's signal. This architecture significantly decreases the ADT output quantization noise, improving its $C_{eff}$ and flexibility by relaxing the output filter quality factor requirements.

The proposed method, shown in Fig. 3.38, uses a mixed-domain digital-analog combination filter with delayed versions of the transmitter's output. Using multiple delayed outputs that are weighted and combined, it is possible to cancel the signal's power at particular frequencies. By adjusting the delays $d_n$ and the coefficients $a_n$, the filter's frequency response can be changed. This type of filter has a finite impulse response; however, the signals are combined in the analog domain using a power combiner. Similar techniques were used in [17, 199] to improve digital transmitter's Adjacent Channel Leakage Ratio (ACLR) and dynamic range. However, in both works the noise removed was in a very narrow bandwidth, which does not improve the transmitter $C_{eff}$ and still requires a narrow band output filter.

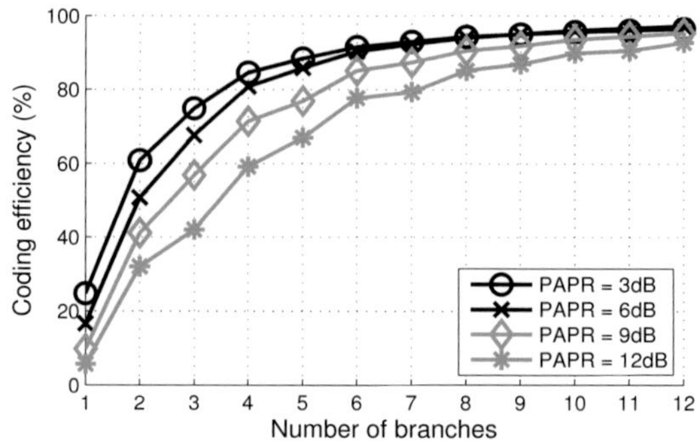


**FIGURE 3.39:** Coding efficiency depending on the number of branches and signal PAPR.

Using the structure of Fig. 3.38, any frequency response can be approximated with an error that depends on the filter order, i.e., number of delayed branches. For a narrow band frequency response, a high order is required, making this filter impractical. Despite this, for reasonably narrow band responses, it is possible to use this filter technique with a low number of combination branches using interpolated filter responses.

Combining several RF branches to create interference using analog circuitry is challenging, as it is difficult to guarantee a constant delay over a large bandwidth, especially if that delay needs to be adjustable. On the other hand, with an ADT architecture, the signal can be delayed using digital shift registers to handle the time delays (Fig. 3.38). Since the RF signal is digital, controlling the delays is simple and can be used to adjust the filter for different situations.

The weight gain stage, which for analog implementations typically requires a set of variable amplifiers or attenuators, can be implemented by regulating the supply voltage of the output buffers. This is possible because the output wave from the ADT is a two-level square wave.

The number of combination branches used in the combination filter (shown in Fig. 3.38) depends on the desired $C_{eff}$ or noise attenuation. A higher number of branches leads to higher noise rejection and $C_{eff}$; however, for $\Delta\Sigma$M, the $C_{eff}$ and potential filtering gains are conditioned on the signal's PAPR. Since the modulator's quantization noise hinges on the signal, higher PAPR signals require additional noise attenuation to achieve the same $C_{eff}$.

Fig. 3.39 shows simulation results for the ADT $C_{eff}$ variation with the number of combination branches for different PAPR values. While, as expected, for a higher number of branches the $C_{eff}$ approaches 100 %, the hardware complexity also increases, which should be taken into account. For each branch, a delay shift register and a high-speed digital output buffer are required. A compromise between hardware complexity and $C_{eff}$ was found with 8 branches guaranteeing a $C_{eff}$ above 80 % for signals up to 12 dB PAPR.

The filter's time response can be deduced from the combined signals as shown in Eq. (3.26). The conversion to the frequency domain, as shown in Eq. (3.27), is found using the Fourier transform.

$$y(t) = \sum_{n=1}^{N} a_n x(t - d_n) \tag{3.26}$$

$$Y(f) = \underbrace{\left( \sum_{n=1}^{N} a_n e^{-j2\pi f d_n} \right)}_{H(f)} X(f) \tag{3.27}$$

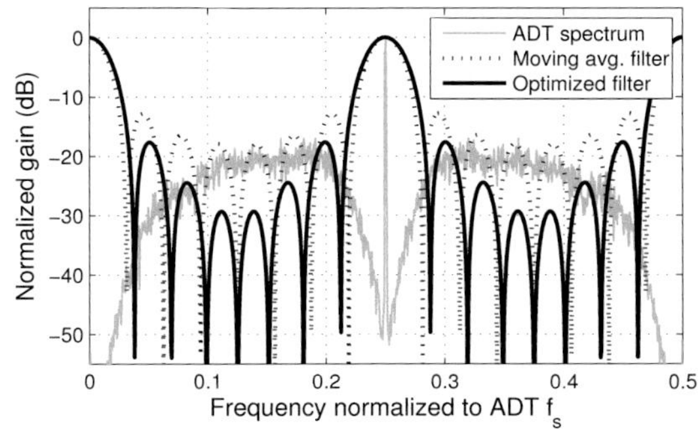


**FIGURE 3.40:** ΔΣM-based digital transmitter spectrum and two proposed ideal frequency responses for the combination filter.

The first term of Eq. (3.27), $H(f)$, gives the equivalent filter frequency response generated from the interference between the combined signals.

Fig. 3.40 shows an example of an ADT frequency power spectrum for the transmitter's first NZ. Since the majority of the quantization noise is located around the center of the delta-sigma side lobes, the filter coefficients, $a_n$ and $d_n$, should be adjusted for higher attenuation at those frequencies.

On the other hand, DC and $f_s/2$ frequencies have quantization noise power nulls, which allows relaxation of $H(f)$ at those points. Considering this simplification, Fig. 3.40 shows two filter magnitude frequency responses of $H(f)$, with ideal $a_n$ and $d_n$ values, to attenuate the quantization noise using 8 parallel branches, according to the architecture in Fig. 3.38. The presented frequency responses were obtained assuming that the delays $d_n$ are fixed step delays, so that $d_n = k_n d_0$, because the delay implementation is done using digital registers. The minimum delay step $d_0$ is the inverse of the ADT sampling frequency $f_s$, $d_0 = 1/f_s$. The first shown filter response (dotted plot in Fig. 3.40) is a moving average interpolated filter with coefficients $a_n = [0.125, 0.125, 0.125, 0.125, 0.125, 0.125, 0.125, 0.125]$ and $k_n = [0, 4, 8, 12, 16, 20, 24, 28]$. This represents the sum of the eight signal phases with the same square wave amplitude.

The optimized filter (black line plot in Fig. 3.40) is a filter optimization to further reduce the quantization noise from the side lobes using different $a_n$ values. The chosen coefficients were $a_n = [0.4375, 0.8125, 0.9375, 1, 1, 0.9375, 0.8125, 0.4375]$ and $k_n = [0, 4, 8, 12, 16, 20, 24, 28]$. This solution was obtained using an adapted Parks-McClellan algorithm, which finds the coefficients that provide the best approximation to the desired frequency response in the minimax sense [151]. The adapted algorithm performs successive weight adaptations to reduce the error in the magnitude response of the filter only above the desired frequency response. By doing this, it is possible to maximize the filter attenuation for frequencies in which the quantization noise is still higher than the desired carrier-to-noise ratio. The real values for the $a_n$ coefficients will depend on the hardware limitations of the buffer's differential voltage. The presented values were the best-case achievable for the current hardware implementation but do not represent the best-case achievable for arbitrary non-quantized values of $a_n$.

### 3.5.3 Hardware implementation

The proposed ADT architecture uses a combination filtering technique that significantly reduces the transmitter quantization noise and improves the transmitter $C_{eff}$. Compared

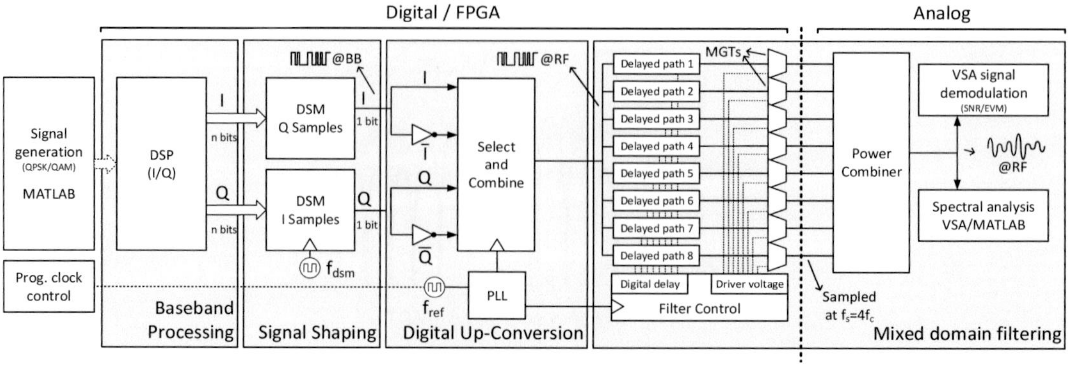


**FIGURE 3.41:** Hardware FPGA-based implementation of the proposed method.

to similar pulse ADTs, it reduces the transmitter's output filtering requirements, thereby improving system flexibility.

The proposed transmitter architecture, presented in Fig. 3.41, shares a similar topology to the previous architecture, Fig. 3.37, up to the digital up-conversion stage.

The baseband signal was previously generated off-line and loaded into memory to emulate a typical transmitter chain. The DSP stage of the transmitter increases the sampling rate up to $f_{dsm}$. This frequency may vary, but it must be guaranteed that the oversampling ratio is enough for the $\Delta\Sigma$M to provide a good SNR. Hardware frequency limitations constrain the NTF to a maximum achievable value.

The $\Delta\Sigma$M stage converts the I and Q signals to a 1-bit equivalent at $f_{dsm}$. The modulator used in this work is a time-interleaved first-order low-pass delta-sigma modulator presented in [36]. This architecture is a parallel polyphase $\Delta\Sigma$M with 4 phases capable of achieving the combined sampling frequency ($f_{dsm}$) of 1.6 GHz.

The digital up-conversion is done using an integrated high-speed serializer, the FPGA's MGT, that multiplexes in time the I and Q signals from the $\Delta\Sigma$M with their inverted version. This is done accordingly to the $\Delta\Sigma$M sample order defined in the Select and Combine block. The carrier frequency is defined by setting the ADT sampling frequency $f_s$ so that $f_s = 4 \times f_c$.

The mixed domain filter was implemented by adding the hardware block shown in Fig. 3.38 to process the RF signal from the select and combine block. The digital RF signal is delayed into 8 different delay paths, each one used as input to an embedded MGT. These delays are integer multiples of the minimum delay step $d_0$. Each delayed path feeds a high-speed digital output driver (up to 6.4 Gbps). The voltage swing of these drivers is digitally controlled, acting as an adjustable gain $a_n$, for each filter path. The voltage swing had a limited number of possible values; hence, the optimized values of $a_n$ were quantized. The filtering effect is achieved by combining the 8 different paths in the analog domain using a power combiner. The power combiner is wide-band to allow for noise cancellation over the whole bandwidth of the delta-sigma modulated signal. For a higher level of integration, power combining can be achieved using fully integrated Complementary Metal–Oxide–Semiconductor (CMOS) combiners [10], resulting in a compact transmitter architecture.

### 3.5.4 Experimental setup and results

This subsection presents the main experimental results for the implementation of the proposed ADT. The experimental scheme setup is detailed in Fig. 3.41. For the proposed

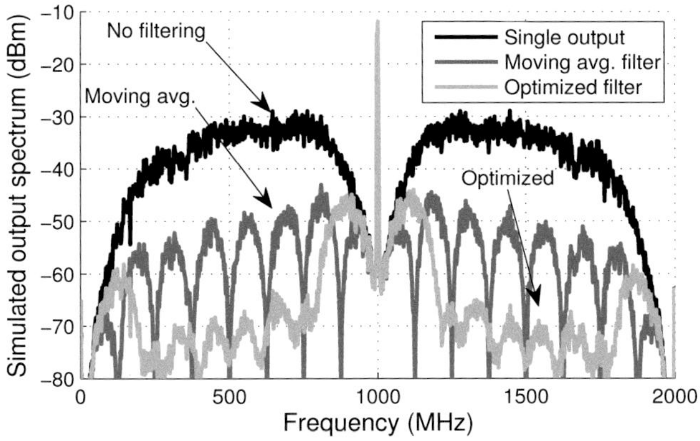


**FIGURE 3.42:** Simulated transmitter spectrum with 500 kHz resolution bandwidth for the two proposed ideal frequency responses of the combination filter.

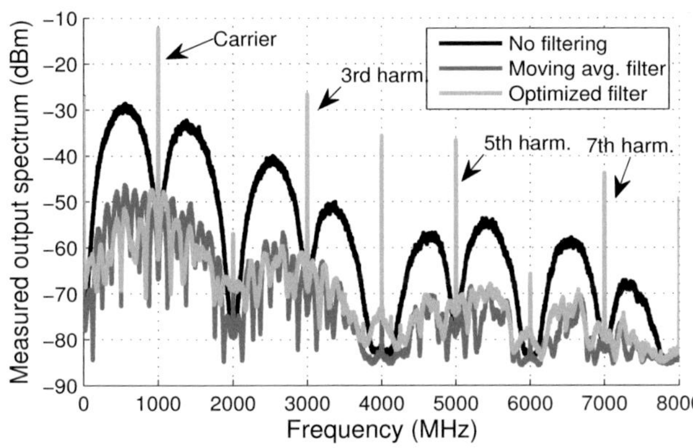


**FIGURE 3.43:** Measured spectra, with 500 kHz resolution bandwidth and 5 kHz video bandwidth, captured with a 2 GHz span. Signal at 1 GHz carrier frequency.

prototype, an ML628 board from Xilinx was used. The digital hardware project was developed for a Virtex6 HX380T FPGA as a proof-of-concept. The used power combiner is an 8:1 RF power combiner, model ZN8PD1-53+, from Mini-Circuits. A set of fixed delay lines is used to guarantee the correct time alignment of the 8 delayed outputs. This is done because the development board layout does not guarantee a perfect time alignment for different MGT outputs. The signals were aligned using microstrip lines with the appropriate length.

The transmitter output was connected to a VSA as shown in Fig. 3.41. From the VSA, the EVM, SNR, coding efficiency, and spectrum were obtained. The input signals, used to test the transmitter, were Quadrature Phase-Shifting Keying (QPSK) and 64-QAM modulations with symbol rates ranging from 1.5625 Msps to 15.625 Msps.

The transmitted spectrum is shown in Fig. 3.43 for three different cases and can be compared with the simulation results in Fig. 3.42. The first plot, in black, shows the original ADT signal without any filtering. The dark gray plot represents the output signal obtained using 8 branches with moving average filtering. Finally, the last plot in light gray is the output spectrum using the optimized filtering coefficients.

Comparing the simulation with the measured results, it is possible to see that for the optimized filter, the obtained results are quite different from the expected in the simulation.

**TABLE 3.6:** Transmitter Coding Efficiency and Carrier to noise ratio

| Carrier frequency | | 0.5 $GHzZ$ | 1 $GHz$ | 1.6 $GHz$ |
|---|---|---|---|---|
| | | Transmitter Coding Efficiency | | |
| QPSK 1.5625 Msps | No filtering | 19.03 % | 19.16 % | 20.05 % |
| | Moving avg. | 88.29 % | 90.89 % | 93.25 % |
| | Optimized | 91.28 % | 94.18 % | 96.71 % |
| QAM 15.625 Msps | No filtering | 11.66 % | 11.14 % | 11.02 % |
| | Moving avg. | 85.22 % | 88.09 % | 89.08 % |
| | Optimized | 89.87 % | 92.89 % | 93.84 % |
| | | Carrier to noise ratio | | |
| QAM 15.625 Msps | No filtering | 18.67 dB | 19.21 dB | 19.47 dB |
| | Moving avg. | 35.17 dB | 37.12 dB | 37.13 dB |
| | Optimized | 39.93 dB | 40.89 dB | 41.49 dB |
| QAM 15.625 Msps | No filtering | 7.21 dB | 6.05 dB | 5.70 dB |
| | Moving avg. | 24.74 dB | 23.77 dB | 22.61 dB |
| | Optimized | 28.49 dB | 28.27 dB | 28.46 dB |

**TABLE 3.7:** Comparison with other all-digital transmitter approaches

| | Ghan10 [72] | Pods12 [158] | Silv13 [183] | Ebra13 [55] | Chun15 [30] | This work | | | |
|---|---|---|---|---|---|---|---|---|---|
| Signal | WiMAX | QPSK | 64-QAM | OFDM | LTE | QPSK | 64-QAM | 64-QAM | 2×64-QAM* |
| Bandwidth | 1.25 MHz | 0.625 MHz | 1.5 MHz | 1.4 MHz | 5+10 MHz | 3.5 MHz | 3.5 MHz | 19 MHz | 2×9.5 MHz* |
| PAPR | 3.8 dB - 11.9 dB | 8 dB | - | 7.8 dB | 11.7 dB | 3.8 dB | 6.8 dB | 6.9 dB | 9.5 dB |
| Cod. Eff. | 25.8 - 3.8 % 21 % | 79.3 % | 18.7 % | 35.4 - 47.1 % | 94.1 % | 91.9 % | 93.8 % | 88.5 % | |
| SNR | 45 dB | - | 50 dB | 41 dB | - | 39.5 dB | 42.6 dB | 35.2 dB | 32.9 dB |
| Modulation | 2-level DSM | 2-level DSM | Multi-path PWM | QNRIF DSM | BPDSM and RFPWM | 2-level DSM w/ mixed-domain filtering | | | |

*Two adjacent channels with simultaneous transmission.

This is due to the time misalignment at the outputs of the parallel delay paths and the amplitude mismatch between the different drivers. While these errors are small for very low gain frequencies in the filter frequency response, any small mismatch in the coefficients $a_n$ or $d_n$ results in considerable changes on the logarithmic scale.

The carrier-to-noise power ratio without filtering is very low, which reduces $C_{eff}$ and further increases the need for an analog filter with a high quality factor. Table 3.6 shows the $C_{eff}$ and the carrier-to-noise ratio for the first Nyquist zone results using two signals for the worst and best tested scenario in terms of CE, QPSK at 1.5625 Mbps and 64 QAM at 15.625 Mbps respectively.

From these measures, it is possible to see that this method allows for achieving a $C_{eff}$ of more than 89 % for both scenarios within the transmitter's band (500 MHz to 1600 MHz) using the optimized filtering. Table 3.6 also shows the noise peak power reduction from the carrier to noise measure, which improves from about 19 dB to more than 40 dB in the best case and from 6 dB to 28 dB in the worst case.

The results show a high carrier-to-noise power ratio improvement as well as a significant advance in the ADTs state-of-the-art for the achievable $C_{eff}$. A comparison between several state-of-the-art all-digital transmitters is shown in Table 3.7. As can be seen, the proposed transmitter with mixed-mode filtering provides a very high $C_{eff}$ and high SNR, while improving the flexibility of ADT by eliminating the need for an output narrow-band filter.

### 3.5.5 Key Remarks

In this section, a new filtering technique for all-digital transmitters was presented to extend their flexibility to the filtering stage. It utilizes signal interference to cancel the signal at the designated frequencies, thereby reducing the quantization noise common to all-digital transmitter radios.

With the proposed filtering technique, the design of the output narrow band filter can be avoided. Furthermore, the filtering characteristics can be easily digitally controlled to match the transmitter carrier frequency, thereby enhancing the overall transmitter flexibility.

## 3.6 Out-of-band Noise Reduction Through the Use of PAM4

Several solutions to reducing the out-of-band noise generated by these transmitters have been presented. To better evaluate the performance of each implementation, it is important to know the FoM named $C_{eff}$, which is given by:

$$CE = \frac{P_s}{P_t} \tag{3.28}$$

where $P_s$ is the in-band signals' power and $P_t$ is the total power transmitted. This FoM is very important for the characterization of an ADT because it allows for a quick understanding of how much total transmitted energy is useful energy that is being put into the signal and not into the noise.

A summary of the state of the art of out-of-band noise reduction for ADTs reveals implementations that range from using high-order modulators [65], cascaded modulators [52], subtracting the out-of-band noise in the analog domain [31, 49], or implementing a mixed domain filters [35]. Starting with [52], the author presented an implementation for relaxing analog filtering requirements using cascaded modulators. This implementation utilized two consecutive ΔΣMs, where the first one is a multi-bit one, followed by a faster single-bit one. Due to the higher output levels of the first modulator, it generates less noise. After that, the second modulator, which operates at a higher frequency, sends its noise further away from the signal. This results in a lower noise floor near the signal. However, the $C_{eff}$ is still low due to the fact that the last modulator is a single-bit one. This means that an analog filter with a high attenuation is still required.

Another approach to the reduction of ADTs' noise is presented in both [49] and [31]. It consists of the subtraction of the out-of-band noise, in the analog domain, using a power combiner. Despite the potential for great results in terms of $C_{eff}$, generating the out-of-band noise canceler signal requires a multi-bit DAC. This not only sharply increases costs but also increases energy consumption and complexity of the system, negating any energy efficiency improvements gained from the lower out-of-band noise. Furthermore, the use of a multi-bit DAC invalidates the advantages of using an ADT, such as their scalability and maximum carrier frequency.

In [35], a mixed-domain combination filter is presented. Such a filter is implemented using several MGTs. Each MGT sends a copy of the same signal with a specific delay and gain, where the gain is controlled by the voltage of the MGT. By combining all outputs using a power combiner, this implementation results in an FIR filter. While this implementation achieves good results in terms of $C_{eff}$, the utilization of multiple MGTs and a power combiner results in a more complex and costly implementation.

From this review of the state of the art, it is evident that a new approach to reducing out-of-band noise is necessary. In this paper, a solution based on PAM-4 for ADTs will be presented and tested in an FPGA implementation. The main contribution of this paper is a novel technique that reduces out-of-band noise without requiring excessive digital resources or introducing significant complexity.

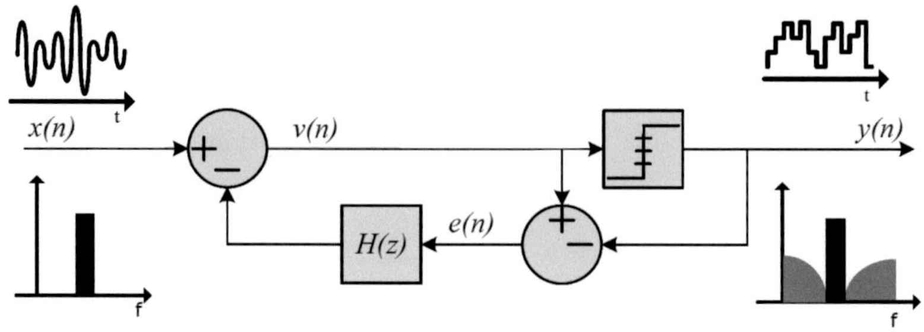


**FIGURE 3.44:** Block diagram of the delta-sigma modulator.

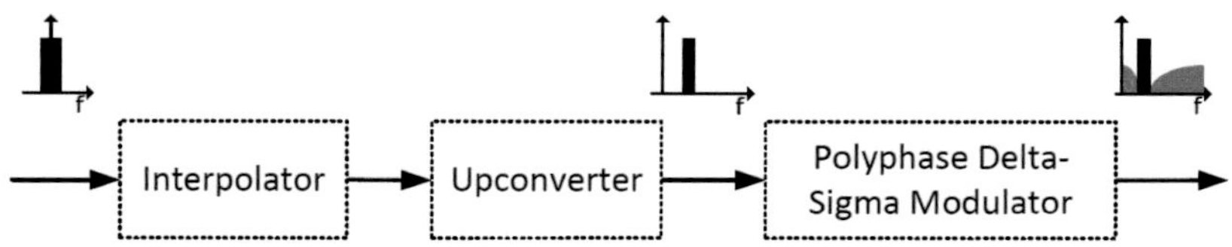


**FIGURE 3.45:** Block diagram of the architecture of the transmitter.

### 3.6.1 Delta-Sigma Modulator Design

For the modulators' architecture, the error feedback structure, depicted in Fig. 3.44, was chosen due to its simplicity, while also achieving good performance through the increase in the order of its filter, $H(z)$. As can be seen from the image, changing the number of output levels of such a modulator is straightforward and, in theory, requires only modification of the quantization block. In practice, this change also requires an optimization of the filter, $H(z)$. Furthermore, the levels of decision need to be symmetrical, equidistant, and adjusted to the values of the input signal. This is not a problem for a one-bit $\Delta\Sigma$M because there are only two levels, and the decision value can be set to zero. However, for a two-bit modulator, there are four levels, which means that there are three decision values that must be specified accordingly to the input signal.

The block diagram of the final transmitter is shown in Fig. 3.45. There, due to the baseband signal having a lower sampling rate when compared to that of the $\Delta\Sigma$M, the first block is the interpolator, followed by the upconverter. Here, a polyphase digital direct synthesis (DDS) is used to generate sine and cosine waves at the desired final carrier frequency. The final block is the polyphase $\Delta\Sigma$M, where multiple $\Delta\Sigma$M cores are used in parallel, allowing each of them to run at a lower clock frequency. This enables the overall sampling rate of the entire system to be higher than it would otherwise be. A technique presented in [46], known as state propagation, is also implemented here. It allows for the reduction of time discontinuities in the final modulated signal, reducing the overall noise.

### 3.6.2 Hardware Implementation

One of the criteria for choosing the FPGA kit was the presence of high sampling rate MGTs capable of delivering a PAM-4 signal. The VCU129 kit possesses GTMs transmitters capable of operating at 28 Gsps, which means that, in PAM-4 mode, it transmits up to 56 Gbps. With the FPGA kit chosen, a proof of concept was implemented in it. The frequency of

**FIGURE 3.46:** Photograph of the experimental setup.

operation for the MGTs was chosen to be 25.6 Gsps, which also defines the sampling rate of the ΔΣM.

As described in the previous section, to achieve higher sampling rates, in this case in the tens of GHz range, a parallel approach is necessary. In this work, we chose $M$, the number of parallel phases, to be 128. This means that the FPGA clock is set to 200 MHz, a value that can be reached in the current generation of FPGAs.

For comparison with the proposed PAM-4 implementation, a traditional ADT with two-level Non-Return-to-Zero (NRZ) output was implemented. This second transmitter used the same frequencies for the MGT and value of $M$ as the PAM-4 version to ensure a fair comparison.

### 3.6.3 Experimental Results

Fig. 3.46 depicts the experimental setup. In the figure, highlighted in blue is the PC, in green the clock generator, in red an R&S FSW-8 VSA, and in yellow the FPGA. The resource usage of both the PAM-4 and NRZ implementations can be seen in Table 3.8. From it, it is possible to see that both implementations use around the same number of resources, with the PAM-4 implementation using a slightly higher number of LUTs and flip-flops. However, this increase in resource usage is relatively small, and, as the results will show later, it is well justified by the increase in performance.

**TABLE 3.8:** Resource usage of the implemented all-digital transmitters (PAM-4 vs NRZ).

| Implementation | PAM-4 | | NRZ | | |
|---|---|---|---|---|---|
| Resources | Used | % | Used | % | Total |
| Flip Flops | 495712 | 14.34 | 463008 | 13.4 | 3456000 |
| LUTs | 512924 | 29.68 | 476013 | 27.55 | 1728000 |
| BRAM | 199 | 7.38 | 199 | 7.38 | 2688 |
| DSP48 | 1461 | 11.89 | 1461 | 11.89 | 12288 |
| GTM | 1 | 4.17 | 1 | 4.17 | 24 |

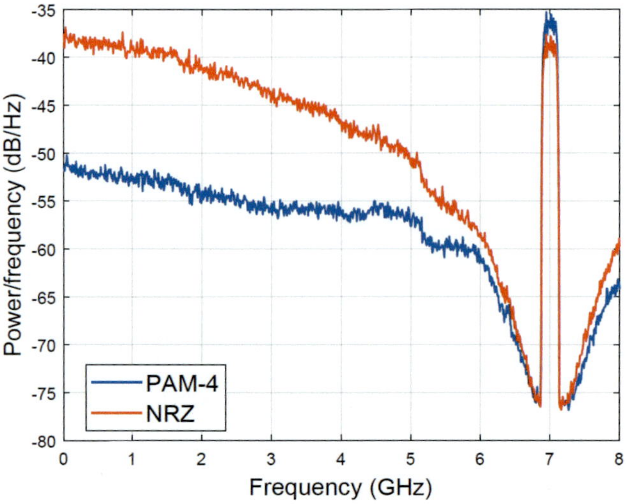


**FIGURE 3.47:** Comparison of the output spectrum of the ADT with PAM-4 (black), and with NRZ (gray).

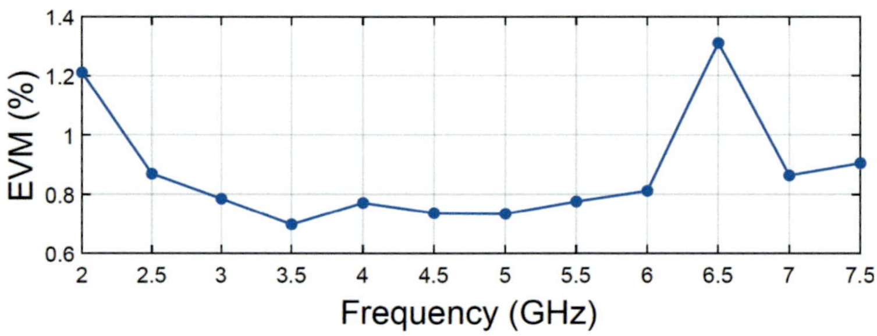


**FIGURE 3.48:** EVM sweep from 2 GHz up to 7.5 GHz using a 200 Msps, 16-QAM signal.

A depiction of both output spectra is shown in Fig. 3.47. The signal being transmitted is a 200 Msps, 16-QAM at 7 GHz. In the figure, the difference between the output spectrum of the PAM-4 and NRZ transmitters is clearly visible, with the first one (PAM-4) presenting a much lower out-of-band noise. This difference in the level of the noise outside of the signal band can reach more than 10 dB (as is the case at 2 GHz for the example in Fig. 3.47). Using Eq. (3.28), an increase in $C_{eff}$ was measured to go from 5.8%, in the NRZ case, to 63.2%, in the PAM-4 transmitter. Such sharp improvement is primarily due to the reduction in out-of-band noise level, with an increase in in-band signal power of only around 2 dB.

To demonstrate the integrity of the transmitted signal alongside the flexibility of the transmitter, an EVM sweep was realized from 2 GHz up to 7.5 GHz. The results of this sweep can be seen in Fig. 3.48, and, as it is possible to see in the image, the EVM never goes above 1.4%. Which, accordingly to 3GPP guidelines for 5G, is enough to transmit a 256-QAM [2]. These results prove that not only is the implemented PAM-4 transmitter capable of transmitting high-order modulations, but it can also do so in a wide range of frequencies.

### 3.6.4 Key Remarks

In this work, a new PAM-4-based ADT is implemented and compared with a similar ADT that uses only a one-bit output. For the implementation of such a PAM-4 transmitter, the quantizer levels were optimized to cover the range of the input signal. Further filter optimization was also done to ensure the best possible results for this new transmitter.

The final results of this work clearly demonstrate a significant decrease in out-of-band noise, accompanied by a measurable increase in in-band signal power. These two points resulted in a sharp improvement in the $C_{eff}$ figure of merit, where a typical one-bit ADT presents a value of 5.8% and the PAM-4 transmitter presented in this paper showed a value of 63.2%. This improvement in $C_{eff}$ not only makes the transmitter potentially more energy-efficient but also increases its practical flexibility, as a lower-quality analog filter is required.

---

## 3.7 Cascading Delta-Sigma Modulators

The stringent requirements for the output reconstruction filter impose several constraints on the deployment of the ADTs in realistic scenarios. This is essentially due to the need for a high-Q analog reconstruction filter, which ultimately hinders the flexibility and agility of the digital radio transmitter. Thus, in this section, this research line is addressed, and a promising architecture that relaxes the requirements for the reconstruction filter is proposed.

This section is supported by the conference paper [52]. The section presents a BB-stage ADT architecture that combines cascaded ΔΣMs with different sampling rates, with the underlying idea of relaxing the analog filtering requirements.

### 3.7.1 Contextualization

The relevant ADTs presented in the SoA typically relies on the use of a single-stage modulator with a single-bit quantizer in the output. However, by doing this, a huge amount of quantization noise is generated and shaped according to the modulator's NTF, with a considerable part located around the BB signal. The typical way to improve the available processing bandwidth is to increment the modulator's order by including more zeros and poles [170]. However, this tends to increase the modulator's critical path, resulting in lower sampling frequencies. Furthermore, instability problems may also appear [179].

### 3.7.2 Proposed Architecture

In this work, to relax the analog filtering requirements, a cascaded 2-stage ΔΣ architecture with different quantization levels and sampling frequencies is proposed. As the modulator's order is maintained, stability and critical path issues are no longer a problem.

The cascaded ΔΣM equivalents involve the same architecture divided into two modulators: a multi-bit modulator, followed by a faster single-bit ΔΣM. The illustration of this conceptual idea is depicted in Fig. 3.49. On the one hand, using a multi-bit modulator as the first stage results in lower quantization noise. In fact, the quantization noise decreases by 6 dB for each added bit to the quantizer [179]. On the other hand, the ΔΣM NTF's processing bandwidth is proportional to the sampling rate. Thus, a faster 1-bit

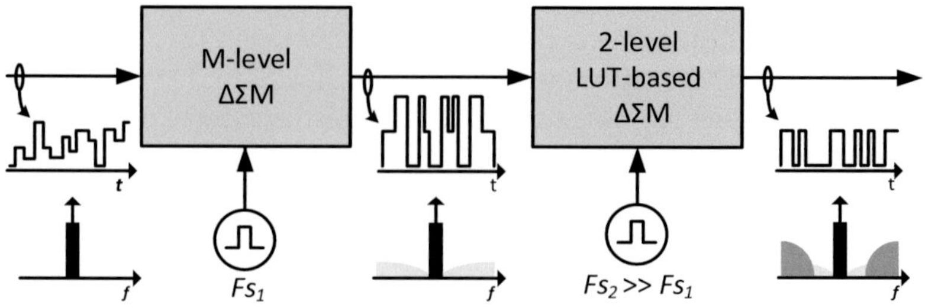


**FIGURE 3.49:** Conceptual idea of the proposed architecture.

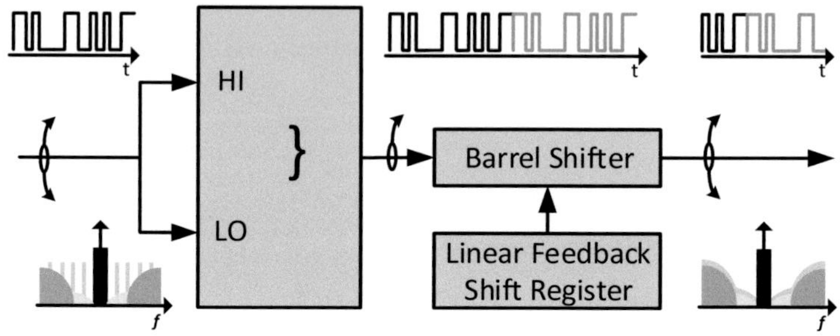


**FIGURE 3.50:** Illustration of the implemented dithering mechanism.

modulator as the last stage implies that the BB signal, together with the low quantization noise generated by the first modulator around the signal carrier, will be transmitted with less distortion, thereby alleviating the constraints of the analog filter. Nonetheless, the apparent simplicity of this concept masks the fact that designing modulators with higher sampling rates is quite challenging. This difficulty stems from the fact that the sampling rate is inversely proportional to the maximum critical path. As pipeline stages cannot be introduced in $\Delta\Sigma$M without the risk of corrupting the transfer functions, the sampling rates are always bounded by the system designer's expertise and the technological process.

For the purpose of this proof-of-concept, the chosen architectural solution for the last stage relied on the use of LUT-based approximations to $\Delta\Sigma$Ms, as proposed in [167]. These try to emulate the typical $\Delta\Sigma$M's behavior using offline-computed LUTs, avoiding feedback loops, with the inherent gains in terms of sampling rates. However, the modulator's finite memory imposes several discontinuities in the output signal, which worsen FoMs, such as in-band signal integrity. These discontinuities will affect the performance of the modulator; however, as will be seen, their effect on the FoM will not be as dominant.

Another drawback arising from the use of LUTs (either in $\Delta\Sigma$M or in PWM) is the spurious tones that tend to corrupt the entire spectrum at well-defined frequencies. This appears to be due to the implicit determinism inherent in the chosen patterns that are transmitted.

To overcome this issue, a methodology similar to the one proposed in [189] for PWM randomization is followed to smooth the spurious out using dithering. The underlying idea is the random circular shifting of each LUT's output word. In this way, the energy of each word is maintained without the deterministic pattern, spreading the spurious tones' energy across the out-of-band spectrum. To accomplish this in a practical way, a 20-bit Linear Feedback Shift Register (LFSR) was connected to a Pipelined Barrel-Shifter Module. The block diagram is depicted in Fig. 3.50. The incoming data is concatenated

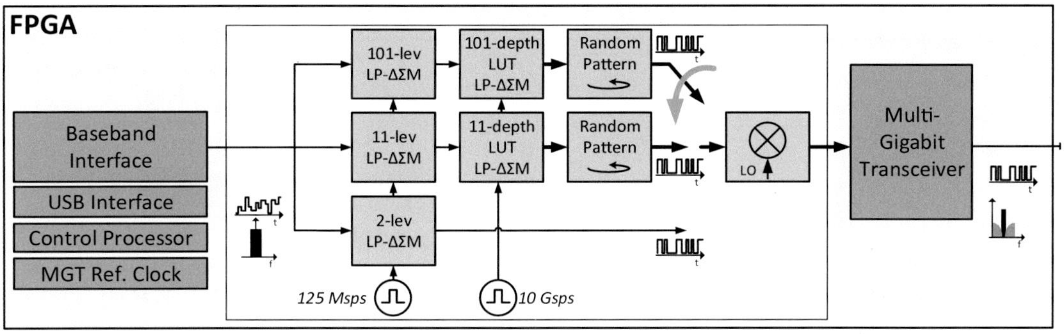


**FIGURE 3.51:** Block diagram of the implemented All-Digital Transmitter used to explore different combinations of cascaded modulators (conventional single-stage [*2-lev*], cascaded with 11 levels at the first stage [*11-lev*] and with 101 levels [*101-lev*]).

with itself, generating a word with the double of the input length. Then, it is forwarded to the Pipelined Barrel Shifter Module, which shifts the data word by a specified number of bits. The number of bits is in the range of 0 to the length of the input data and is given by the LFSR.

Overall, it may be said that the first stage provides an adequate SNR and reduces the number of output combinations. The second stage should be as fast as possible to relax the analog filtering requirements. The maximum achievable bandwidth is limited by the combination of the two stages, but primarily by the quantization noise introduced by the first stage.

### 3.7.3 Prototype Validation

To demonstrate the potential and flexibility of the proposed design, two different cascaded ΔΣMs and a conventional single-stage modulator were embedded into a single FPGA. The block diagram is depicted in Fig. 3.51. In the cascaded topologies, 11 (2.8 bit) and 101 (6.65 bit) quantization levels are used at the first modulator, defining the LUT's depth of the second stage. As the typical input values from the baseband I/Q samples have 10 bits, it was decided to choose these values to uniformly cover the range from 1 to 10 bits.

The samples are continuously transmitted from the BB Interface to the different modulators. Thereafter, they are combined and replicated by the *Select&Combine* module to generate a parallel output word that contains the four components $[I, \overline{Q}, Q, \overline{I}]$ and inverted versions of the desired signal. This output word is transmitted by the MGT. The LUT-based Low-Pass Delta-Sigma Modulator (LP-ΔΣM) has an equivalent sampling rate equal to the MGT's bit-rate.

The implemented LP-ΔΣM is based on an Error-feedback structure. It was chosen due to its short critical path, leading to higher sampling rates. Its block diagram is shown in Fig. 3.52. The modulator's transfer function can be computed by modeling the quantizer by a noise source ($E_q(z)$):

$$Y(z) = STF(z)X(z) - NTF(z)E_q(z) \tag{3.29}$$

where $STF(z) = 1$ and $NTF(z) = 1 - 2z^{-1} + z^{-2}$. This NTF is a high-pass filter, resulting in a ΔΣM with a low-pass response.

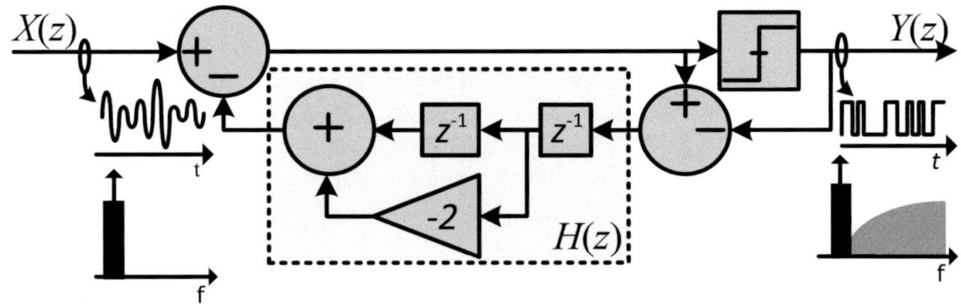


**FIGURE 3.52:** Representation in z-domain of the single-bit LP-ΔΣM.

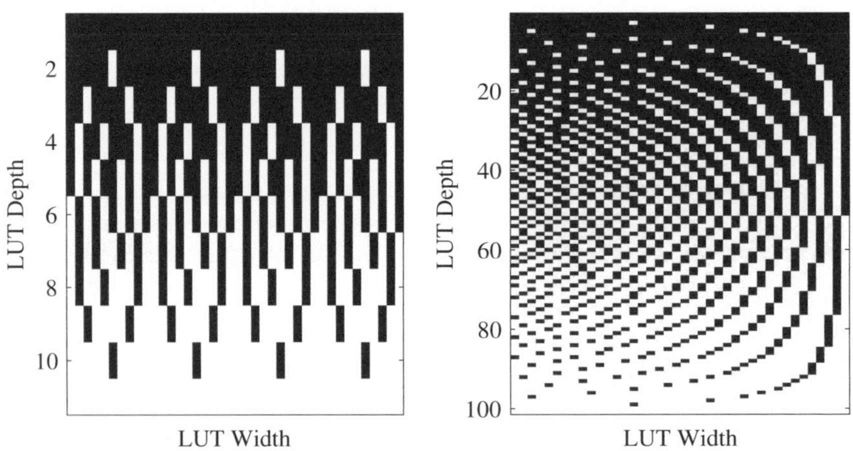


**FIGURE 3.53:** ΔΣMs LUT bitmap representations, with 11 (left) and 101 levels (right).

The same modulator was used for the generation of the LUT's bitmap. In particular, the ΔΣM processes each input word, whose result is written in the respective LUT row. The offline generated LUTs are depicted in Fig. 3.53.

To validate the proposed idea, an ADT was fully implemented into a Virtex Ultrascale XCVU095 FPGA using the Xilinx VCU1287 Characterization Kit. The experimental setup is depicted in Fig. 3.54.

The MGT line rate was configured to 10 Gbps to obtain a carrier frequency at 2.5 GHz, and was then connected to the VSA R&S FSW 8 for signal analysis and demodulation. In particular, spectrum and EVM measurements were obtained for the three architectures: the *2-lev*, the *11-lev*, and the *101-lev*.

The obtained spectra illustrating the difference between the conventional and the proposed architectures are shown in Fig. 3.55, where a 16-QAM was transmitted with a Symbol Rate (SR) of 3.125 Msps. It can be concluded that for the proposed modulators, the LUT's NTF has a large notch, which preserves the BB signal, along with the quantization noise (around the signal) from the first stage. Thus, the higher the number of levels in the first stage, the lower the analog filtering requirements will be.

The importance of dithering in the output spectrum is depicted in Fig. 3.56. The pattern randomization effectively removes the spurious tones, resulting in an inherent increase in the noise floor, but without degrading the integrity of the in-band signal.

Table 3.9 summarizes the comparison between the conventional LP-ΔΣM and the two cascaded modulators. Different SRs were tested, with a 16-QAM signal. As can be seen,

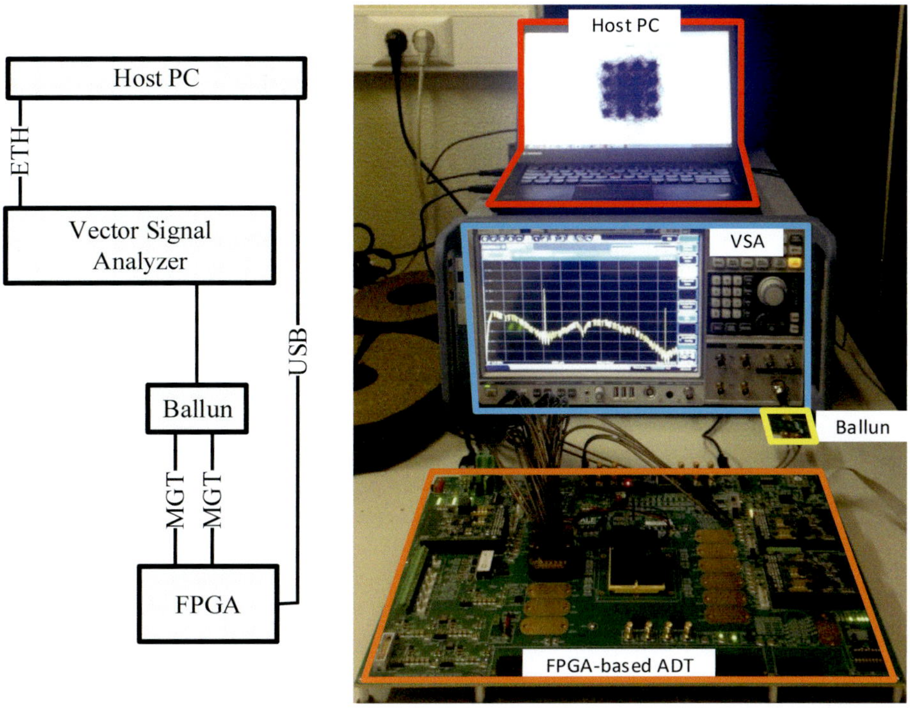


**FIGURE 3.54:** Photo of the experimental setup.

**TABLE 3.9:** Comparison between a single-stage *2-lev* vs two cascaded modulators with dithering, with a 16-QAM signal. Symbol rate is in Msps, EVM in % and MER in dB.

| | Conventional | | 11 Levels | | 101 Levels | |
|---|---|---|---|---|---|---|
| SRate | EVM | MER | EVM | MER | EVM | MER |
| 3.125 | 1.08 | 39.3 | 0.96 | 40.3 | 1.10 | 38.5 |
| 6.25 | 1.35 | 37.3 | 1.18 | 38.5 | 1.46 | 36.7 |
| 12.5 | 4.53 | 26.8 | 2.04 | 33.8 | 2.37 | 32.5 |
| 15.625 | 7.78 | 22.1 | 2.46 | 32.2 | 2.84 | 30.9 |
| 31.25 | n/a | n/a | 2.94 | 30.6 | 3.30 | 29.5 |

reducing the quantization noise around the carrier frequency allows for higher bandwidths to be accommodated and demodulated in the same spectral notches, which is another advantage arising from the proposed architecture. In particular, all the experimental results present EVM values below 3.3%, even for the case of 31.25 Msps (37.5 MHz of signal's bandwidth).

Finally, the resources required for both architectures are presented in Table 3.10. The major conclusion that can be derived is that the increase in hardware complexity for choosing the cascaded $\Delta\Sigma$Ms instead of the conventional one is almost negligible, even when considering the specifications of low- and mid-range FPGAs available on the market.

### 3.7.4 Key Remarks

In this subsection, a novel architecture based on cascading two $\Delta\Sigma$Ms with different sampling frequencies and quantization levels was presented as a possible core for the

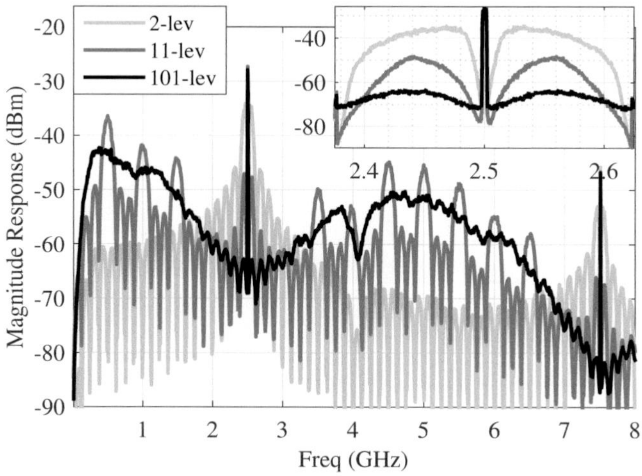


**FIGURE 3.55:** Comparison between the different spectra: single-stage *2-lev*, cascaded *11-lev* and cascaded *101-lev*.

**TABLE 3.10:** Comparison between the different ΔΣM occupied resources: the single-stage *2-lev*, the cascaded *11-lev* and the cascaded *101-lev*.

| | | Conventional | | 11 Levels | | 101 Levels | |
|---|---|---|---|---|---|---|---|
| Logic Resources | Total | Used | % | Used | % | Used | % |
| Flip Flops | 1075200 | 261 | 0.02 | 1519 | 0.14 | 1519 | 0.14 |
| LUTs | 537600 | 328 | 0.06 | 1235 | 0.23 | 4286 | 0.80 |
| Memory LUT | 76800 | 141 | 0.18 | 167 | 0.22 | 167 | 0.22 |
| GTHs | 28 | 1 | 3.57 | 1 | 3.57 | 1 | 3.57 |

development of novel BB-stage ADTs. It was shown that instead of designing a pulse encoder based on a single-stage 1-bit LP-ΔΣM, the system's FoMs are greatly improved if a cascaded 2-stage ΔΣM architecture is chosen instead. In particular, the quantization noise near the transmitted carrier frequency can be highly reduced, avoiding the need for a high-quality factor analog output filter.

This novel concept has been successfully implemented and validated on an FPGA-based transmitter and compared with the conventional architectures that perform a single-bit quantization into a single stage. The proposed technique is relatively simple and requires only minor changes to the modulator's hardware.

All the experimental results attest to the validity of the proposed architecture and clearly demonstrate that the RF reconstruction filter's quality factor requirements can be relaxed with only minor modifications to the hardware. Moreover, the proposed technique is completely independent of the modulator's topology, and thus, stability and critical path issues are no longer a concern. At last, even though LUT-based ΔΣMs were used in this proof-of-concept, another type of modulator can be used. To effectively take advantage of this proposed architecture, the focus should always be on increasing the sampling rate of the cascaded modulator's last stage. That, together with the number of output levels from the first stage, will dictate the final output spectrum's shape and, therefore, the analog filtering constraints.

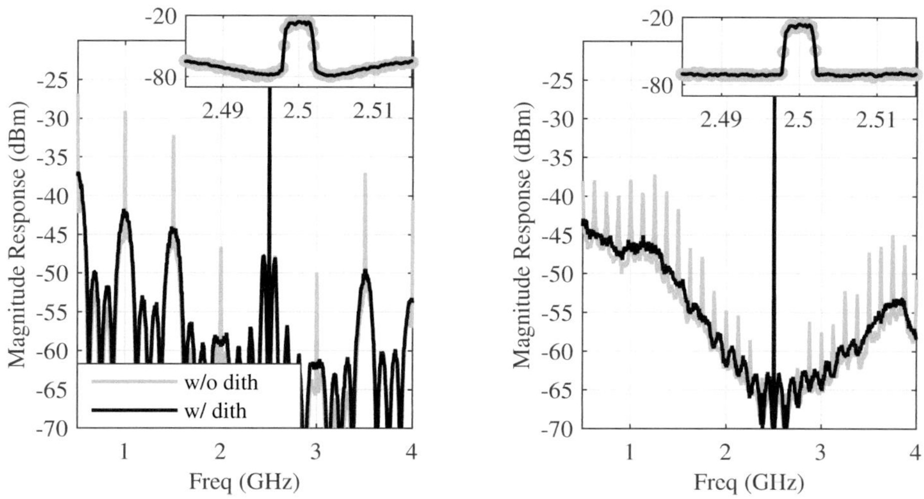


**FIGURE 3.56:** Impact of pattern randomization (dithering) in the spectrum: cascaded with *11-lev* (left) and with *101-lev* (right).

## 3.8 Conclusion

The out-of-band noise generated by ADTs is one of the most significant challenges that is still preventing these transmitters from becoming mainstream in the telecommunications market. In this chapter, several architectures that tackle this problem were presented.

The first one uses polyphase multipaths, where the signal's phase is kept constant, but the noise's phase is changed to cancel the out-of-band noise generated by the $\Delta\Sigma$M. This technique was able to significantly increase the coding efficiency of the output signal; moreover, EVM measurements even show an increase in the signal's quality as more paths were used. All this proves the feasibility and potential of this polyphase multipath architecture.

The second architecture changes the waveform of the PWM signal in order to spread the noise harmonics to frequencies away from the signal. Due to this implementation being LUT-based, this technique requires no additional hardware when compared to a traditional LUT-based PWM transmitter. A reduction of up to 52 dB was achieved for the first harmonic, enabling the use of a much lower quality factor filter.

The third architecture proposed a LUT-based transmitter where it precomputes $\Delta\Sigma$ modulated waveforms that are stored and addressed by the baseband signal, reducing FPGA resource usage while enabling carrier agility up to 7.62 GHz and bandwidths of 150 MHz. A "Half-LUT Implementation" further halves memory requirements without degrading performance, proving the feasibility of resource-efficient ADTs for future 6G massive-MIMO systems.

The forth architecture implements a mixed-domain combination filter that uses interference to cancel the out-of-band noise. The implemented filter is equivalent to an FIR-filter where the delays are implemented in the digital domain and the gains are controlled by changing the voltage of each MGT. After that, a power combiner is used to combine all signals. This architecture greatly increases both the coding efficiency and the SNR of the output signal.

The fifth one implements a PAM-4-based ADT, extending the modulator to four levels in order to reduce out-of-band noise. This technique achieves a substantial improvement in coding efficiency, with $C_{eff}$ increasing from 5.8% in the traditional NRZ case to 63.2%. Experimental results confirm more than 10 dB reduction of out-of-band noise and EVM values below 1.4% up to 7.5 GHz, enabling compliance with 3GPP requirements for 256-QAM. Consequently, this approach not only enhances spectral efficiency but also relaxes analog filtering requirements, improving both practicality and energy efficiency of ADTs.

Finally, the last architecture employs a cascading ΔΣM where the first modulator is a multi-bit one, operating at a lower sampling frequency, and the second modulator is a one-bit one, operating at a higher frequency. This allows for less out-of-band noise to be generated by the first modulator. Then, the second modulator, due to its higher sampling frequency, will have its out-of-band noise further away from the signal. Ultimately, this enables a higher SNR, which is reflected in a lower EVM, and also allows for the use of a lower-quality analog filter.

# Chapter 4

## Fully Agile Modulators

The rapid diversification of wireless standards and spectrum usage demands transmitters that can retune carriers, bandwidths, and waveforms in real time without sacrificing efficiency. This chapter advances that goal by presenting fully agile, ADT architectures that decouple carrier agility from serializer clocking.

### 4.1 Introduction

The unparalleled proliferation of wireless standards operating at different frequencies, using dissimilar coding and modulation schemes, and targeted for various purposes, fosters the development of new, flexible radios capable of adapting to different communication scenarios. However, a key challenge for achieving this new communications paradigm involves the development of a highly flexible physical layer that supports the simultaneous transmission of multiple, multi-rate, multi-band, and multi-standard signals, which is very difficult to achieve using conventional approaches in practice.

Nonetheless, recent advances in this field include the development of novel all-digital transmitters [86, 188, 204, 226], where the datapath is entirely digital up to the RF stage, as shown in Fig. 4.1. The inherent high flexibility of such a concept represents a significant step toward the development of new radios and shortens the gap toward the ideal SDR transmitter [130]. However, current all-digital transmitters still have a few key limitations that prevent their wider dissemination. In fact, almost all the above-cited transmitters do not support multichannel operation, others require expensive external multiplexers for the RF up-conversion, and most of them require very high-quality filters to remove out-of-band noise.

This chapter is supported by the conference papers [7,190] and by the journal article [45]. The chapter starts by showing a solution to frequency agility based on IF-placement in BB-stage modulators, where the final RF frequency is controlled by moving the IF carrier frequency of the signal prior to the final upconversion to RF. The second part of the chapter then introduces the concept of RF-placement, where the signal is shifted to the final RF carrier frequency before the quantizer, whether it is a PWM modulator or a $\Delta\Sigma$M. At the end the chapter, presents a new approach for implementation of multiple less complex ADT in a combinatorial approach for a more agile and flexible modular-ADT, capable of adjusting to the network needs.

DOI: 10.1201/9781003225713-4

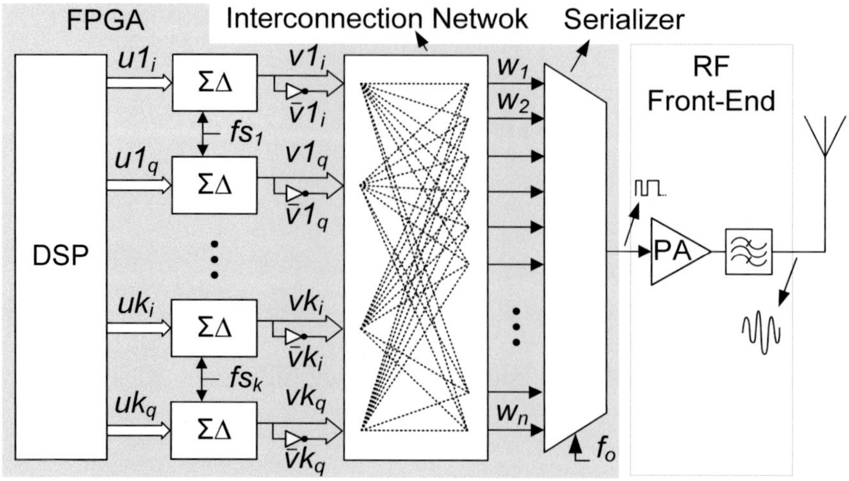


**FIGURE 4.1:** Block diagram illustrating the architecture of a conventional multichannel agile transmitter, adapted from [184].

## 4.2 Tunable IF-placement in BB-stage Modulators

In this section, we address the frequency agility limitations presented in old implementations by proposing a new transmitter architecture that allows the fine frequency tuning for each carrier, where the RF output operating in the microwave frequency range is generated inside a single FPGA. Additionally, the proposed RF transmitter explores the use of polyphase datapaths for improving the quality of the transmitting signal in terms of SNR and out-of-band noise.

### 4.2.1 Proposed All-Digital Transmitter

This subsection details the design and improvement stages of the proposed fine-tunable multichannel transmitter. In this architecture, fine frequency tuning is achieved by using a two-stage up-conversion approach, where IF image rejection is employed to maximize the usable bandwidth of the RF transmitter. The first up-conversion stage uses a digital SSB signal generation approach based on the Weaver modulator [53], where a set of two multipliers shift the baseband components ($b_i$ and $b_q$) into an IF signal ($u_i$) and two other multipliers generate a 90° phase-shifted IF signal centered at the same frequency ($u_q$), as can be seen in Fig. 4.2.

After the first up-conversion stage, the IF signal is converted into a 2-level representation using $\Delta\Sigma$ modulation. The $\Delta\Sigma$ coefficients were precomputed using the $\Delta\Sigma$-toolbox [177] for a sampling rate of $f_s = 150$ MHz and a bandpass of 12 MHz, resulting in an effective OSR of 12.5. These coefficients were then rounded to powers of two, or if not suitable, to a two's-complement binary representation using up to eight bits. This rounding approach contributes to reducing the area and, most importantly, shortens the critical path. The resulting highpass NTF can be given by:

$$\text{NTF}(z) = \frac{z^2 - 2z + 1.039}{z^2 - 0.875z + 0.3145}. \quad (4.1)$$

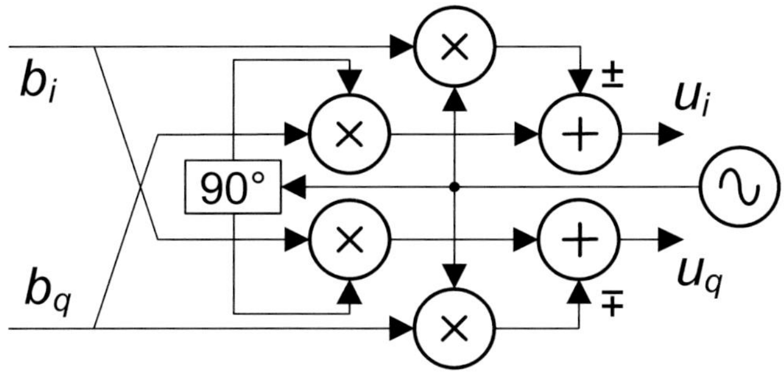


**FIGURE 4.2:** Block diagram illustrating the implemented quadrature single-sideband up-converter for IF image rejection.

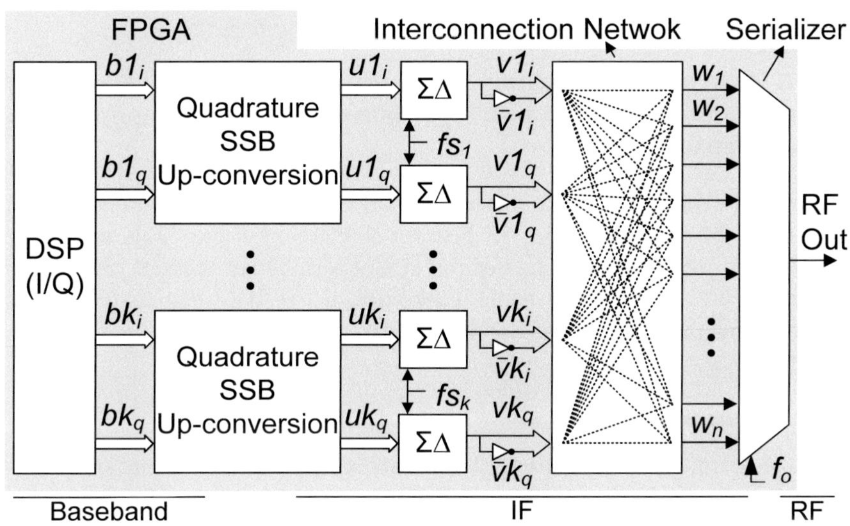


**FIGURE 4.3:** Proposed agile transmitter detailing the fine frequency tuning of the multichannel carriers by using a two-stage up-conversion approach combined with SSB signal generation.

A block diagram detailing the architecture of the proposed fine-tunable multichannel transmitter is shown in Fig. 4.3. In this architecture, each channel uses a $\Delta\Sigma$ pair to process the IF components ($u_i$ and $u_q$) generated by the first up-conversion stage. The resulting signals of the $\Delta\Sigma$Ms and their inverted versions are then connected to a multi-gigabit serializer through an interconnection network. This network combines and replicates the $\Delta\Sigma$ output signals to generate a parallel output word ($w$) that contains, for each channel, the four components of the desired signal ($v_i v_q \bar{v}_i \bar{v}_q$).

The serialization of this parallel word will generate a multichannel RF SSB signal, where the order given to the four components of the desired signal determines, for each channel, which sideband is generated. For instance, serializing $v_i v_q \bar{v}_i \bar{v}_q$ will generate an RF carrier centered at the Lower SideBand (LSB), and serializing $v_i \bar{v}_q \bar{v}_i v_q$ will result in an RF carrier centered at the Upper SideBand of the spectrum.

The proposed architecture, illustrated in Fig. 4.3, was further enhanced by using a polyphase datapath approach to improve the quality of the transmitting signal in terms of SNR and out-of-band noise emission, as shown in Fig. 4.4.

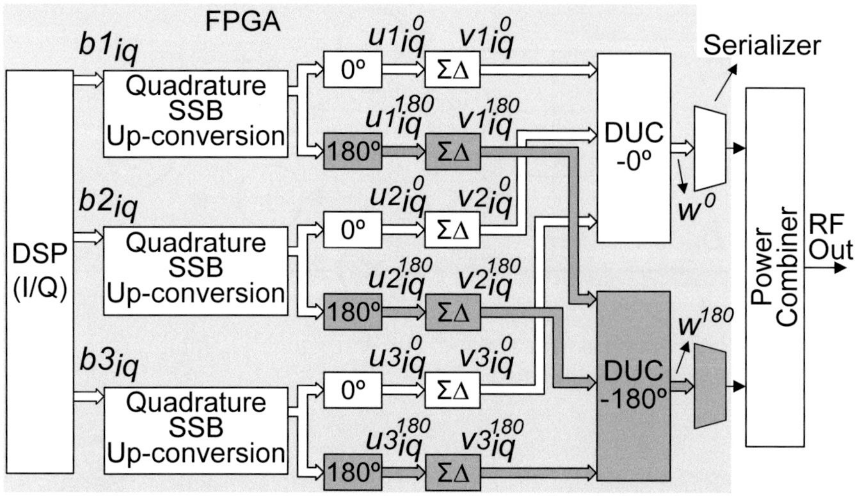


**FIGURE 4.4:** High-level block diagram illustrating the implemented fine frequency tuning multichannel transmitter combined with a polyphase datapath for improving the quality of the transmitting signal.

In this approach, the phase of the desired signal is shifted before entering the ΔΣMs and shifted back to the original phase in the last up-conversion stage. This way, each serializer will generate an RF output where the desired signal retains its original phase, while the ΔΣ quantization noise has a different phase. Therefore, if the RF outputs from each serializer are combined, the in-phase energy of the desired signal will increase more than the unwanted noise.

## 4.2.2 Experimental Results

The proposed fine-tunable multichannel transmitter was prototyped in an *ML628* development board containing a *Virtex6 HX380T* FPGA. For the purpose of this proof-of-concept, the datapath shown in Fig. 4.4 was implemented, and its digital RF outputs were connected to a *Mini-Circuits* 2:1 RF power combiner model *ZN2PD2-50-S+*. Finally, the power combiner was directly connected to a Vector Signal Analyzer (VSA). The baseband test signals used in this setup are one WiMAX (OFDM-based) and two 64-QAM signals, all stored inside the FPGA's internal memory.

As a first experiment, the implemented RF transmitter was configured to simultaneously transmit three independent carriers, one centered at 745.5 MHz containing a Narrow Band (NB) 64-QAM ($b1_{iq}$), a second one centered at 748.5 MHz containing a wider band 64-QAM ($b2_{iq}$) and a third one centered at 753.0 MHz, containing a WiMAX signal ($b3_{iq}$).

For this implementation example, the first up-conversion stage was configured to shift each one of the three baseband signals to a different IF band. The wider 64-QAM signal was up-converted to a 1.5 MHz IF, while the narrow band 64-QAM was shifted to 4.5 MHz and the WiMAX to 3 MHz.

Next, all ΔΣMs were set to operate at a sampling frequency of 187.5 MHz, and each serializer was configured to generate a line rate of 9 Gbps. For the polyphase datapath containing the *DUC-0°* shown in Fig. 4.4, the interconnection network was configured to generate the following 48-bit word:

$$w^0 = [v1_i\ v2_i\ v3_i\ \bar{v}1_q\ \bar{v}2_q\ v3_q\ \bar{v}1_i\ \bar{v}2_i\ \bar{v}3_i\ v1_q\ v2_q\ \bar{v}3_q\ v1_i\ v2_i\ v3_i\ \bar{v}1_q\ \bar{v}2_q\ v3_q\ \bar{v}1_i\ \bar{v}2_i\ \bar{v}3_i\ v1_q\ v2_q\ \bar{v}3_q\ v1_i\ v2_i\ v3_i\ \bar{v}1_q\ \bar{v}2_q\ v3_q\ \bar{v}1_i\ \bar{v}2_i\ \bar{v}3_i\ v1_q\ v2_q\ \bar{v}3_q\ v1_i\ v2_i\ v3_i\ \bar{v}1_q\ \bar{v}2_q\ v3_q\ \bar{v}1_i\ \bar{v}2_i\ \bar{v}3_i\ v1_q\ v2_q\ \bar{v}3_q].$$

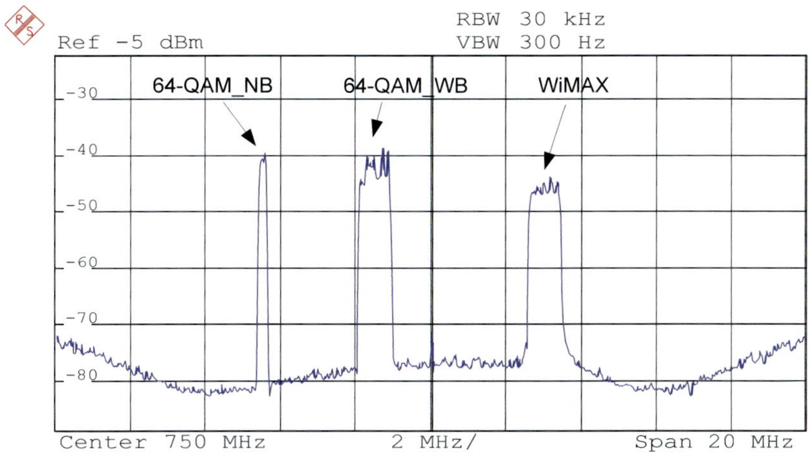


**FIGURE 4.5:** Measured multichannel spectrum when the fine tunable RF transmitter is configured to simultaneously generate three independent carriers, one centered at 745.5 MHz, a second one centered at 748.5 MHz and a third one centered at 753.0 MHz.

The serialization of $w^0$ at a rate of 9 Gbps will produce a multichannel RF signal with a 750 MHz central frequency, given by the line rate / (number of channels × 4). Moreover, the order given in $w^0$ to the four components $(v_i v_q \bar{v}_i \bar{v}_q)$ of the three channels will impose the two 64-QAM signals appearing at the LSB of the central frequency, while the WiMAX signal will appear at the Higher SideBand (HSB), as shown in Fig. 4.5. For the second path, the *DUC-180°* was configured to generate the following 48-bit output word:

$$w^{180} = [\bar{v}1_i\ \bar{v}2_i\ \bar{v}3_i\ v1_q\ v2_q\ \bar{v}3_q\ v1_i\ v2_i\ v3_i\ \bar{v}1_q\ \bar{v}2_q\ v3_q\ \bar{v}1_i\ \bar{v}2_i\ \bar{v}3_i\ v1_q\ v2_q\ \bar{v}3_q\ v1_i\ v2_i\ v3_i\ \bar{v}1_q\ \bar{v}2_q\ v3_q\ \bar{v}1_i\ \bar{v}2_i\ \bar{v}3_i\ v1_q\ v2_q\ \bar{v}3_q\ v1_i\ v2_i\ v3_i\ \bar{v}1_q\ \bar{v}2_q\ v3_q\ \bar{v}1_i\ \bar{v}2_i\ \bar{v}3_i\ v1_q\ v2_q\ \bar{v}3_q\ v1_i\ v2_i\ v3_i\ \bar{v}1_q\ \bar{v}2_q\ v3_q].$$

This second word is an inverted version of $w^0$ so that the desired signals are shifted back to their original phases. This way, the RF output of both serializers can be combined, and the desired signals will add constructively. In fact, using the VSA to demodulate the wider 64-QAM signal, it was possible to quantify the MER gain when two polyphase datapaths are used. The obtained 35.2 dB for the two polyphase datapaths versus 32.3 dB for the single-path approach shows a ≈3 dB improvement in the desired signal.

As a second experiment, the implemented RF transmitter was configured to perform the simultaneous transmission of three carriers, one centered at 373.5 MHz, a second one centered at 378.0 MHz and a third one centered at 753.0 MHz.

In this implementation example, all frequencies and data rates are equal to the first experiment, with the exception of the narrow band 64-QAM signal that was shifted to a 3 MHz IF in the first up-conversion stage. For the polyphase datapath containing the *DUC-0°*, the interconnection network was configured to generate the following 48-bit output word:

$$w^0 = [v1_i\ v2_i\ v3_i\ v1_i\ v2_i\ v3_q\ v1_q\ \bar{v}2_q\ \bar{v}3_i\ v1_q\ \bar{v}2_q\ \bar{v}3_q\ \bar{v}1_i\ \bar{v}2_i\ v3_i\ \bar{v}1_i\ \bar{v}2_i\ v3_q\ \bar{v}1_q\ v2_q\ \bar{v}3_i\ \bar{v}1_q\ v2_q\ \bar{v}3_q\ v1_i\ v2_i\ v3_i\ v1_i\ v2_i\ v3_q\ v1_q\ \bar{v}2_q\ \bar{v}3_i\ v1_q\ \bar{v}2_q\ \bar{v}3_q\ \bar{v}1_i\ \bar{v}2_i\ v3_i\ \bar{v}1_i\ \bar{v}2_i\ v3_q\ \bar{v}1_q\ v2_q\ \bar{v}3_i\ \bar{v}1_q\ v2_q\ \bar{v}3_q].$$

The serialization of $w^0$ at a rate of 9 Gbps will generate a multichannel RF signal with a lower central frequency of 375 MHz and a higher central frequency of 750 MHz. Once again, the order given in $w^0$ to the four components $(v_i v_q \bar{v}_i \bar{v}_q)$ of the three channels will

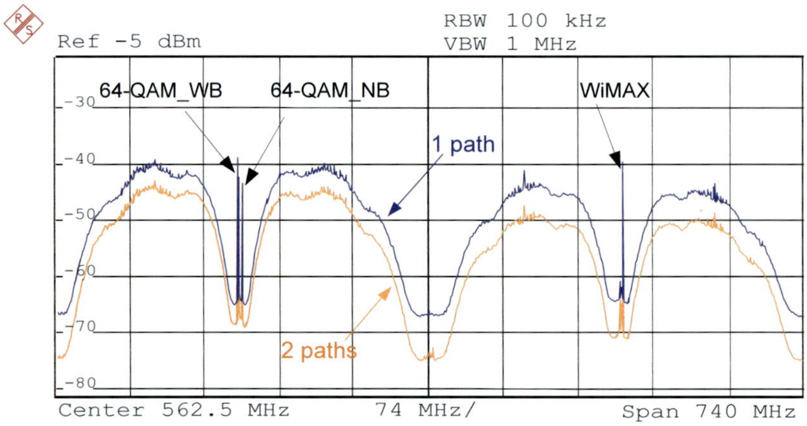


**FIGURE 4.6:** Measured multichannel spectrum when the proposed agile transmitter is configured to transmit three carriers and where the spectrum obtained by using two independent paths is overlapped with a single path approach with a gain normalized to the 2 paths.

impose the narrow band 64-QAM and the WiMAX signals to appear at the HSB of their central frequencies, while the wider 64-QAM will appear at the NB. For the second path, the *DUC-180°* was configured to generate the following 48-bit output word:

$$w^{180} = [\bar{v}1_i\ \bar{v}2_i\ \bar{v}3_i\ \bar{v}1_i\ \bar{v}2_i\ \bar{v}3_q\ \bar{v}1_q\ v2_q\ v3_i\ \bar{v}1_q\ v2_q\ v3_q\ v1_i\ v2_i\ \bar{v}3_i\ v1_i\ v2_i\ \bar{v}3_q\ v1_q\ \bar{v}2_q\ v3_i\ v1_q\ \bar{v}2_q\ v3_q\ \bar{v}1_i\ \bar{v}2_i\ \bar{v}3_i\ \bar{v}1_i\ \bar{v}2_i\ \bar{v}3_q\ \bar{v}1_q\ v2_q\ v3_i\ \bar{v}1_q\ v2_q\ v3_q\ v1_i\ v2_i\ \bar{v}3_i\ v1_i\ v2_i\ \bar{v}3_q\ v1_q\ \bar{v}2_q\ v3_i\ v1_q\ \bar{v}2_q\ v3_q].$$

As illustrated in Fig. 4.6, the use of polyphase datapaths allows for combining the energy of the desired signal in a proportion that is higher than the unwanted noise. This way, it is possible to increase the SNR of the transmitting signal while simultaneously reducing out-of-band noise.

### 4.2.3 Key Remarks

A new fine-tunable all-digital transmitter architecture was presented in this paper. The proposed RF transmitter supports the concurrent transmission of multi-rate, multi-band, and multi-standard signals, utilizing polyphase datapaths to enhance the quality of the transmitted signal.

The obtained results demonstrate the feasibility and potential of FPGA-based RF transmitters, where a two-stage up-conversion approach enables fine frequency tuning for each transmitting channel, and combining two polyphase datapaths significantly improves the SNR of the desired signals.

---

## 4.3 Tunable RF-placement in RF-stage Modulators

Real-time frequency agility is one of the key limitations of current ADTs. Thus, in this section, this research line is addressed, and a promising architecture is proposed.

This section is supported by the journal article [45]. In particular, the section proposes a fully parallel and digital architecture that can be embedded into an FPGA. This

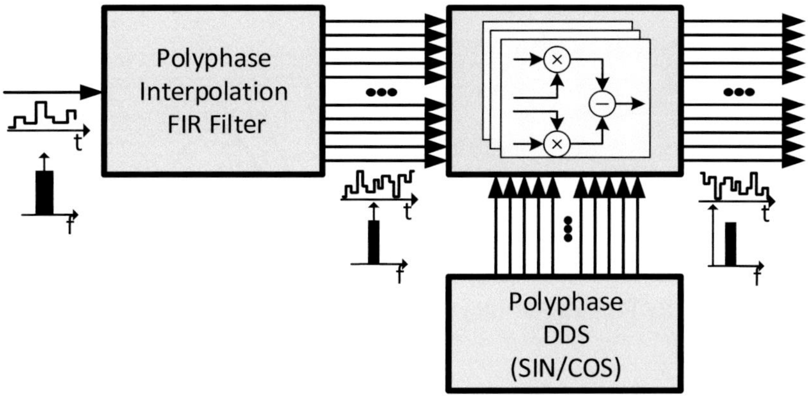


**FIGURE 4.7:** Block diagram of the proposed DUC.

architecture was the first one reported to enable the design of RF-stage modulators within an FPGA.

Wideband frequency agility requires a high sampling rate at the pulse encoder, and only RF-stage modulators can claim it. These sampling rates must be at least twice the desired carrier frequency. As the programmable logic subsystems from the FPGAs have a limited sampling rate (typically below 300 MHz), the SoA was unanimous to assume that FPGAs were not suitable to design RF-stage modulators. Thus, all the ADT techniques from the SoA fall in the implementation of BB-stage modulators by placing the DUC stage after the Pulse Encoder [37, 126, 200]. As aforementioned, this specific DUC is entangled with the serializer's sampling rate. Thus, the frequency agility in those architectures is only accomplished through the real-time reconfiguration of the serializers, which may jeopardize the overall FoM.

### 4.3.1 Proposed Architecture

In order to embed an RF-stage modulator into an FPGA, the proposed idea falls into the exploration of highly-parallelized architectures. In these architectures, high sampling rates can be aggregated while maintaining lower individual sampling rates. Thus, as a high global sampling rate at the pulse encoder can be maintained, the RF waveforms can be synthesized in the digital domain before the pulse encoder, allowing for unprecedented frequency agility.

#### 4.3.1.1 Digital Upconversion Stage

To ensure a high level of carrier agility, the signal must be upconverted before the pulse encoding.

Considering the strict timing constraints of the commercial FPGAs, common digital circuits designed within the programmable logic can typically be clocked up to 300 MHz. Due to this, a single-rate upconversion from BB up to the RF stage is not a feasible option. To overcome this limitation, multi-rate techniques were applied to the DUC stage. In particular, a polyphase equivalent is designed with $N$ different paths to achieve a global sampling frequency $N$ times higher than the BB sampling rates.

The proposed DUC is depicted in Fig. 4.7 and consists of a polyphase interpolation FIR filter, a polyphase I/Q DDS, working as a digital LO, and parallel digital mixers. The polyphase interpolation FIR filter increases the sampling rate of the BB signal from $Fs_{BB}$ to $Fs_{RF}$, while dividing the samples in $N$ different phases [205]. The polyphase I/Q DDS

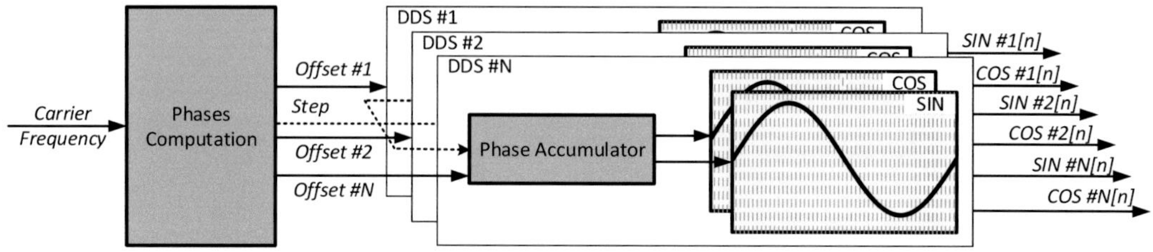


**FIGURE 4.8:** Functional block diagram of the polyphase DDS module.

is the core of this DUC stage, and it is illustrated in Fig. 4.8. In view of the aforementioned FPGA's timing constraints, a classic single-rate DDS cannot be used in this particular case because it is not possible to achieve a sampling rate equal to the serializer's sampling frequency ($Fs_{RF}$). To overcome this problem, a polyphase equivalent was designed. Since the DDS is a feedforward system, polyphase techniques do not significantly impact the critical path. In its essence, it combines $N$ conventional and single-rate DDS modules. Each DDS has a specific phase accumulator and two Read-Only Memory (ROM)s containing the sine and the cosine waveforms. The DDS modules work in parallel to simultaneously generate $N$ samples of the required carrier frequency ($\omega_c$) waveform. Each of the DDS is actually working at a rate $N$ times lower than the equivalent global sampling rate. Thus, it is possible to achieve a global sampling rate equal to the serializer's sampling frequency $Fs_{RF}$ with $N$ different phases being clocked at $Fs_{BB}$. The Phase Computation block, depicted in the same figure, performs the mapping between the desired carrier frequency, the phase step, and the respective DDS phase offset. The minimum frequency resolution ($\Delta f$) is computed as in the classical single-rate DDS, but now with a sampling rate $N$ times higher:

$$\Delta f = \frac{Fs_{BB}\,N}{2^L} = \frac{Fs_{RF}}{2^L} \tag{4.2}$$

where $L$ is the number of bits from each phase accumulator. The last module of the proposed DUC is a digital mixer, which performs the element-wise multiplication and subtraction between the interpolated samples and the different samples from the cosine and sine waveforms, according to Eq. (2.15).

It is worth mentioning a few notes regarding the efficient design of the polyphase DDS modules. The block diagram of Fig. 4.8 aims to provide a simplistic and conceptual overview of the polyphase DDS. However, depending on the system requirements, several optimization techniques and novel subsystems could be included. For instance, efficient memory usage can be achieved by exploiting the symmetry of sinusoidal waveforms. Interpolation or phase dithering can also be applied to increase the Spurious-Free Dynamic Range (SFDR). Moreover, the required SFDR is also dependent on the system application, as well as on the pulse encoding mechanism. In fact, generally speaking, the SFDR will be imposed by the pulse encoding mechanism, which will generate a substantially higher amount of noise than the DDS. These optimization techniques are typically embedded into the DDS Intellectual Property (IP) cores from the FPGA's vendors [219]. For this reason, their implementation is out of the scope of this work.

#### 4.3.1.2 RF Pulse Encoder

To validate the proposed architecture, an RF Pulse Encoder to perform the single-bit quantization must be designed and integrated after the DUC. This RF Pulse Encoder must be

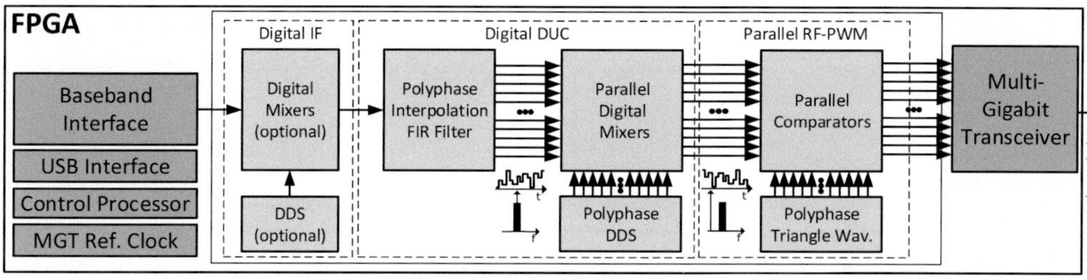


**FIGURE 4.9:** Block diagram of the implemented FPGA-based RF-PWM Digital Transmitter.

embedded into an FPGA, with no specialized application-specific circuit-level techniques. In this sense, this topic will be addressed next, and two parallel architectures will be proposed and implemented: a quadrature RF-PWM and a tunable RF-$\Delta\Sigma$M.

### 4.3.1.3 RF-PWM Architecture

The design proposed in this work is based on the traditional PWM generation mechanism, introduced in Section 2.2.3.1. Nonetheless, instead of synthesizing the two signals ($x[n]$ and $r[n]$) at low sampling frequencies ($Fs_{BB}$) (as performed in the literature [173]), the signals are generated in a multi-rate approach with a global sampling frequency of $Fs_{RF}$. In addition to this, the reference waveform's frequency $F_{ref}$ is lower than the RF carrier frequency, but higher than the signal's bandwidth (at least twice the bandwidth). An optimization procedure for the frequency of the reference waveform must be done to ensure a clean spectrum in the band of interest. To guarantee a global sampling frequency of $Fs_{RF}$, this reference waveform is also synthesized using a polyphase DDS with $N$ different phases (using a triangular waveform stored in the ROM).

The overall architecture is illustrated in Fig. 4.9. For the purpose of this proof-of-concept, the BB samples are continuously transmitted from the Baseband Interface to the Polyphase Interpolation FIR filter. This module increases the sampling rate by a factor of $N$, distributed into $N$ different phases, which are then upconverted according to Eq. (2.15). The frequency resolution from the whole modulator is given by Eq. (4.2). However, depending on the system requirements, a fine frequency resolution may be needed. While increasing the phase accumulators' resolution seems to be the most intuitive way to accomplish this, it results in unnecessary high resource usage. To avoid this, a single-rate Digital-IF can be included together with a high-resolution DDS before the Polyphase Interpolation FIR filter, as depicted in Fig. 4.9. In this way, the resolution of the phase accumulators from the polyphase DDS can be significantly reduced, leading to savings in the usage of memory-primitive resources, while the single-rate digital-IF provides fine frequency resolution. The trade-off is the bandwidth of the Polyphase Interpolation FIR filter, which must accommodate the IF-modulated signal.

After having the desired signal interpolated and upconverted to a given carrier frequency, it will be directly compared with the reference waveform, according to:

$$y_{RF}[n] = \text{sign}(r[n] - x_{RF}[n]) \tag{4.3}$$

At last, the resultant 1-bit signals from each branch are concatenated to generate a parallel output word that will be serialized by the MGT, running at $Fs_{RF}$.

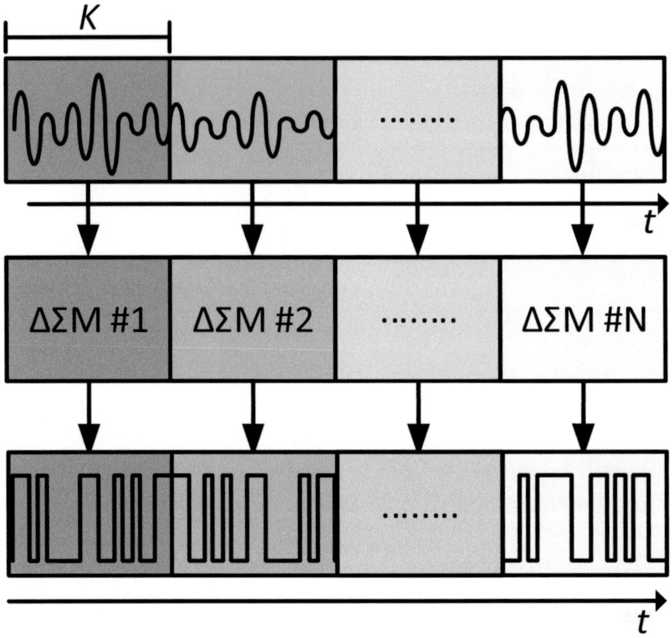


**FIGURE 4.10:** Illustration of the multicore ΔΣ modulators proposed in [37].[1]

#### 4.3.1.4 RF-ΔΣM Architecture

One of the main advantages of the use of ΔΣM is the inherent noise shaping that preserves the in-band integrity. Thus, there is a strong motivation for designing a parallel RF-ΔΣ architecture. The architecture proposed in this work relies on the replication and combination of multiple ΔΣ modulator cores running in parallel, as proposed in [37]. However, instead of modulating an incoming BB signal, several tunable bandpass modulators are included in parallel to enable the encoding of an upconverted RF signal, with a global sampling frequency equal to $Fs_{RF}$.

This multicore-based architecture is chosen because, to date, it is the closest approach to an ideal parallel representation of a ΔΣ modulator. The underlying idea is illustrated in Fig. 4.10 and is based on slicing the input signal into blocks of size $K$, which will feed each of the $N$ modulators in parallel. Even though all the modulators are running at a lower clock, the resultant behavior approaches the ideal modulator running at a global sampling frequency $N$ times higher. In addition to that, in a careful design, the resultant critical path (that binds the maximum global sampling rate) is imposed by the modulator itself and not by the system as a whole. This is of paramount importance to achieve higher sampling rate gains.

To implement this architecture, two new subsystems—Deinterleaving and Interleaving modules—were introduced. A graphical illustration of the Deinterleaving and the Interleaving modules is depicted in Fig. 4.11. As the temporal progression in a multi-rate system unfolds between the different phases with every clock tick, the Deinterleaving module is required to rearrange the incoming samples to maintain temporal continuity within each phase. This is of paramount importance to make sure that the ΔΣMs cores are processing contiguous samples. The Interleaving module is responsible for performing the inverse

[1]The concept of multicore ΔΣ will be further introduced in Chapter 5.

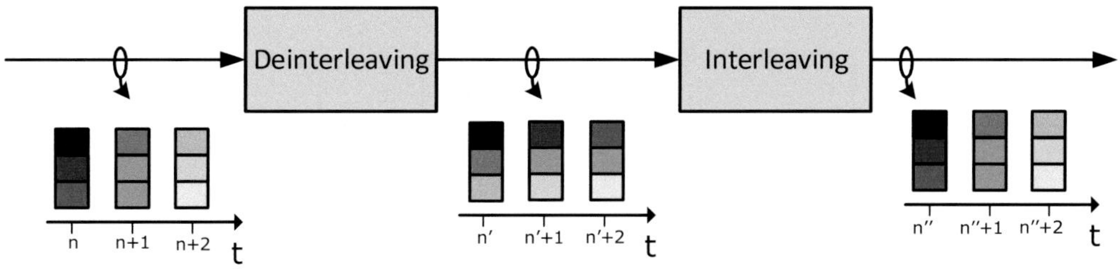


**FIGURE 4.11:** Graphical illustration of the Deinterleaving/Interleaving functions for the case of $N = 3$ phases and $K = 3$ samples. In the case where $N$ equals $K$, one can see that the Deinterleaving/Interleaving modules are simply matrix transposers. In the figure, $n'$ and $n''$ refer to the clock instant $n + L$, where $L$ is the latency of each module.

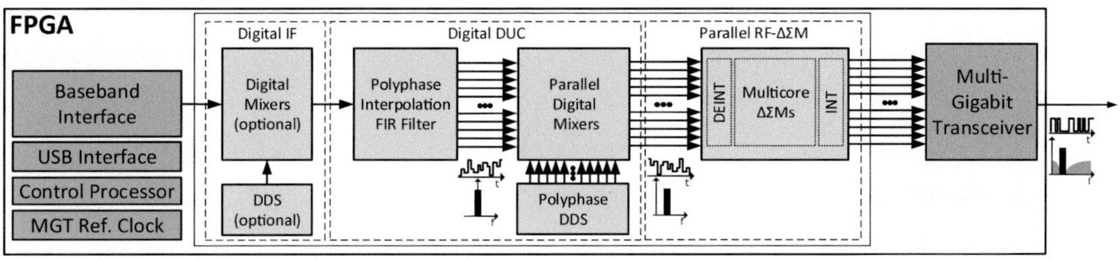


**FIGURE 4.12:** Block diagram of the implemented FPGA-based RF-ΔΣM Digital Transmitter.

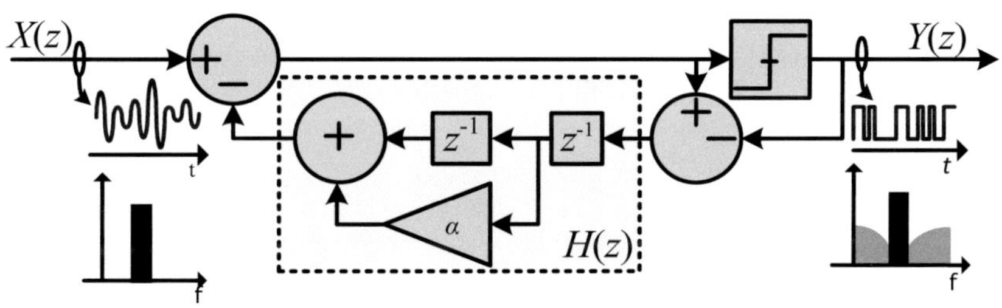


**FIGURE 4.13:** Z-domain representation of the tunable Error-feedback ΔΣM.

operation: converting temporal continuity within a phase to temporal continuity between different phases.

Nonetheless, it must be noted that this is not an error-free architecture, as the discontinuities introduced in every $K$ samples degrade the system's performance. Hence, the higher this value is, the better will be the modulator's FoM. However, this increases the amount of resources required to temporarily store all the samples in the Deinterleaving/Interleaving modules.

The overall architecture is illustrated in Fig. 4.12, and as can be seen, the block diagram is quite similar to the RF-PWM case (Fig. 4.9). The main difference lies in the last stage, the RF Pulse Encoder, which is based on multicore ΔΣ modulators. In particular, Deinterleaving and Interleaving modules must be designed to accommodate and re-arrange the input/output data from $N$ tunable bandpass ΔΣ modulators. Similar to the RF-PWM case, the resultant 1-bit signals from all phases are concatenated to generate a parallel word that will be serialized by the MGT, running at $Fs_{RF}$.

Focusing on the ΔΣ topology, the design has fallen into the Error-feedback structure, whose illustration is depicted in Fig. 4.13. This architecture presents a high level of

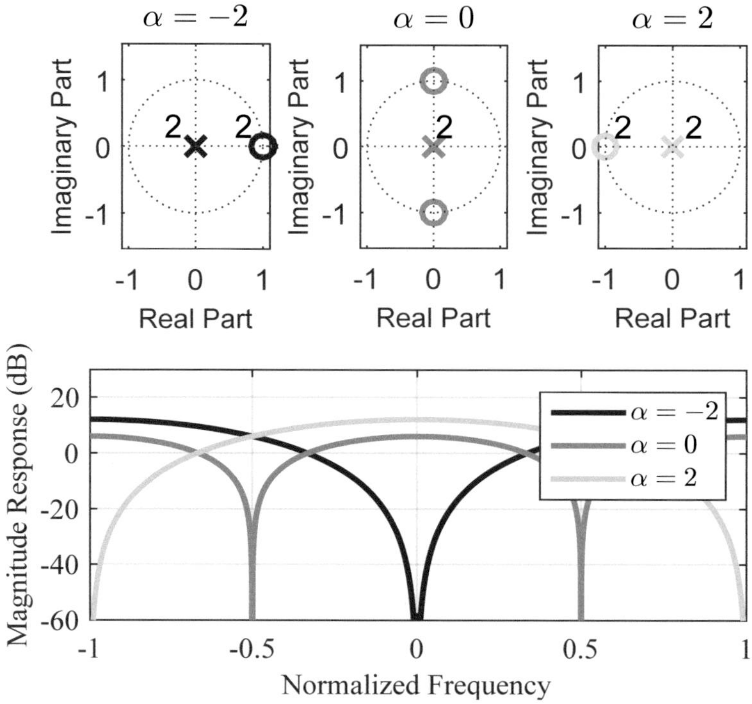


**FIGURE 4.14:** NTF's frequency response according to different $\alpha$ values.

simplicity and stability, with a short critical path, which eases the implementation process and leverages the use of higher sampling rates. Modeling the quantizer by a noise source ($E_q(z)$), one can derive:

$$\begin{aligned} Y(z) &= X(z) - (1 + H(z))E_q(z) \\ &= STF(z)X(z) - NTF(z)E_q(z) \end{aligned} \tag{4.4}$$

where $STF(z) = 1$, $NTF(z) = (1 + H(z))$ and $H(z) = \alpha z^{-1} + z^{-2}$. . To determine the range of the gain $\alpha$, the frequency response's magnitude of the transfer function must be assessed, which can be easily done by evaluating the $NTF(z)$ on the unit circle, $z = e^{j\theta}$.

$$\begin{aligned} |NTF(e^{j\theta})| &= |1 + \alpha e^{-j\theta} + e^{-2j\theta}| \\ &= |2\cos(\theta) + \alpha| \end{aligned} \tag{4.5}$$

The notch frequency(ies) can be found when the magnitude of the frequency response is zero, which leads to $\alpha = -2\cos(\theta)$. In particular, as depicted in Fig. 4.14, the NTF's frequency response can be fully reconfigured from a low-pass filter ($\alpha = -2$) up to a high-pass filter ($\alpha = +2$), passing through a band-pass filter with tunable center frequency ($-2 < \alpha < +2$).

### 4.3.2 Prototype Validation

To validate the proposed parallel direct-conversion architecture, the two different ADTs (depicted in Figs. 4.9 and 4.12) were implemented in a Virtex Ultrascale XCVU095 FPGA using the Xilinx VCU1287 Characterization Kit. For the sake of a proof-of-concept, the polyphase DDS modules were designed with custom phase accumulators and custom LUTs,

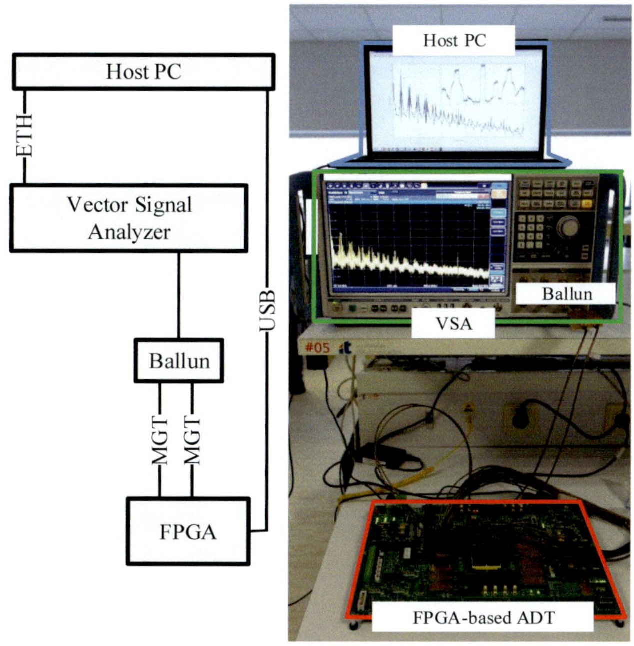


**FIGURE 4.15:** Block diagram and photo of the experimental setup.

and not with Xilinx-provided IP cores. For the same reasons, the optional Digital-IF modules were also not included. In both projects, the number of phases is equal to 64. In the ΔΣM case, an implementation with $K$ set to 256 samples was chosen, considering both maximum resource usage and system performance.

In the proposed architectures, the MGT bitrate defines the modulator's first NZ (from 0 to $Fs_{RF}/2$). This means that the carrier frequency can be located in any bin from the frequency grid given by the DDS' frequency resolution within this NZ. Nonetheless, it is therefore clear that the higher the MGT bitrate, the better will be the system's FoM due to the increased OSR. However, this will impact the carrier frequency resolution, as shown in Eq. (4.2). For the RF-PWM case, the maximum timing constraints imposed a maximum $Fs_{BB}$ of 200 MHz, and consequently, the bitrate of 12.8 Gbps was chosen. For the RF-ΔΣM, the feedback caused a reduction of the maximum BB sampling rate down to 160 MHz, leading to the serializer's bitrate of 10.24 Gbps. The frequency resolutions can be computed by using Eq. (4.2). In particular, the different bitrates together with the 13-bit resolution of the DDS' phase accumulator lead to 1.5625 MHz and 1.25 MHz for the RF-PWM and RF-ΔΣM, respectively.

The MGT differential outputs were directly connected through a balun to the VSA R&S FSW 8 for signal analysis and demodulation. The experimental setup is depicted in Fig. 4.15. The host computer controls and manages the FPGA-based ADT through a Universal Serial Bus (USB) connection. At the same time, it controls and requests the data captured from the VSA. This automated testbed can sweep different carrier frequencies, modulations, and bandwidths; and record such FoM as spectrum, MER, and EVM.

#### 4.3.2.1 RF-PWM-based ADT

As also introduced in Section 2.2.3.1, in the discrete PWM encoder, several expanded replicas or quantization noise from other NZs may fold back into the first one. The distribution and placement of this noise are dependent on the choice of the reference waveform frequency.

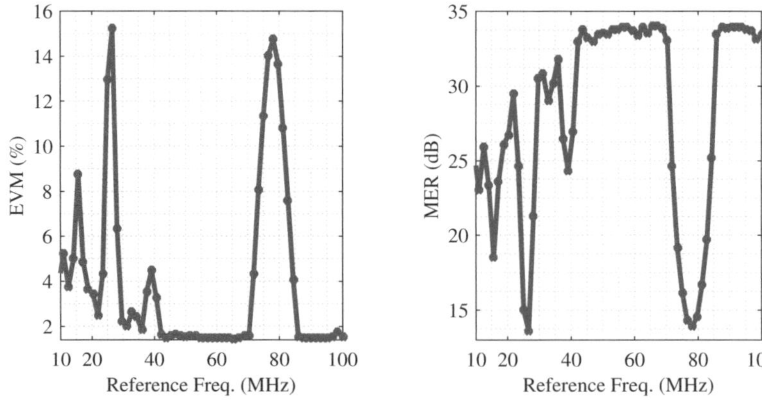


**FIGURE 4.16:** Measured results of a sweep in the reference waveform frequency for a carrier frequency of 4240.625 MHz with a 16-QAM 6 MHz modulated signal.

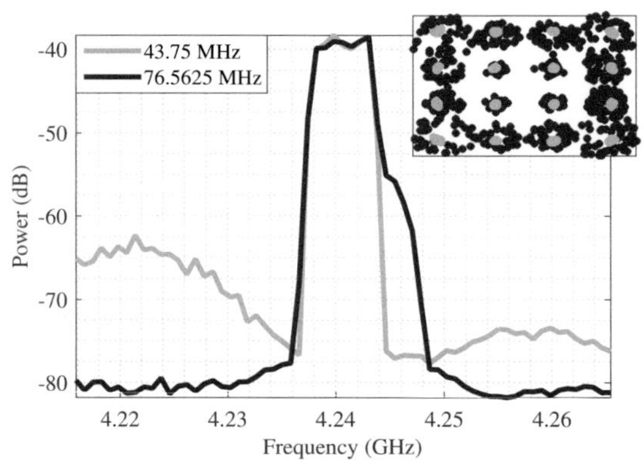


**FIGURE 4.17:** Obtained spectra for two of the points from Figs. 4.16. 43.75 MHz corresponds to EVM 1.48% and 76.5625 MHz to EVM 14.03% (RBW=100 kHz, VBW=2 kHz, SPAN=50 MHz).

To validate this assumption, a given carrier frequency was fixed, and the reference frequency was changed from 43.75 MHz to 76.5625 MHz. The results are depicted in Figs. 4.16 and 4.17, in the form of MER/EVM pairs and obtained spectrum, respectively. The results successfully demonstrate that, depending on the frequency of the reference waveform, harmonic distortion and quantization noise can interfere with the desired signal, negatively impacting the signal's in-band quality. Thus, one can conclude that the frequency of the reference waveform must be carefully chosen to ensure optimum performance for all the carrier frequencies.

In this system, the frequencies of the reference waveforms were *a priori* defined as 29.6875 MHz and 40.625 MHz, at two different symbol rates of 5 Msps and 10 Msps, respectively. These values were chosen to allow a considerable distance between the in-band signal and the first expanded harmonics. Then, to ensure a high-performance system and to minimize the amount of in-band distortion from higher NZs, a simple optimization methodology was designed. This low-complexity methodology involves sweeping the reference waveform frequency, within a small set of frequencies around the *a priori* defined ones, while the main FoM are measured in the VSA. Then, the reference frequencies that provide reduced

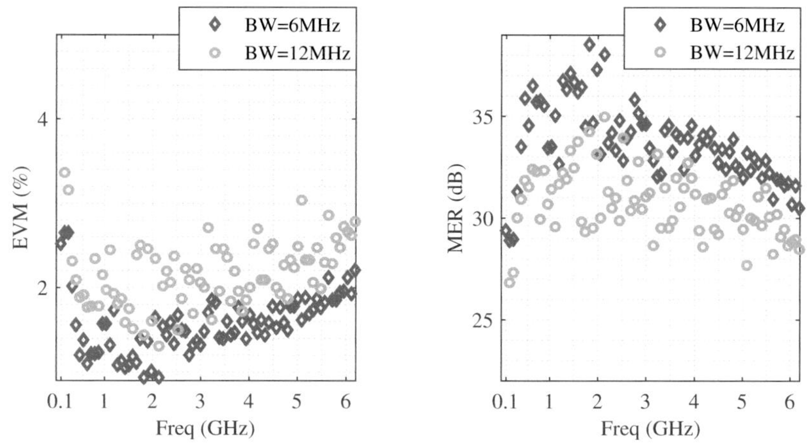


**FIGURE 4.18:** Measured results of a sweep in the carrier frequency with 16-QAM 5 Msps and 10 Msps modulated signals, which correspond to an available transmission bandwidth of 6 MHz and 12 MHz (roll-off factor of 0.2).

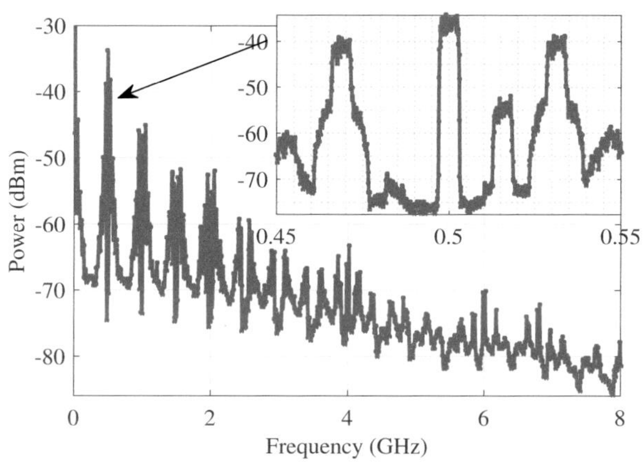


**FIGURE 4.19:** Obtained spectrum with a full span of 8 GHz for a carrier frequency of 500 MHz (RBW=100 kHz, VBW=2 kHz).

in-band distortion for each carrier frequency are stored in the FPGA. This is a one-time procedure for a given signal bandwidth.

Fig. 4.18 shows the measured results of a sweep of carrier frequencies in the range of 100 MHz up to 6.4 GHz, with a step of 78.125 MHz, for two different symbol rates: 5 Msps and 10 Msps. Focusing on the specific FoM, the maximum obtained MERs were 39 dB and 36 dB for the 5 Msps and the 10 Msps symbol rates, respectively. As expected, the reduction of the signal power as the frequency increases results in a lower MER. Nonetheless, even at 6.4 GHz, the average MER is close to 30 dB, resulting in well-defined constellations.

The spectrum in a span of 8 GHz for a carrier frequency of 500 MHz is depicted in Fig. 4.19. This is the spectrum resultant from the *sign* operation in the PWM.

Finally, the occupied resources in the FPGA for the proposed architecture are presented in Table 4.1. A high percentage of DSP slices ($\simeq$ 49%) are being used in the polyphase interpolation FIR filters. These filters were designed with a 3-tap transposed direct-form

**TABLE 4.1:** Occupied resources for the proposed RF-PWM transmitter design

| Logic Resources | Used | Total | % |
|---|---|---|---|
| Flip Flops | 15256 | 1075200 | 1.42 |
| LUTs | 20677 | 537600 | 3.85 |
| Memory LUT | 640 | 76800 | 0.83 |
| BRAM | 96 | 1728 | 5.56 |
| DSP48 | 372 | 768 | 48.44 |
| GTHs | 1 | 28 | 3.57 |

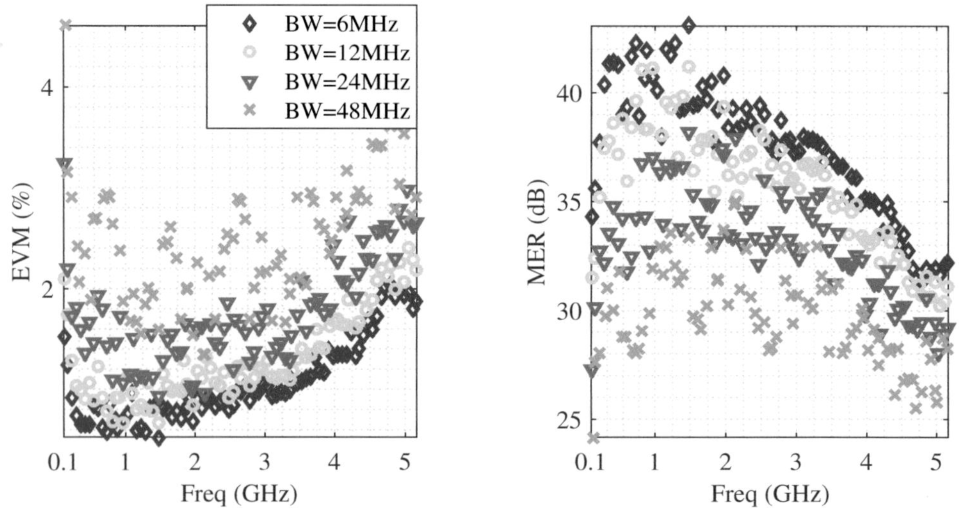


**FIGURE 4.20:** Measured results of a sweep in the carrier frequency with 16-QAM 5 Msps, 10 Msps, 20 Msps and 40 Msps modulated signals, corresponding to available transmission bandwidths of 6 MHz, 12 MHz, 24 MHz and 48 MHz (roll-off factor of 0.2).

architecture per phase. Nonetheless, this percentage can be lowered with the resort to multiplier-less FIR or Cascaded Integrator-Comb (CIC) filters.

#### 4.3.2.2 RF-ΔΣM-based ADT

Fig. 4.20 shows the measured results of a sweep of carrier frequencies in the range of 100 MHz up to 5.12 GHz, with a step of 62.5 MHz for four different symbol rates: 5 Msps, 10 Msps, 20 Msps, and 40 Msps. Focusing on the specific FoM, the maximum obtained MERs were 43 dB, 42 dB, 38 dB, and 35 dB, respectively. Similarly to the PWM case, the signal power tends to decrease as the frequency increases, resulting in lower MERs. Nonetheless, even in the 5.12 GHz of carrier frequency, the average MER is close to 25 dB (for 40 Msps of symbol rate), resulting in well-defined constellations.

The spectrum in a span of 8 GHz for a carrier frequency of 2.475 GHz is shown in Fig. 4.21. Additionally, Fig. 4.22 shows the comparison between the different bandwidths in terms of spectrum and modulated constellation. The signal replicas in the spectrum presented in Fig. 4.21 are spaced by $Fs/2$. This evidences a global sampling rate of 10.24 Gsps. The noise shaping capabilities from the ΔΣM can also be observed in Fig. 4.21. Moreover, the image in the second NZ also presents a reasonable MER, meaning it could be used to achieve higher carrier frequencies [66]. The spurs around the desired carrier (and around

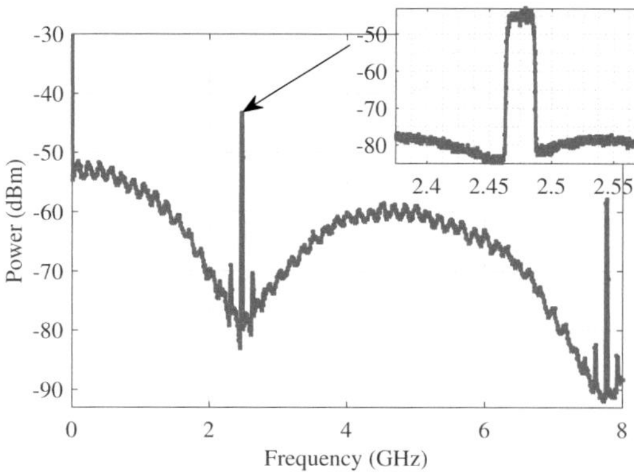


**FIGURE 4.21:** Obtained spectrum with two different spans (8 GHz and 100 MHz) for a 16-QAM modulated signal with 24 MHz of bandwidth at 2.475 GHz (RBW=100 kHz, VBW=2 kHz).

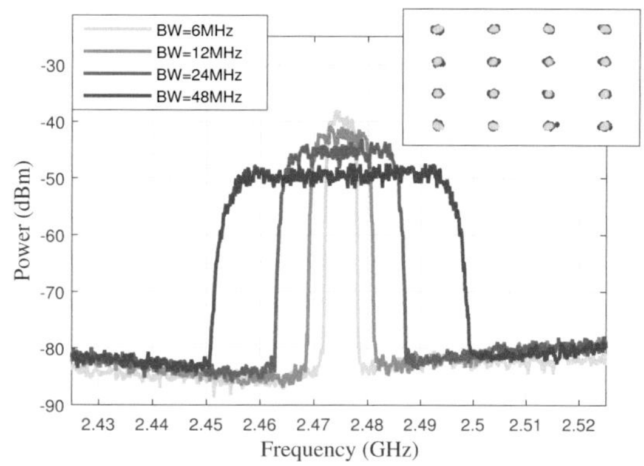


**FIGURE 4.22:** Different obtained spectra and respective constellations for bandwidths of 6 MHz, 12 MHz, 24 MHz and 48 MHz in a carrier frequency of 2.475 GHz (RBW=100 kHz, VBW=2 kHz).

**TABLE 4.2:** Occupied resources for the proposed RF-ΔΣM transmitter design

| Logic Resources | Used | Total | % |
|---|---|---|---|
| Flip Flops | 25222 | 1075200 | 2.35 |
| LUTs | 54516 | 537600 | 10.14 |
| Memory LUT | 2438 | 76800 | 3.17 |
| BRAM | 1443 | 1728 | 83.51 |
| DSP48 | 256 | 768 | 33.33 |
| GTHs | 1 | 28 | 3.57 |

the image) are caused by the inefficient interpolation FIR filter based on ZOH [163]. This interpolation technique was chosen due to its simplicity, efficiency in terms of occupied resources (no arithmetic operations are required), and low complexity, making it ideal for application in a proof-of-concept. Nonetheless, other topologies could have been selected, depending on the system's requirements.

Finally, the occupied resources for the RF-ΔΣM case are shown in Table 4.2. In this case, the Deinterleaving and the Interleaving modules are responsible for using a high percentage of BRAM primitives. These modules are required to temporarily accommodate all the incoming/outgoing data, before performing the required rearrangements. Furthermore, a high percentage of DSP slices is used in the parallel ΔΣ modulators. Contrasting with the PWM case, the interpolation filter does not occupy a significant amount of resources. This is due to the use of the ZOH topology.

#### 4.3.2.3 Comparison to the SoA

Table 4.3 presents an overview of the relevant modulators presented in the literature. The comparison focuses on the frequency agility (Min. $F_c$ and Max. $F_c$), together with the modulator's specifications: output sampling frequency (Fout), signal's BW, frequency resolution, modulator's type (Mod. Type), and modulator's implementation (HW Impl.). By agility (last column), one is referring to implementations or techniques that allow the adjustment of the carrier frequency without changing the output's sampling rate. Although all published works exhibit a certain degree of frequency agility, achieved by reconfiguring the

**TABLE 4.3:** Comparison of the proposed architectures with some relevant modulators from the SoA[1]

| Reference | Min. Fc (MHz) | Max. Fc (MHz) | Max. Fout (Gsps) | Max. BW (MHz) | Freq. Res. (MHz) | Mod. Typ | HW Impl. | Agility |
|---|---|---|---|---|---|---|---|---|
| **This (Impl. 1)** | **100** | **6200** | **12.4** | **12** | **1.5625** | **RF-PWM** | **FPGA** | **Yes** |
| **This (Impl. 2)** | **100** | **5120** | **10.24** | **48** | **1.25** | **RF-ΔΣM** | **FPGA** | **Yes** |
| [126] (Impl. 1) | - | 983 | 63 | 5 | - | BB-ΔΣM-PWM | ASIC | No |
| [126] (Impl. 2) | - | 2620; 5240 | 24.5 | 20 | - | BB-ΔΣM-PWM | FPGA | No |
| [201] | - | 5200 | 20.8 | 20 | - | BB-ΔΣM | FPGA | No |
| [43] | 2375 | 2625 | 10 | 3.75 | 0.05 | BB-ΔΣM | FPGA | Yes |
| [37] | - | 1600 | 6.4 | 122 | - | BB-ΔΣM | FPGA | No |
| [30] | - | 856; 1450 | 25 | 20 | n/a | RF-PWM | AWG | Yes |
| [230] | - | 1905 | 7.9 | 20 | - | BB-PWM | FPGA | No |
| [195] | - | 900; 1950 | n/a | 5 | - | RF-PWM | ASIC | No |
| [36] | - | 900; 1000 | 4 | 15 | - | BB-ΔΣM | FPGA | No |
| [145] | 2100 | 2200 | 5 | 5 | (Analog) | RF-ΔΣM | ASIC | Yes |
| [66] | - | 650; 950 | 4 | 5 | n/a | RF-ΔΣM | ASIC | No |
| [144] | 2100 | 2200 | 9 | 5 | (Analog) | RF-ΔΣM | ASIC | Yes |
| [175] | 1550 | 2450 | 7.5 | 20 | (Analog) | RF-ΔΣM | ASIC | Yes |

[1]In the table, "-" and "n/a" stand for "not-applicable" and "not-available", respectively.

output sampling frequency, this solution is not ideal. As first introduced in Section 2.2.5, this reconfiguration is limited, or not feasible at all, and will impair the system's performance by unintentionally varying the modulator's OSR. This is why the frequency agility of the solutions presented in this work is considerably higher than that of their SoA counterparts. Moreover, this work has demonstrated the generation of signals centered around a maximum carrier frequency of $F_s/2$, showing that the carrier frequency and the output sampling frequency do not need any OSR (generation is possible up to the Nyquist Limit). Ultimately, this will relax the final system requirements and, at the same time, will minimize the switching losses from the amplifier stage. In addition to this, the work also presents the first design of RF Pulse Encoders with no dedicated, custom, application-specific devices. In particular, the fully parallel approach enables the integration of pulse encoders operating at RF frequencies into an FPGA, with a maximum sampling rate of less than 200 MHz in the programmable logic part.

Finally, when considering the two different implementations proposed in this work, it is worth noting that there is a trade-off between performance and resource utilization. On the one hand, the RF-PWM case is a minimalist architecture, with a low degree of resource occupation, but with a number of drawbacks: the dependence on the reference waveform's frequency, the low available bandwidth, and the high requirements for the analog filtering. On the other hand, the multicore-based RF-ΔΣM is a more robust solution, with relaxed analog filtering requirements, with a reasonable MER and a scalable solution (in the sense that any topology can be chosen for the modulator itself). However, to achieve superior performance, a reasonable number of contiguous samples is required, which leads to excessive resource occupation.

One final note regarding the scalability of this architecture is worth highlighting. The main focus of this work was to demonstrate a parallel RF-stage transmitter that could be synthesized and embedded into programmable logic. This limits the sampling rate to be significantly inferior to the required RF carriers. Thus, to validate the aforementioned architecture, two different RF pulse encoders were designed and integrated as a proof of concept. Consequently, the amount of logic resources reported in both implementations should not be taken as the minimum required values for designing and implementing this architecture. Several different optimization techniques can be applied in the critical subsystems, depending on varying requirements. For instance, a recent proposal for relaxing the resource usage in multicore-based $\Delta\Sigma$M was presented in [46].

### 4.3.3 Key Remarks

This section presented a new architecture for designing agile and real-time reconfigurable RF-stage ADTs. The technique was proposed and validated in an FPGA-based transmitter. By enabling the synthesis of a fully parallel RF-PWM/$\Delta\Sigma$M transmitter, unprecedented frequency agility was achieved (in the range of 100 MHz to 6 GHz, with a maximum frequency resolution of 1.5625 MHz).

This represents a considerable improvement over conventional single-rate counterparts that operate directly in the RF stage. These single-rate architectures require sampling rates much higher than the carrier frequencies. The proposed architecture exceeds the SoA in terms of simplicity and flexibility. The system's behavior, in terms of carrier frequency, symbol rate, and modulation type, can be updated in real-time by adjusting the set of variables that control the entire system.

In the end, it can be concluded that the proposed parallel architecture has significant potential for designing real-time, reconfigurable digital RF Transmitters. The experimental results demonstrate that it is possible to generate signals with high MER at the desired carrier frequency, which is reconfigurable in real-time. The real-time reconfiguration involves a minor adjustment to the set of variables that control the polyphase DDS and the polyphase reference waveform generator (in the RF-PWM case). Thus, the MGT's bitrate can be optimized and maintained constant during the system's operation to provide adequate OSR and, ultimately, the ideal FoM.

---

## 4.4 Multi-Channel Modular All-Digital Transmitter

With every next generation of mobile telecommunications, the previous generation capabilities will be further pushed and more demanding with regard to higher data rates, reduced latency, and improved spectral latency, among others. Massive MIMO is a promising technology to address these needs by efficiently focusing energy and reducing inter-cell interference.

However, realizing its full potential presents several challenges. Overcoming these hurdles is essential to advance next-generation communication and fully leveraging the advantages that Massive MIMO offers. ADTs present as a key piece for unlocking the full capabilities of SDR systems, as they convert multi-bit data into single-bit streams using $\Delta\Sigma$Ms.

This approach enables the use of simpler, more efficient, and higher-speed one-bit DACs, which is advantageous for the expected next generation of large-scale Massive MIMO systems with many antenna elements. However, $\Delta\Sigma$-based ADTs face challenges in balancing carrier flexibility and bandwidth. Recent research has explored parallel $\Delta\Sigma$ core designs to increase operating frequencies, carrier flexibility, and bandwidth.

Although these designs can enhance performance, they demand substantial resources, which can limit scalability and adaptability in systems where resource sharing and dynamic allocation are critical. Thus, in this section, a new flexible tile-based multi-channel digital architecture of modular ΔΣ-ADTs aimed at Multi-User RUs is presented.

The section is supported by the conference paper [7]. The section presents a novel flexible tile-based multi-channel digital architecture of modular ΔΣ-ADTs aimed at Multi-User RUs. The proposed architecture displays key improvements compared to the more traditional approach to the ΔΣ-ADTs by combining less complex ΔΣ-ADTs units operating in parallel as individual "tiles" which can be dynamically combined.

This tile-based approach enables improvements in internal filtering and quantization processes within the ΔΣ cores, achieving better SNR, higher bandwidth signals, center carrier frequencies and a more scalable, modular solution that supports a wide range of configurations. The design allows for dynamic adjustments in response to network demands, offering greater flexibility in resource allocation and supporting future advancements in 6G systems.

### 4.4.1 Contextualization

ADTs are a key technology for achieving the true potential of SDR systems, as envisioned in [130]. These transmitters convert the original multi-bit data into a single-bit stream using ΔΣ Modulators. This simplified signal allows them to leverage special one-bit DACs. Compared to traditional RF-DACs, these one-bit versions are cheaper, require less power, and can handle higher sampling rates. Also, considering that massive MIMO systems with a large number of radiating elements are one of the enablers of 6G [4], ADT present as a scalable and flexible option to these types of systems [51].

Nevertheless, ΔΣ-ADT systems suffer from a design trade-off in terms of carrier flexibility and wide bandwidths [41, 45]. Some works have proposed new approaches to achieve ADTs that are more agile, carrier flexible, and with higher bandwidths [46, 154]. By using the parallel approach to the ΔΣ-Cores, as shown in [46], it is possible to achieve higher operating frequencies in the FPGA, leading to higher carrier frequencies and bandwidths.

Nonetheless, the required resources for implementing these types of architectures can be problematic in some systems, thus the need for a new approach where the available resources are limited. Also, this type of fixed approach can present as a challenge to a scalable and dynamic vision of the RU, where the resources can be shared and assigned dynamically.

This section presents a novel flexible tile-based multi-channel digital architecture of modular ΔΣ-ADTs aimed at Multi-User RUs. The proposed architecture utilizes multiple, less complex ΔΣ-ADT operating in a tile-based approach. The design offers flexibility and reconfigurability of the system by enabling various modular configurations of digital tile-based ΔΣ-ADTs channels that can be combined to achieve better system performance by trading off the number of available channel outputs.

Also, combining multiple tile-based ΔΣ-ADTs, it results in improving the internal filtering and quantization of the ΔΣ-Cores and enables higher carrier frequencies and bandwidths, as similar works are performed in a more fixed solution [154]. This approach allows the RU to dynamically adjust the available resources to meet the needs of the network.

### 4.4.2 Architecture Description

The proposed flexible tile-based multi-channel digital architecture of modular ΔΣ-ADTs channels is based from the typical architecture of a ΔΣ-Core with a feedback loop shown in Fig. 4.23. Building upon this architecture as a foundation, and applying parallelization with the aim to use multiple individual less complex ΔΣ-ADT, a new architecture was designed and proposed, as seen in Fig. 4.24.

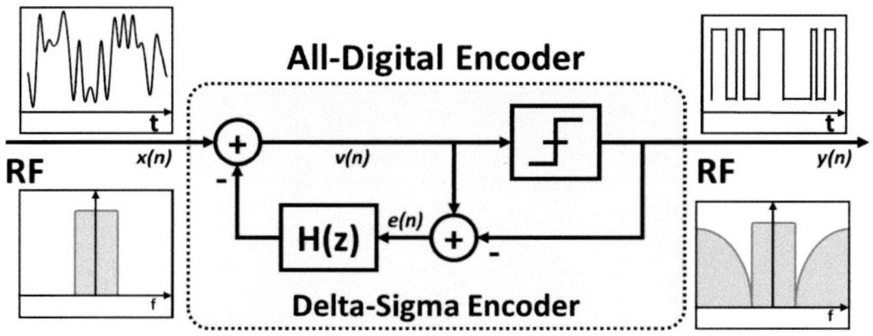


**FIGURE 4.23:** ΔΣ feedback ADT.

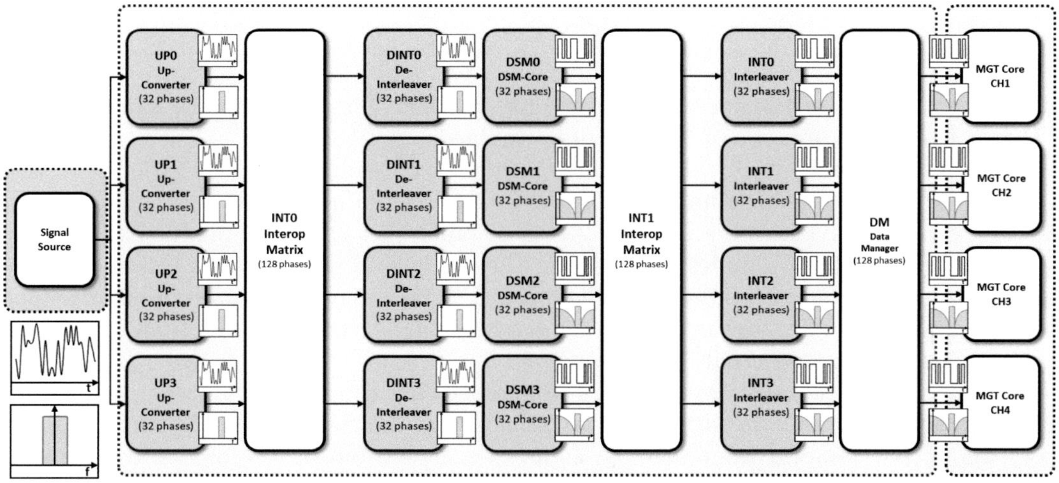


**FIGURE 4.24:** Proposed flexible tile-based multi-channel digital architecture of modular ΔΣ-ADT channels.

The proposed architecture consists of four individual tile-based ΔΣ-ADT channels composed of four main blocks each: an Up-Converter, a De-Interleaver, a ΔΣ-Core, and an Interleaver. As in its name, the Up-Converter performs an up-conversion on the baseband source signal, using a DDS approach, to the desired output carrier frequency. The Interleaver reorganizes the samples to adjust to the ΔΣ-Core. The ΔΣ-Core is responsible for the quantization and filtering of the up-converted signal, similar to the process shown in Fig. 4.23. Finally, the Interleaver reorganizes the output samples from the ΔΣ-Core to align them temporally and attain a full RF conversion of the source signal.

Each of these individual blocks was designed for a 32-phase internal parallelization, thus capable of processing 32 samples simultaneously in a polyphase architecture, which reduces the required FPGA sampling frequency. Also, the ΔΣ-Core block internally has 32 ΔΣ-Cores to process the 32 input samples. In addition to these main blocks of each of the individual channels, there are three other main blocks, named “INT0”, “INT1” and “DM” as demonstrated in Fig. 4.24. These blocks are responsible for interacting with each of the connecting blocks to reconfigure them according to the required modular configuration setup for the system.

This architecture is designed to perform under three different modular configurations. “Single”, maintains the channels individually running, “Dual”, combines two channels to achieve double the maximum carrier frequency compared to “Single”, but loses one of the MGT outputs. Lastly, “Quad”, combines the 4 available individual channels, capable of

**TABLE 4.4:** Resource Utilization Report

| | ADT Single Channel | ADT Quad Channel | Additional Control Logic | Total |
|---|---|---|---|---|
| **LUTs** | 29.730 | 118.920 | 159.242 | 278.162 |
| **FFs** | 6.713 | 26.852 | 16.815 | 43.667 |
| **CARRYs** | 710 | 2.840 | 0 | 2.840 |
| **MUXs** | 1.536 | 6.144 | 0 | 6.144 |
| **CLBs** | 3.011 | 12.044 | 10.048 | 22.092 |
| **BRAMs** | 32 | 128 | 0 | 128 |
| **DSPs** | 353 | 1.412 | 0 | 1.412 |

achieving 4 times the maximum carrier frequency of the "Single" configuration. Still, as in "Dual", it will only use one of the outputs combined.

One of the benefits of digitally combining different individual channels is that it reduces the noise introduced by the internal ΔΣ-Core filter on the quantization process. By connecting various individual tile-based ΔΣ-ADT channels to form a new modular channel, we increase the number of samples processed in the modular channel, by the same amount of samples processed on each individual tile-based ΔΣ-ADT channel connected. This reduces the number of discontinuities introduced on the quantized signal, resulting from the filtering process of each of the internal ΔΣ-Cores.

The "INT0" is responsible for generating and setting the individual step used for each of the tile-based Up-Converter blocks to achieve the desired carrier frequency and also sort the output samples from them to supply to the De-Interleaver blocks. The "INT1" is responsible for setting the filter coefficients on each of the tile-based ΔΣ-Cores internal filters and sorting the output samples from them to supply to the interleaver blocks.

The "DM" performs the final alignment of the output samples and selects the appropriate MGT output depending on the modular configuration setup for the system. Using this approach we can achieve multiple configurations on this simple 4 tile-based individual channels system: 4 individual ΔΣ-ADT channels with different waveforms each; 2 individual ΔΣ-ADT and 1 modular ΔΣ-ADT channel, resulting in 3 different output waveforms, with one of them with double the capability; and 4 modular ΔΣ-ADT channels outputting a single waveform with higher performance. All these configurations can select each available MGT-Core block as its output.

### 4.4.3 Prototype Validation

The proposed ADT architecture implementation was done in a *Virtex UltraScale+ 56G PAM4 VCU129 FPGA Evaluation Kit* from AMD ideal for deploying ADTs due to the high number of available MGTs in various interfaces, with sampling rates up-to 56 Gb/s in a PAM with 4 levels configuration.

The implementation is composed of the proposed ΔΣ architecture presented in Fig. 4.24, as discussed previously, and additional logic required for supplying the source signal for encoding. Any type of modulation protocol could be used in the source signal, but with the available laboratory equipment at the time of writing this paper, the selected protocol was the QAM.

Since this work is an exploratory approach to building a multi-channel modular ΔΣ-ADT, the implemented project did not use the available PAM4 MGT and focused on normal functionality of the MGT at a sampling frequency of 20 Gb/s. Table 4.4 shows the utilization

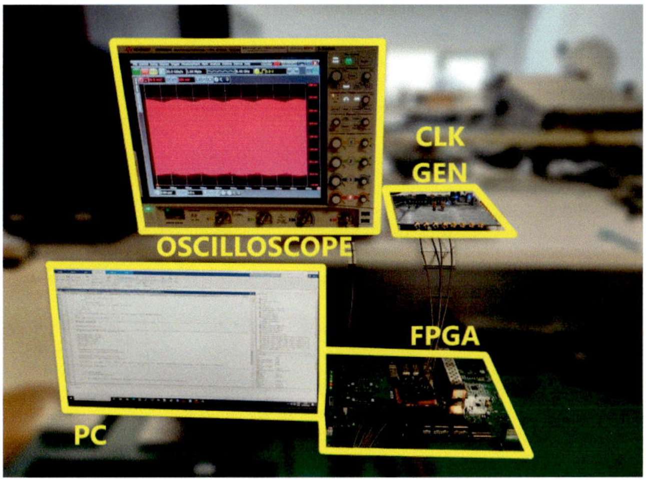


**FIGURE 4.25:** Laboratorial setup.

report for a single ADT, and all the 4 individual ADT implemented plus the additional logic required for control and interaction between the multiple channels depending on the selected configuration of the system.

Table 4.4 shows all the BRAMs, DSPs, MUXs, and CARRYs that are used in a single tile-based $\Delta\Sigma$-ADT channel and also the total amount required for the proposed system in Fig. 4.24. At the same time, the remaining resources are approximately split in half between the implemented $\Delta\Sigma$-ADT channels and the additional logic required for the control of the system and the modular configuration setup. Referring to the selected FPGA kit (VCU129), the implemented system uses, on average, 11% of the available LUTs, CLBs, and DSPs while the other primitives are well below the 5%. This demonstrates the possibility of having multiple individual or modular channels capable of generating RF signals with carriers up to half the maximum sampling frequency of the MGT technology used.

A picture of the laboratory setup used can be seen in Fig. 4.25. As mentioned, an AMD VCU129 FPGA kit was used. A clock generator was used to supply the reference clock to the MGT bank in the FPGA. A Keysight DSOS804A Oscilloscope was used to record each of the output waveform channels. And, the PC was used to control the system and store the recorded waveforms from the oscilloscope.

To evaluate the system performance and capabilities, a sweep was performed through a range of frequencies and modular configurations of the available $\Delta\Sigma$-ADT channels, using four 16-QAM signals with different bandwidths, 100 MHz, 200 MHz, 400 MHz, and 800 MHz. Fig. 4.26 shows the results of that sweep. In that figure, the various types of lines refer to a different bandwidth signal and the color to a specific modular configuration between the individual tile-based $\Delta\Sigma$-ADT channels.

Looking at Fig. 4.24 as a reference of the implemented design, we can have four individual channels in the "Single" modular setup, which means four types of different waveforms and center carriers. Also, we can have two "Dual" configurations, or two "Single" and one "Dual" or one "Quad" setup as mentioned before.

In this laboratory setup, each of the individual $\Delta\Sigma$-ADT channels works at a sampling rate of 5 Gb/s capable of generating RF signals with a carrier frequency up to 2.5 GHz. By combining multiple tile-based $\Delta\Sigma$-ADT channels in a modular configuration, it is possible to achieve 10 Gb/s sampling rate and 5 GHz center frequency in the "Dual" configuration and 20 Gb/s sampling rate and 10 GHz in "Quad" configuration. This is why the maximum frequencies change between the various modular configuration setup results.

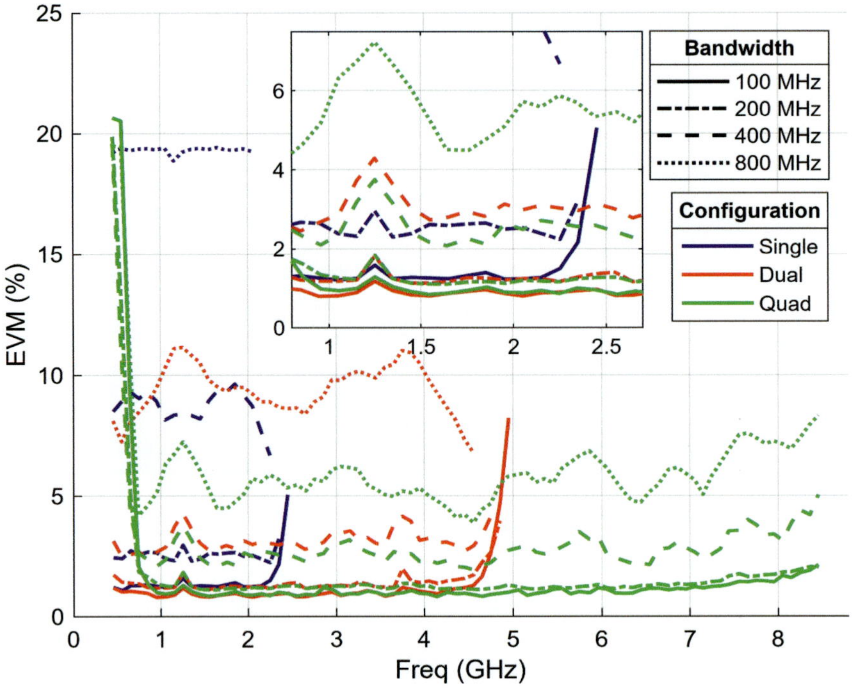


**FIGURE 4.26:** Frequency sweep EVM results.

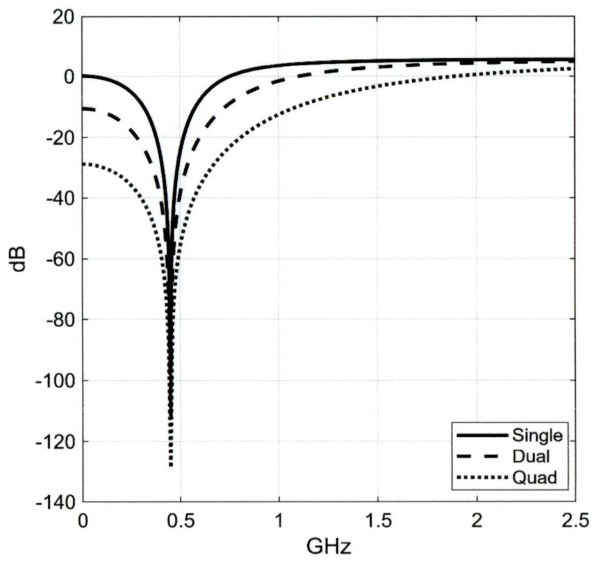


**FIGURE 4.27:** $\Delta\Sigma$-core internal filter response.

As expected, by switching from the "Single" to the "Dual" and "Quad" modular configuration, we see a better response of the system in terms of EVM and subsequently SNR. Also, the closer the center frequency is to the maximum of each configuration, the worse the EVM response.

The initial EVM values on the "Quad" configuration are a result of the $\Delta\Sigma$-Core internal filter response, as shown in Fig. 4.27. Since the filter for the "Quad" configuration is designed

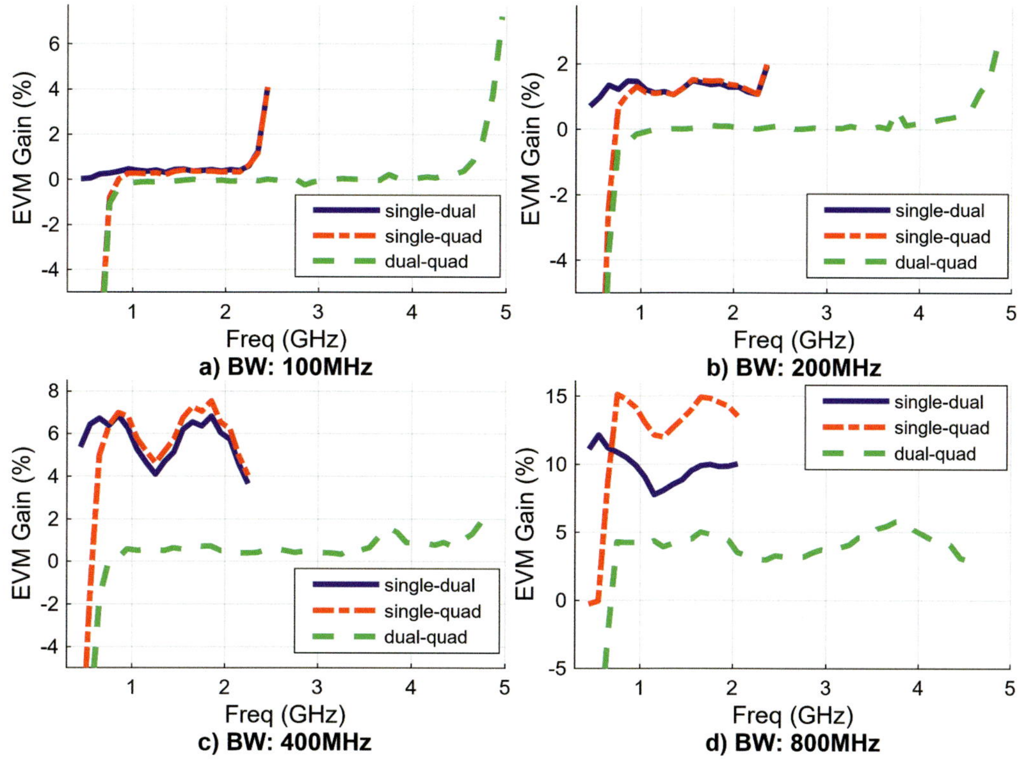


**FIGURE 4.28:** EVM gains in different system configurations.

for a higher sampling rate, we can see that the filter response is not as "symmetrical" as in the other two configurations for the same carrier frequency, but with lower ΔΣ-ADT sampling rates, even though the "Dual" configuration already shows some distortion, in the "Quad" it is very evident. This filter response leads the internal feedback loop of the ΔΣ-Core to become unstable, thus resulting in the destruction of the output signal of the ΔΣ-ADT. Furthermore, as the carrier frequency increases, the filter response becomes more "symmetrical", which stabilizes the internal feedback loop of the ΔΣ-Core, hence, resulting in a properly quantized output signal.

Fig. 4.28 shows the results for the four signals with different bandwidths tested. For the 100 MHz bandwidth signal, it can be seen that, although the "Dual" and "Quad" configurations have slightly better EVM, for lower center frequencies (below 2.25 GHz), the "Single" configuration is the better option, while the "Dual" and "Quad" are for the remaining spectrum. For the 200 MHz, 400 MHz, and 800 MHz bandwidth signals, we can see the better option is the "Quad" configuration.

### 4.4.4 Key Remarks

In this section, we presented a new flexible tile-based multi-channel digital architecture of modular ΔΣ-ADTs channels. This new approach to how ΔΣ-ADT is designed, is capable of providing multiple individual RF channels in a single system or, by combining multiple tile-based individual channels, provide better EVM and SNR, as well as, to achieve higher carrier frequencies. Three modular configurations of the system were presented, "Single", "Dual" and "Quad", representing a specific modular setup of the available individual tile-based

$\Delta\Sigma$-ADTs channels. Four different bandwidth signals were tested in these configurations, achieving an EVM lower than 1.5% for 100 MHz, 1.83% for 200 MHz, 4.01% for 400 MHz, and 8.30% for 800 MHz in the case of the "Quad" configuration. These results demonstrate the capability of this approach to combine less complex implementations of tile-based $\Delta\Sigma$-ADTs to achieve higher carrier frequencies and higher bandwidth depending on the required system configuration. This leads to an increase of flexibility and reconfigurability of the system, allowing the RU to dynamically adjust the available resources to meet the needs of the network.

In summary, the multi-channel modular ADT presented here offers extensive adaptability and reconfigurability, featuring reduced quantization noise, higher carrier frequencies (capable of generating waveforms up to 12 GHz, limited by the chosen FPGA kit's maximum rate), high-bandwidth signal generation (up to 800 MHz as demonstrated), and dynamic system reconfiguration. These capabilities support adaptable, efficient multi-user SDR and MIMO systems, essential for future mobile communication networks.

---

## 4.5 Conclusion

This chapter provided an overview of the evolution of frequency agility in ADTs. Prior to the works presented here, the SoA of carrier agility was very limited and sometimes only possible through the changing of the serializer's frequency. To solve this problem IF solutions were the first to be proposed. These solutions, such as the one presented in Section 4.2, already allow for a good degree of flexibility. However, bandwidth was still limited. Then RF-placement solutions were proposed to introduce an even greater degree of flexibility and wider bandwidths, with the first of these solutions being introduced in Section 4.3. Section 4.4 introduced a multi-channel modular ADT designed for enhanced adaptability and reconfigurability, featuring reduced quantization noise, higher carrier frequencies and high-bandwidth signal generation, enabling efficient and flexible multi-user SDR and MIMO systems.

# Chapter 5

# Wideband Delta-Sigma Modulators

## 5.1 Introduction

With the new telecommunications standards, the increase in BW is a constant. However, a key limitation of conventional ADT architectures has always been the low bandwidth (typically up to hundreds of kHz) achieved in the signal shaping stage due to the low OSR from the $\Delta\Sigma$M or PWM. This problem is directly connected to the frequency limitations of the typical platform used, the FPGA, and its logic, which cannot easily achieve frequencies high enough for the modulators to work with reasonable bandwidths. In this chapter, the history of the evolution toward larger BWs will be told. With each section, the reader will follow along the steps that, at the time, contribute to an improvement in the state of the art of ADTs. This chapter is supported by the conference papers [36, 37, 46] and by the journal articles [48, 153].

Low OSR in $\Delta\Sigma$M not only minimizes the transmitted bandwidth but also deteriorates the SNR for the output signal. Due to these limitations, some effort has already been made to overcome the disadvantages of the low SNR factor. In [183], a multi-path approach, combined with a bandpass $\Delta\Sigma$M, is presented to enhance both the SNR and the transmitter bandwidth. However, this represents a fairly complex architecture that requires multiple serializer outputs. In [82], a parallel $\Delta\Sigma$M is presented that performs delta-sigma modulation with a smaller OSR, but the high hardware complexity limited the hardware validation to signals with low bandwidths. In the following sections, several techniques aimed at increasing the available BW of the ADT will be analyzed.

The first architecture, referred to as Time-interleaved Delta-sigma (TIDS), was also addressed in [54] using a parallelization technique that allowed the improvement of the OSR. Although this technique enables the generation of simultaneous delta-sigma output samples within a single clock period, to the best of our knowledge no prior work has considered the impact of complexity scaling and hardware frequency limitations in such approaches, which constrains their applicability in practical systems. This issue will be addressed in Section 5.2. The second architecture to be presented is referred to as Multicore-based Architecture and aims to address the scalability issues presented by TIDS. Then, Section 5.4 refines this architecture by incorporating a new block in the $\Delta\Sigma$M that addresses some performance issues exhibited by the Multicore-based Architecture. Section 5.5 presents an architecture that moves the $\Delta\Sigma$M forward, after the Upconversion stage, creating an RF-stage modulator (see Section 4.3), which can deliver up to 1.25 GHz BW with an EVM lower than 2.3%. This is achieved through the use of 8 MGTs. Finally, Section 5.6 switches from an FIR filter to an IIR filter, which with some optimization reaches 1 GHz bandwidth and carrier frequency agility from 1 to 12 GHz, using only one MGT.

DOI: 10.1201/9781003225713-5

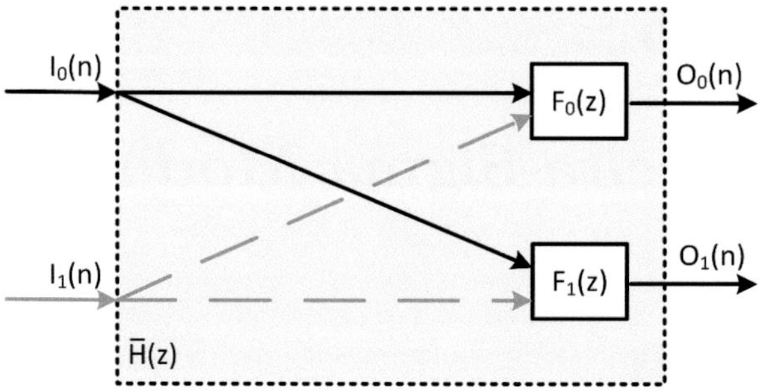


**FIGURE 5.1:** Interdependence of each output phase to multiple input phases at present or delayed samples.

## 5.2 Time-Interleaved Architecture

In this section, we show several optimizations to increase the OSR on a ΔΣM modulator. Analysis is conducted in terms of hardware scaling, complexity, and critical path delay to maximize the hardware sampling rate. This results in a polyphase modulator working at an equivalent frequency of 1 GHz in a fully integrated FPGA-based transmitter, representing a considerable increase in the OSR when compared to other FPGA-based ADTs.

The following subsections demonstrate how conventional polyphase equivalents can be identified, how they can be adapted to accommodate the FPGA hardware characteristics, and, in the case of a ΔΣM, how to minimize the impact of increased complexity.

### 5.2.1 Polyphase Decomposition

Time-interleaved delta-sigma architectures leverage polyphase representations of the modulator's internal filter H(z), enabling relaxed hardware speed requirements and achieving higher OSR. A more in-depth theoretical analysis of this decomposition can be seen in Subsection 2.2.4.

#### 5.2.1.1 Hardware Implementation Issues

Polyphase decomposition seems to be an interesting approach for a hardware implementation of a high-speed ΔΣM. By changing the ΔΣM from a SISO block to a MIMO equivalent, the data rate can be potentially increased by the number of used paths, $M$. However, depending on the modulator's filter function $H(z)$, this might not be as trivial as expected.

From Eq. 2.10, one can understand that each system output will now depend on a combination of all the input phases. Moreover, this dependence may be due to other phases in the present sample, rather than a delayed one, as shown in Fig. 5.1, where $F_x(Z)$ represents the interdependence between the different phases. In a common MIMO system, this would not represent an implementation problem; however, ΔΣM uses a feedback path. This gives rise to a time dependence of some input phases relative to the outputs of others. If the resulting transfer function does not introduce a register delay, then the outputs cannot be computed independently. For this reason, polyphase decomposition does not create a truly parallel system because some paths are still in series with other ones without a delay breaking them. Furthermore, pipeline techniques cannot be used within a feedback loop because the resulting transfer function would be altered.

Another problem that decreases the gains achieved by a polyphase equivalent representation can also be derived from the fact that now each output depends on all inputs. To

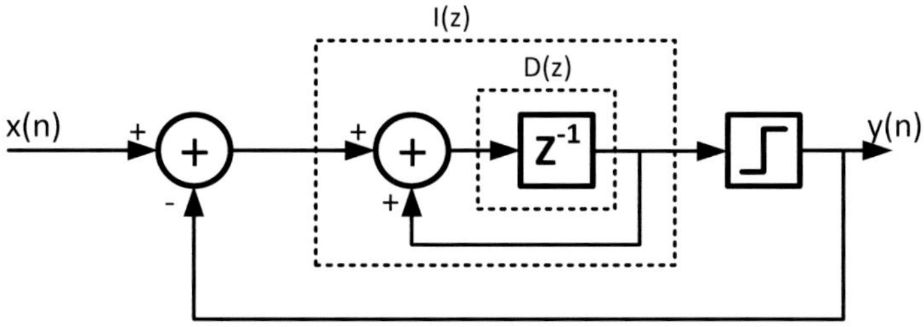


**FIGURE 5.2:** DSM with delayed integrator. I(z) is the delayed integrator transfer function and D(z) is the delay transfer function.

combine all the effects from all the phases, additional logic must be added. In fact, because an output phase is a sum of all the input phases, logic adders are necessary. For each doubling of the number of paths $M$, one additional level of adders is needed.

#### 5.2.1.2 Hardware Implementation and Optimization

In the $\Delta\Sigma$M case, the polyphase decomposition can be applied to the internal filter $H(z)$ on the $\Delta\Sigma$M loop. For a practical example, and without any loss of generality, we will use the example of a first-order $\Delta\Sigma$M modulator with a delayed integrator as the function $H(z)$ (see Fig. 5.2).

$\Delta\Sigma$M's transfer functions $H(z)$ is in Eq. (5.1).

$$H(z) = I(z) = \frac{z^{-1}}{1 - z^{-1}} \tag{5.1}$$

Rewriting the integrator function, using the Eqs. (2.11) and (2.9), the polyphase representation with $M = 4$ parallel paths is found in (5.2).

$$\overline{H}(z) = \begin{bmatrix} \frac{z^{-1}}{1-z^{-1}} & \frac{1}{1-z^{-1}} & \frac{1}{1-z^{-1}} & \frac{1}{1-z^{-1}} \\ \frac{z^{-1}}{1-z^{-1}} & \frac{z^{-1}}{1-z^{-1}} & \frac{1}{1-z^{-1}} & \frac{1}{1-z^{-1}} \\ \frac{z^{-1}}{1-z^{-1}} & \frac{z^{-1}}{1-z^{-1}} & \frac{z^{-1}}{1-z^{-1}} & \frac{1}{1-z^{-1}} \\ \frac{z^{-1}}{1-z^{-1}} & \frac{z^{-1}}{1-z^{-1}} & \frac{z^{-1}}{1-z^{-1}} & \frac{z^{-1}}{1-z^{-1}} \end{bmatrix} \tag{5.2}$$

The result is a sum of four integrators and delayed integrators for each phase. This is a more complex system, but in theory it can output $M$ parallel paths at a time, leading to a $M$ times gain in the OSR.

The four-phase polyphase equivalent for an integrator was found to be a $4 \times 4$ transfer function. Regarding hardware complexity, this solution introduces additional levels of logic to the critical path, thereby decreasing the maximum operating frequency. This occurs because every phase output depends on all the system input phases. As the number of phases increases, the number of sum levels also increases, thereby affecting the delay in the critical path.

Consider now, as in Eq. (5.3), that the transfer function to find the polyphase equivalent is not the integrator but only the delay $D(z)$ in the integrator (Fig. 5.2). Applying the polyphase decomposition to this transfer function, we obtain the result presented in Eq. (5.4).

$$H(z) = D(z) = z^{-1} \tag{5.3}$$

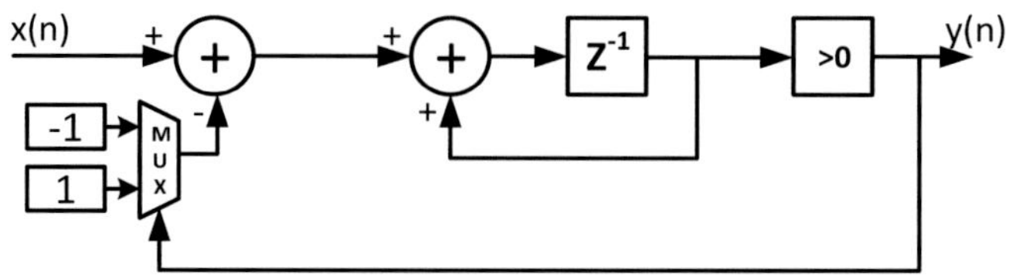


**FIGURE 5.3:** First-order modulator.

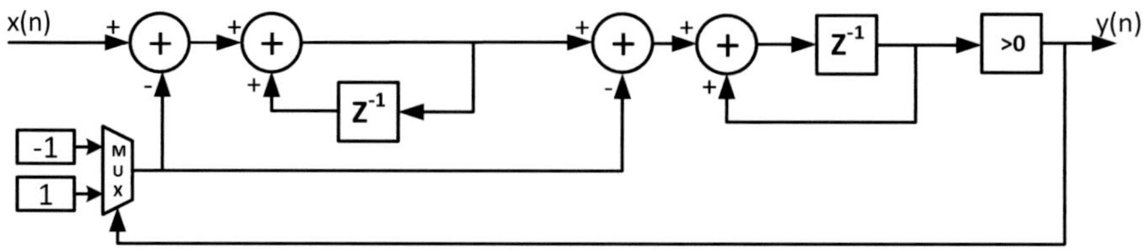


**FIGURE 5.4:** Second-order modulator.

$$\overline{H}(z) = \begin{bmatrix} 0 & 1 & 0 & 0 \\ 0 & 0 & 1 & 0 \\ 0 & 0 & 0 & 1 \\ z^{-1} & 0 & 0 & 0 \end{bmatrix} \tag{5.4}$$

Each of the output phases is a function of only one input phase, and no additional sums are needed. By reducing the number of logic elements needed for each output path, the data path delay will also be decreased. This greatly reduces the degree of hardware complexity in the parallel system and can be easily scaled for higher values of $M$.

An analysis of the maximum clock frequency was made for the hardware implementation of two ΔΣM architectures, a first-order low-pass ΔΣM (Fig. 5.3) and a second-order low-pass ΔΣM (Fig. 5.4).

For each architecture, the maximum achievable frequency was found for different numbers of parallel paths. Also, for these two architectures, both the delayed integrator decomposition and the delay decomposition were tested. The clock frequency analysis was done using Xilinx synthesis tools (ISE Design Suite 14.4). For each different ΔΣM implementation mapping, place and routing routines were performed to maximize the operation frequency. For each implementation, the highest path delay was considered as the frequency limitation.

Fig. 5.5 shows the maximum equivalent frequency of the modulators for the specified cases. From Fig. 5.5 one can see that the frequency gains are far from increasing linearly with $M$. In fact, for the integrator decomposition, there is no gain in using multiple parallel paths for the ΔΣM. This occurs because, with the integrator decomposition, the logic levels in the critical path have additional logic for the sum operation, thereby hampering the potential gains from a polyphase equivalent representation.

Fig. 5.6 illustrates how the number of logic levels, and consequently the critical path delay, evolves for both decomposition methods. Unlike the conventional approach, the delay decomposition does not introduce additional adder logic, resulting in fewer logic levels, a shorter critical path, and thus higher achievable clock frequencies for $M$ parallel paths.

The highest increase in frequency occurs for 4 parallel paths, achieving an equivalent frequency of more than 1 GHz for the first-order ΔΣM. This architecture with four paths

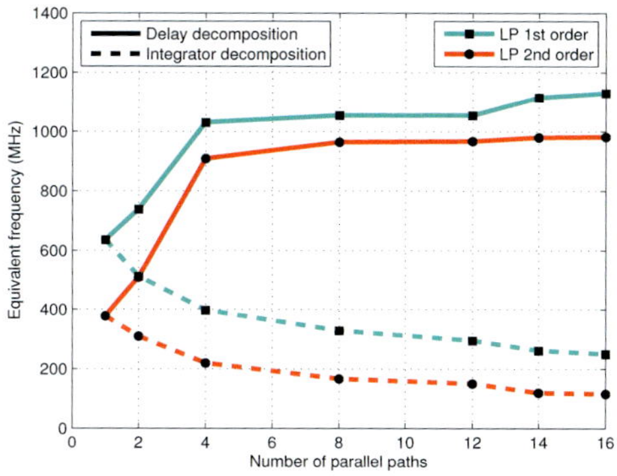


**FIGURE 5.5:** Maximum equivalent frequency for the modulator depending on the number of parallel paths $M$.

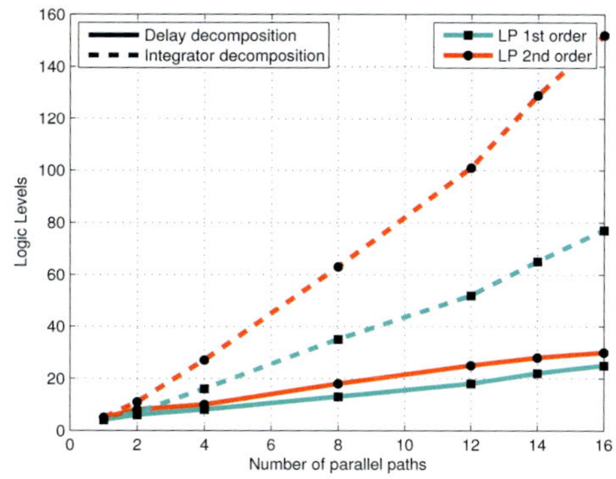


**FIGURE 5.6:** Logic levels at the critical path delay depending on the number of parallel paths $M$.

was used to implement the ADT that will be analyzed below. The following sub-section shows the implemented architecture for the ΔΣM using the delay decomposition approach.

### 5.2.2 Proposed Architecture

The ΔΣM topologies chosen for the transmitter were a first-order low-pass (Fig. 5.3) and a second-order low-pass (Fig. 5.4). A 4-path time-interleaved decomposition was used to achieve an equivalent sampling frequency up to 1 Gsps.

The FPGA implemented 4-path TIDS first-order architecture is depicted in Fig. 5.7. Tables 5.1 and 5.2 show the logic resources for the polyphase interpolation filter and the TIDS modulator implementation in a Virtex-6 HX380T FPGA. This FPGA was used as a proof of concept; however, any other could be used depending on the requirements. A polyphase interpolation FIR filter is used to interpolate the data up to the modulator data rate, with 1 Gsps for the first-order ΔΣM and 900 Msps for the second-order ΔΣM.

Tables 5.1 and 5.2 show a minimal use of resources for the implemented modulators. This means that the increased complexity in a time-interleaved ΔΣM does not come at a high cost in terms of resources for implementation in transmission systems. Moreover, the resource usage difference between the first-order and second-order modulators is minimal.

The use of a polyphase interpolation approach greatly simplifies the generation of a 1 Gsps sample stream, which is actually composed of four paths at 250 Msps each. The parallel polyphase outputs are directly connected to the TIDS in such a way that the correct signal phase is used for the parallel ΔΣM inputs.

The number of filter coefficients must be a multiple of the path number so that the delays between the filter and ΔΣM inputs are correct, or additional delays should be added to correctly align the signal phase.

### 5.2.3 Prototype Validation

The experimental setup, seen in Fig. 5.9, uses the TIDS approach for an ADT architecture as shown in Fig. 5.8. This experimental setup was fully implemented as a proof of concept in a Virtex-6 HX380T FPGA using the Xilinx ML628 development board.

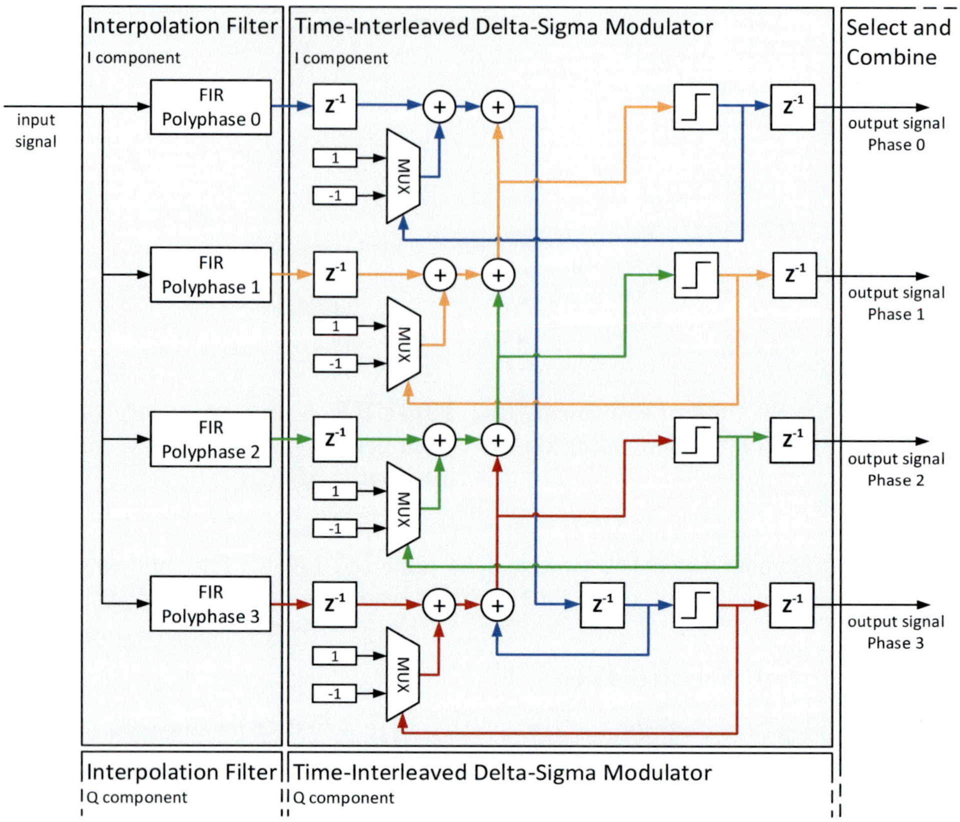


**FIGURE 5.7:** First-order time-interleaved shaper architecture.

**TABLE 5.1:** TIDS first-order modulator and interpolation filter occupied resources on Virtex-6 HX380T FPGA

| Logic resources | Used | Total | Percentage used |
|---|---|---|---|
| Flip Flops | 233 | 478080 | 0.05% |
| LUTs | 108 | 239040 | 0.05% |
| DSP48E1 | 8 | 864 | 0.93% |
| Logic max. frequency - 257.625 MHz | | | |

**TABLE 5.2:** TIDS second order modulator and interpolation filter occupied resources on Virtex-6 HX380T FPGA

| Logic resources | Used | Total | Percentage used |
|---|---|---|---|
| Flip Flops | 245 | 478080 | 0.05% |
| LUTs | 159 | 239040 | 0.07% |
| DSP48E1 | 8 | 864 | 0.93% |
| Logic max. frequency - 227.214 MHz | | | |

Table 5.3 shows the logic resources for the complete transmitter implementation. It is possible to see that a large area is still available for other system functionalities such as baseband processing, higher protocol layers, or user interfaces.

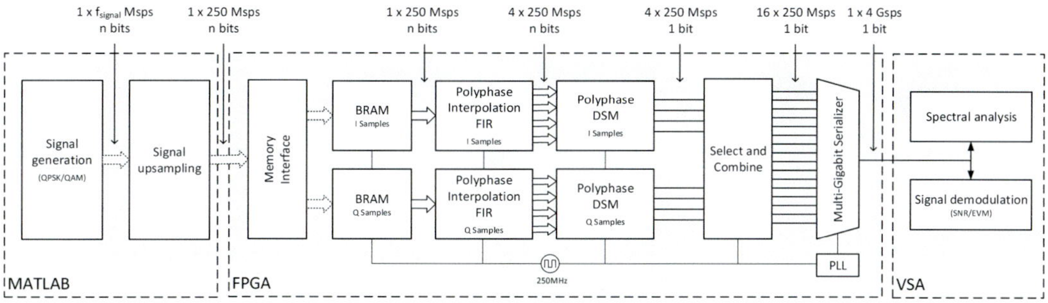


**FIGURE 5.8:** All-digital transmitter architecture implemented for experimental evaluation (frequency values are for the 1 Gsps first-order modulator).

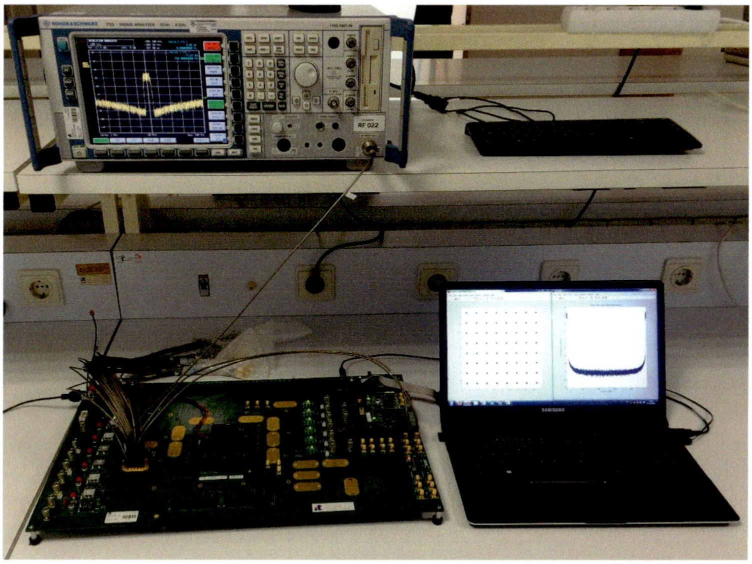

**FIGURE 5.9:** Experimental setup.

**TABLE 5.3:** All-Digital transmitter occupied resources on Virtex-6 HX380T FPGA

| Logic resources | Used | Total | Percentage used |
|---|---|---|---|
| Flip Flops | 919 (943)* | 478080 | 0.19% (0.20%)* |
| LUTs | 685 (787)* | 239040 | 0.29% (0.33%)* |
| DSP48E1 | 16 | 864 | 1.85% |
| RAMB36E | 59 | 768 | 7.68% |
| GTXE1s | 1 | 40 | 2.50% |

*In brackets, values are for the second-order DSM

The transmitter was configured to use a 1 GHz carrier frequency for all the signals, and the $\Delta\Sigma$M logic working sampling rate is 250 MHz.

Several signals were created and pre-processed offline using MATLAB's Communications System and Signal Processing Toolboxes and were loaded into the FPGA's internal memory. The samples are interpolated to a 250 Msps rate with a squared-root raised cosine filter with a roll-off factor of 0.22.

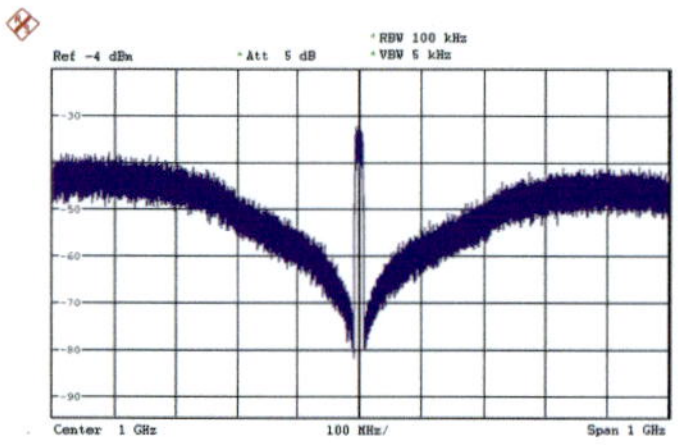


**FIGURE 5.10:** Spectrum of the transmitted signal using a first-order ΔΣM with 12.5 Msps 64-QAM signal.

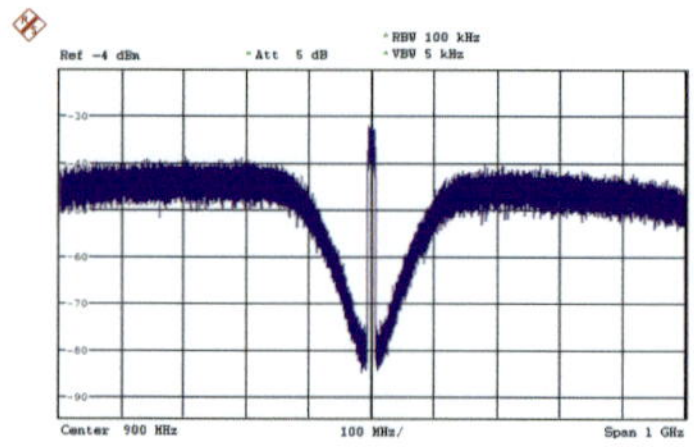


**FIGURE 5.11:** Spectrum of the transmitted signal using a second-order ΔΣM with 12.5 Msps 64-QAM signal.

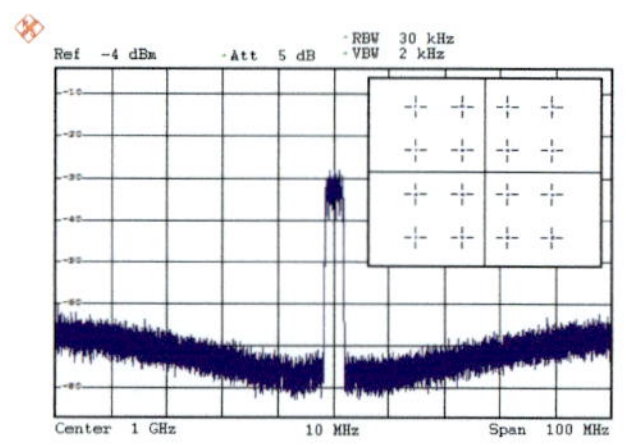


**FIGURE 5.12:** Detailed view of a 16-QAM signal at 3.125 Msps spectrum with 100 MHz span and its received constellation in the VSA.

**TABLE 5.4:** Comparative EVM and SNR values

| | Signal modulation | Symbol rate (Msps) 1.5625 | 3.125 | 6.250 | 12.5 |
|---|---|---|---|---|---|
| | | **First-order ΔΣM at 1 Gsps** | | | |
| SNR (dB) | QPSK | 41.1 | 40.2 | 38.5 | 36.4 |
| | 16-QAM | 40.5 | 39.7 | 38.4 | 35.9 |
| | 64-QAM | 40.4 | 40.0 | 37.8 | 35.2 |
| EVM (%) | QPSK | 0.77 | 0.87 | 1.04 | 1.32 |
| | 16-QAM | 0.82 | 0.90 | 1.05 | 1.41 |
| | 64-QAM | 0.80 | 0.89 | 1.12 | 1.52 |
| | | **Second-order ΔΣM at 900 Msps** | | | |
| SNR (dB) | QPSK | 41.6 | 41.3 | 40.5 | 38.7 |
| | 16-QAM | 41.6 | 41.4 | 40.3 | 38.3 |
| | 64-QAM | 41.8 | 41.3 | 40.1 | 37.9 |
| EVM (%) | QPSK | 0.72 | 0.76 | 0.84 | 1.02 |
| | 16-QAM | 0.72 | 0.73 | 0.85 | 1.05 |
| | 64-QAM | 0.70 | 0.76 | 0.86 | 1.11 |

The input signals used were modulated using QPSK, 16-QAM and 64-QAM with symbol rates of 1.5625 Msps, 3.125 Msps, 6.250 Msps and 12.5 Msps.

The transmitted signal was directly connected to a VSA from Rohde & Schwarz (model FSQ 8). From the VSA demodulation, EVM and SNR were obtained for all the test signals.

Figs. 5.10 and 5.11 show the obtained spectrum directly at the FPGA's serializer output for a 16-QAM modulated signal at 3.125 Msps, resulting in a signal bandwidth of about 4 MHz. Fig. 5.10 uses a first-order TIDS modulator, and Fig. 5.11 uses the second-order TIDS modulator. Fig. 5.12 shows a details view of the signal's channel with a smaller span.

Table 5.4 shows the results obtained with the VSA for the EVM and SNR figures of merit for the used test signals. It is possible to see that the EVM is always under 2%, resulting in a well-defined constellation as the example shown in Fig. 5.12.

The difference between the two presented modulators is more obvious for higher-bandwidth signals. In this case, the second-order modulator exhibits slightly better performance compared to the first-order ΔΣM. This happens because the second-order modulator introduces less noise near the signal's frequency band.

The quality of the results obtained with the presented approach, in terms of SNR and usable bandwidth of the ADT, is similar to that of the state-of-the-art approach presented in [183], but with significantly reduced hardware complexity, requiring only one output.

### 5.2.4 Key Remarks

In this section, a fully integrated SoC ADT was presented and evaluated. What distinguishes this implementation is the fact that both signal processing techniques and hardware limitations were taken into account to maximize the quality of the transmitted signal in terms of SNR and bandwidth. To achieve this goal, a ΔΣM at an equivalent frequency of 1 GHz was designed, implemented, and tested, showing the advantages of parallel processing for embedded FPGA ADTs.

Despite these improvements, this architecture still presents some disadvantages in terms of the long critical path, which limits its final sampling frequency and ultimately bounds the maximum signal bandwidth. In the next section, this problem will be addressed.

---

## 5.3 Multicore-based Architecture

In the previous section, the TIDS architecture was presented and analyzed. Despite several performance improvements, this architecture is still subject to some issues that restrict its use when compared to conventional radio architectures, particularly the transmitter bandwidth. Most of the ADTs presented in the literature have very small bandwidth signals, around a few units of MHz, which limits the achievable data rates in practical application scenarios [30, 36, 56]. The bandwidth limitation is due to the digital implementation of the signal pulsed modulators in ΔΣM or PWM architectures in which the logic switching frequency limit defines the achievable sampling frequency to a maximum value depending on the technology used, constraining either the maximum OSR or the maximum signal bandwidth.

To overcome these limitations, some papers have presented the use of parallel processing, followed by dedicated high-speed logic such as a multiplexer or serializer, to achieve a higher equivalent sampling frequency [36,56,82]. Parallel architectures enable the design of considerably higher bandwidth transmitters operating at reduced logic frequencies for most of the design, allowing for competitive integration with lower-cost technologies. However, even with parallel architectures, the transmitter bandwidth is often limited to a few tens of MHz due to the feedback logic on ΔΣM topologies, which constrain the clock frequency.

### 5.3.1 Multiple DSM Decomposition

In this section, a new scalable wideband ADT architecture for SDR systems is studied. It distinguishes itself from other architectures by improving the achievable bandwidth in

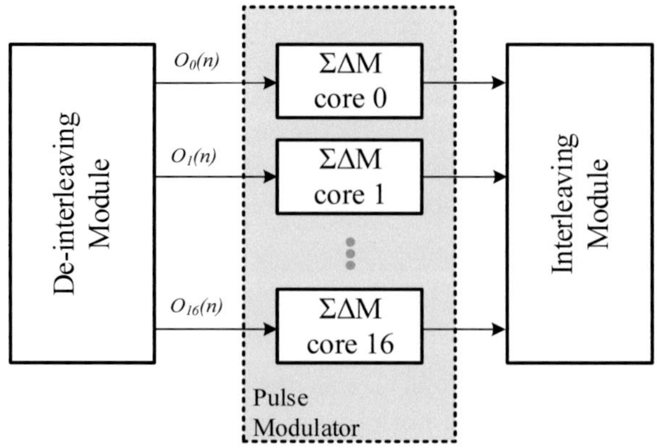


**FIGURE 5.13:** Multicore-based architecture with the de-interleaving and interleaving blocks.

an ADT, compared to other state-of-the-art transmitters, from less than 20 MHz up to a maximum of 122 MHz.

The architecture under study is based on a truly parallel system utilizing multiple ΔΣM cores that operate independently. In particular, it uses 16 parallel paths as shown in the block diagram of Fig. 5.13.

This architecture differs from common parallel ΔΣM implementations that use polyphase decomposition to achieve a time-interleaved ΔΣM. Conventional ΔΣM polyphase architectures do not scale linearly with the number of parallel branches, since the multiple phases are not independent of each other, limiting the achievable frequency gains [36].

The system under study uses a polyphase interpolation filter with 16 phases to achieve a high OSR. Each one of the 16 phases is fed to a de-interleaver block. In this block, the high sampling rate samples from the interpolation filter are reorganized into 16 distinct time blocks, each containing $N$ oversampled samples. By changing the signal sample order accordingly, each one of the independent ΔΣM can process a portion of the baseband signal corresponding to a contiguous time block. Although the processing is done at a lower clock frequency, it is equivalent to the higher sampling frequency imposed by the polyphase filter.

The de-interleaved block input signal is defined as in Eq. (5.5) with 16 parallel phases $I_k(n)$, where $k$ is the signal phase from the interpolation filter.

$$\begin{array}{llllll}
I_0(n) = [ & P_0(0) & P_0(1) & P_0(2) & \cdots & P_0(N)] \\
I_1(n) = [ & P_1(0) & P_1(1) & P_1(2) & \cdots & P_1(N)] \\
I_2(n) = [ & P_2(0) & P_2(1) & P_2(2) & \cdots & P_2(N)] \\
 & & & \ddots & & \\
I_{16}(n) = [ & P_{16}(0) & P_{16}(1) & P_{16}(2) & \cdots & P_{16}(N)]
\end{array} \tag{5.5}$$

In the time domain, these phases are related to each other at each sampling instant $n$, such that $P_0(n)$ is the first sample, then $P_1(n)$, followed by $P_2(n)$ up to $P_{16}(n)$. However, this sequence is in parallel because it would be impossible to process at the equivalent sampling frequency. The de-interleaver stores these parallel samples and, at the cost of a

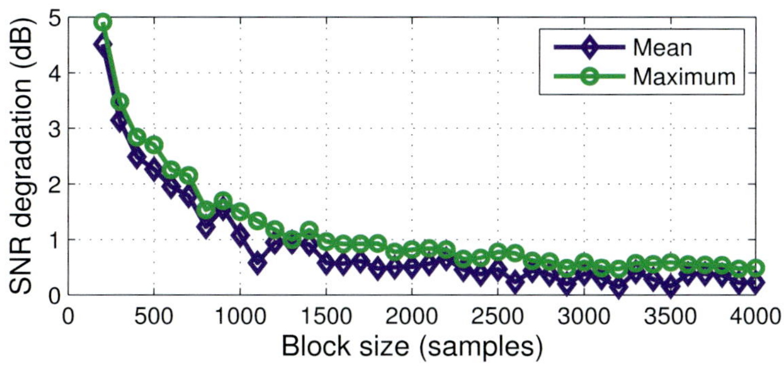


**FIGURE 5.14:** SNR degradation between a single-phase ΔΣM and the parallel multiple-core ΔΣM. The simulation was run using a set of random signals with PAPR values between 3 dB and 12 dB and bandwidths ranging from 5 MHz to 100 MHz.

small amount of added latency, rearranges them in the order presented in Eq. (5.6).

$$\begin{array}{llll} O_0(n) = [P_0(n) & P_1(n) & \cdots & P_{16}(n)] \\ O_1(n) = [P_0(n+N) & P_1(n+N) & \cdots & P_{16}(n+N)] \\ O_2(n) = [P_0(n+2N) & P_1(n+2N) & \cdots & P_{16}(n+2N)] \\ & & \ddots & \\ O_{16}(n) = [P_0(n+16N) & P_1(n+16N) & \cdots & P_{16}(n+16N)] \end{array} \tag{5.6}$$

The output now has the baseband samples arranged into $N$ contiguous sample blocks for each output. The de-interleaver outputs can be processed by independent ΔΣM cores. This means that, although the sampling frequency of the ΔΣM is $f_s$, the equivalent sampling rate is 16 times higher, considerably improving the achievable bandwidth.

The block size in samples, $N$, influences the signal error, as the transition between different time blocks introduces some errors. This occurs because the memory of the ideal ΔΣM is infinite, and by separating the signal into $N$ sample groups, there will be a small transient error in the signal at each $N$-sample transition between blocks. To analyze the introduced error, the ΔΣM outputs of the conventional ΔΣM (operating at a high clock rate) and the parallel ΔΣM architecture under study were compared through simulation, and the signal degradation in terms of SNR was examined. The simulation results in Fig. 5.14 show that for block sizes $N$ above 2000 samples, the maximum SNR degradation is always less than 1 dB. Hence, the chosen value for the block size was 2000 samples.

The modulator cores are the same first-order low-pass ΔΣM replicated 16 times. The choice of a first-order modulator over a higher-order modulator with better noise rejection is due to the fact that the first-order modulator provides a better coding efficiency for the transmitted signal [55]. This has considerable importance in improving the overall transmitter power efficiency.

After modulation, the samples must be rearranged to their original order, corresponding to the polyphase filter output. This is accomplished in the samples interleaving block, which reorganizes the 2-level signals from the ΔΣM cores back to their original configuration, as shown in Eq. (5.5).

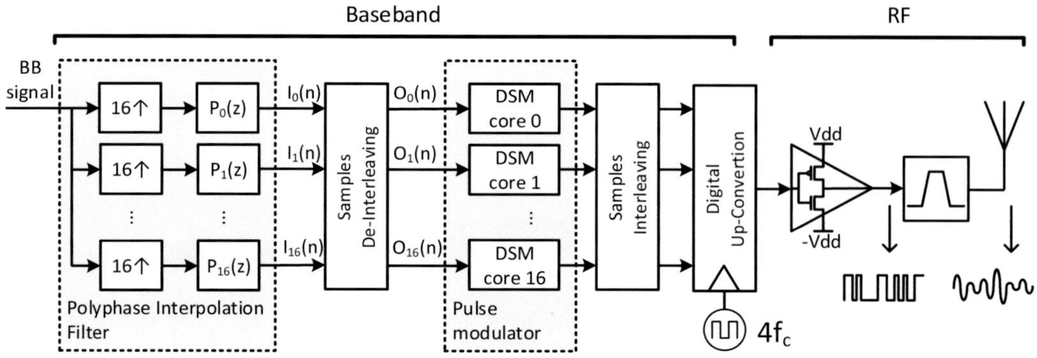


**FIGURE 5.15:** Architecture for the wideband all-digital transmitter. The signal paths are 2-channel paths, comprising in-phase (I) and quadrature (Q) samples.

The digital up-conversion is implemented by combining the $I$ and $Q$ samples as well as their inverted versions $\overline{I}$ and $\overline{Q}$. The vector is defined as a repetition of the ($I\ \overline{Q}\ \overline{I}\ Q$) pattern sampled at a $4 \times f_c$ (carrier frequency). The serialization of this vector with a dedicated high-speed serializer allows a high sampling frequency at the system output while maintaining a lower clock frequency, $f_s$, for the baseband hardware blocks. Once serialized at a frequency $4 \times f_c$, this pattern is equivalent to the multiplication of the I and Q signals with two square waves at $f_c$, with a 90-degree phase shift between them [226]. The ADT-generated output is the baseband ΔΣM signal up-converted to an RF carrier at $f_c$.

### 5.3.2 Proposed Architecture

A proof of concept prototype of the wideband ADT was designed using the architecture presented in Fig. 5.15. The FPGA-based ADT was implemented in an ML628 board from Xilinx with a Virtex-6 VHX280T FPGA. Once again, this FPGA was used as a proof of concept; any other could be used depending on the requirements.

### 5.3.3 Prototype Validation

The RF ADT 2-level output signal was directly connected to a VSA from Rohde & Schwarz (model FSW 8) for signal and spectral analysis. Baseband signals were generated offline and loaded into the FPGA's internal I and Q memories. The generated baseband signals used are QPSK, 16-QAM, and 64-QAM signals with a symbol rate varying from 5 Msps up to 100 Msps with 5 Msps steps. This corresponds to signals with bandwidths ranging from 6.1 MHz to 122 MHz.

The ΔΣM equivalent frequency is 3200 MHz (16 parallel paths at 200 MHz). In Fig. 5.16, three examples of a transmitted 64-QAM signal spectra are shown, the blue plot is the spectrum of the transmitter with a 6.1 MHz signal, the green plot has a 61 MHz bandwidth, and finally the red plot is the 122 MHz bandwidth signal, all centered at a 1.6 GHz carrier.

The SNR and the EVM were measured to evaluate the transmitter signal quality using the VSA demodulator. The SNR and EVM measurements are displayed, respectively, in Figs. 5.17 and 5.18.

The results show that for the tested signals, the achievable SNR is always above 30 dB for all modulations used, including the 122 MHz bandwidth signal. This corresponds to a transmission with less than 3.1% EVM for all modulations in a transmitter capable of

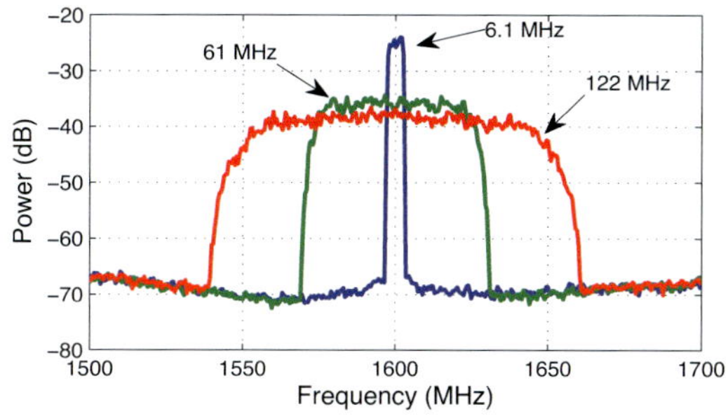


**FIGURE 5.16:** All-digital transmitter spectra for 6.1 MHz, 61 MHz, and 122 MHz bandwidth 64-QAM signals.

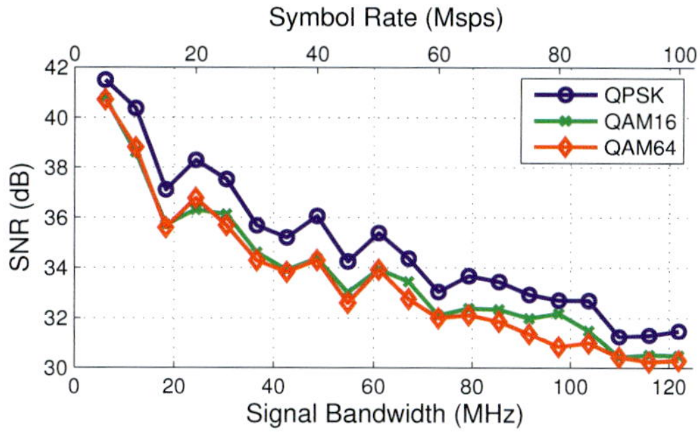


**FIGURE 5.17:** SNR measurement for a bandwidth sweep from 6.1 MHz to 122 MHz with a carrier frequency of 1.6 GHz.

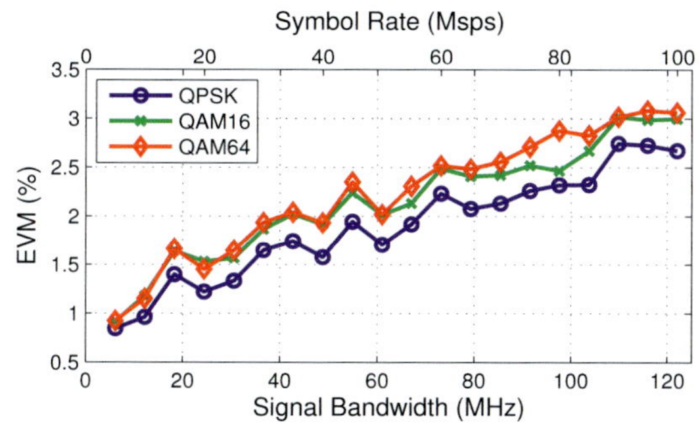


**FIGURE 5.18:** EVM measurement for a bandwidth sweep from 6.1 MHz to 122 MHz with a carrier frequency of 1.6 GHz.

transmitting up to 600 Mbits/s signals. The coding efficiency was also measured for all signals, varying between 11% and 27%, depending on the signal modulation.

Table 5.5 shows a comparison between this work and other state-of-the-art ADT focusing on the transmitter signal bandwidth. As one can see, this work largely exceeds the common signal bandwidths for this category of transmitters.

### 5.3.4 Key Remarks

In this section, a new ADT architecture to improve the digital transmitter's bandwidth was presented and analyzed. The system utilizes a set of parallel $\Delta\Sigma$M cores that are independent of each other, allowing for linear scaling improvements in the sampling frequency of the transmitter modulator. The architecture prototype measurements have shown the

**TABLE 5.5:** Maximum Transmitter Bandwidth State-of-the-Art Comparison.

| | Bandwidth | SNR | Modulation |
|---|---|---|---|
| Ebra11 [56] | 8 MHz | 60 | TIDS |
| Cord13 [36] | 19 MHz | 35.2 dB | TIDS |
| Chun15 [30] | 5 + 10 MHz | - | BPDSM and RFPWM |
| This work | 122 MHz | 31.4 | Parallel DSM |

improvement in terms of transmitter bandwidth compared to other state-of-the-art transmitters. In fact, this architecture enabled the improvement of the state-of-the-art bandwidth from less than 20 MHz to more than 120 MHz. However, a closer analysis reveals the presence of impulsive noise in the time domain; this problem will be the focus of the next section.

## 5.4 Multicore-based Architecture with State Propagation

In this section, new methods for increasing the modulator's sampling frequency using multi-rate methodologies are presented, particularly in the presence of feedback loops, such as in $\Delta\Sigma$Ms. The typical technique to achieve this is based on polyphase decomposition applied to a specific modulator's subsystem [160], or to the entire modulator, as described in Section 5.2. However, the state-of-the-art has also demonstrated that the scalability of this technique is very limited, due to the difficulty of incorporating pipeline stages into $\Delta\Sigma$ architectures [36]. It has been demonstrated that the critical propagation path is proportional to the number of phases, meaning that the maximum global sampling rate is significantly lower than the expected one. A similar alternative is based on deriving novel modulators through the use of node equations, but it has also been demonstrated that, for a high number of phases, the sampling rate gains are still limited [124].

A promising alternative to these techniques was proposed in [37], as shown in Section 5.3, and aims to deploy $N$ multiple modulators running in parallel, each processing independent slices (of size $K$) of the same input signal. In particular, two new subsystems, deinterleaving and interleaving modules, responsible for performing the temporary storage of samples as well as their rearrangement were introduced. However, as a downside, the discontinuities in every $K$ samples cause impulsive noise that degrades the overall system's performance. As a workaround, the length of each slice must be increased. The main drawback is the considerable digital processing resource usage, accompanied by higher latency, which may be problematic for such systems.

Thus, the work analyzed here extends the initial concept presented in [37], where architectural enhancements are proposed to surpass the strict memory requirements without compromising the system's performance. In particular, it will be shown that the impulsive noise in [37] is caused by the initial conditions of each modulator that do not take into account the final state conditions from the subsequent one. To ensure continuity in the initial/final conditions of each modulator, a new subsystem is included that is responsible for propagating the state registers among the different modulators. The idea is similar to the propagation of error values proposed in [3], but instead of propagating the error values in a continuous way, which can lead to instability problems, the state registers from the previous modulators are just taken into consideration during $m$ clock cycles (where $m$ is the order of the $\Delta\Sigma$M) within $K$ clock ticks. By doing this, it is possible to ensure a high level of performance even for small length sizes, ultimately reducing the global system's latency and relaxing the system's requirements in terms of implementation resources.

### 5.4.1 Enhanced Multicore-based Architecture

In this subsection, the architecture under study for designing and implementing parallel $\Delta\Sigma$Ms is presented.

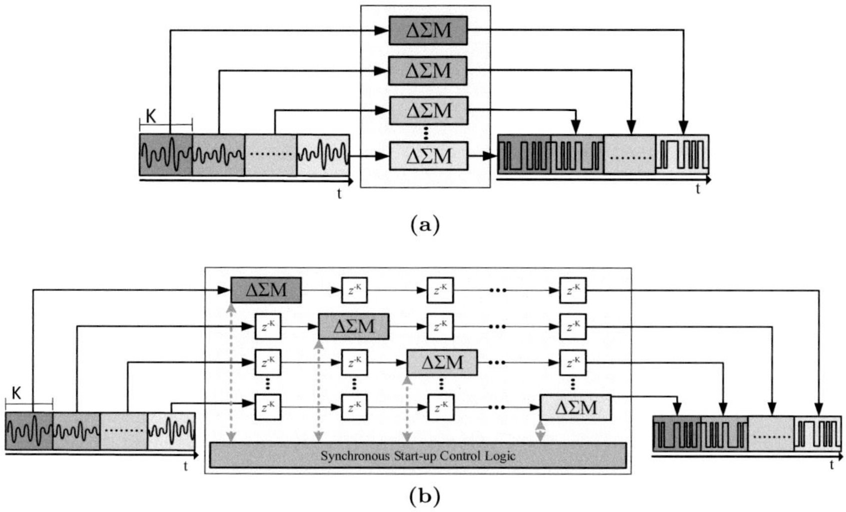


**FIGURE 5.19:** Comparison between the parallel ΔΣM presented in [37] (a) and the one analyzed in this work (b) (herein referred to as "Conventional" and "under study", respectively).

#### 5.4.1.1 Multicore-based Delta-Sigma Modulator

As previously introduced, the parallelization of multiple ΔΣM cores (analyzed in Section 5.3) was chosen as a starting point for this work. This architecture was selected due to its closest proximity to an ideal full parallelization of a ΔΣM, ensuring a high level of scalability with considerable performance. The underlying idea involves slicing an input signal into blocks of length $K$ that are fed to the $N$ modulators in parallel (Fig. 5.19a). However, as also pointed out, this is not an error-free architecture, because the discontinuities cause impulsive noise in the transitions between blocks. In short, this architecture can be subdivided into three main modules: the De-interleaving, the Encoding, and the Interleaving. The first module is in charge of performing the rearrangement of the input signals to ensure a temporal continuity in the order of the samples within each parallel phase. The second block processes and encodes all the samples into a 1-bit representation, through the replication of multiple ΔΣ modulator cores running in parallel. At last, the third block performs the inverse operation of the first one: it rearranges all the samples again to ensure temporal continuity between different phases. To ensure a real-time and online processing of the BB input samples, both the (de)-interleaving modules must be designed with resort to First-In First-Outs (FIFOs), Queues, or other types of memory-based primitives. Thus, the amount of memory resource usage, as well as the final latency, is a function of both the block size of each modulator ($K$) and the number of parallel modulators ($N$). For this reason, a trade-off arises because, for a fixed number of modulators, the higher the $K$, the better the system's performance will be, but with the downside of increased memory usage.

#### 5.4.1.2 Enhancing the Performance

This work aims to present modifications to the previous architecture, with the underlying idea of reducing memory usage and latency constraints. To accomplish this, the architecture was deepened, and it became apparent that the major cause of the performance drop was related to the discontinuities between different blocks, resulting in incorrect initial conditions for each modulator every $K$ ticks. This results in impulsive noise with a limited time duration that must be carefully handled to avoid compromising the system's performance. Thus, to minimize this problem, a new architecture was proposed (illustrated in Fig. 5.19b). Instead of placing all the modulators in parallel, delay registers were inserted before and after them, with the primary objective of propagating the values of the state registers from the previous modulators. To ensure the correctness of the error propagation, a novel block was included into the modulator (referred to as "Synchronous Start-up Control Logic") that is in charge of monitoring the final state registers of every modulator (in every $K$) and to pass the information to the next one. Contrary to [3], this exchange of information is just required in the transitions from blocks. Moreover, this information is used only within a given block for a limited number of clock cycles equal to the modulator's order, thereby not raising concerns about instability issues.

However, two points must be noted. Firstly, the architecture being studied here does not completely eliminate the existence of errors, because maintaining the same BB sampling rate prevents the propagation of the modulator's state registers from the last phase ($N$) to the first one. However, as will be seen below, it is possible to achieve the same performance as the conventional one with a smaller number of contiguous samples per modulator. Ultimately, the reduction of $K$ will be crucial in relaxing the memory and logic resource usage, making this architecture more appealing and efficient. Secondly, the final latency from the whole modulator must take into account three main variables: the latency introduced to perform the de-interleaving process ($L_{DEINT}$), the one introduced in the modulators ($L_{ENC}$), and the one introduced in the interleaving process ($L_{INT}$). At first sight, the extra latency in the modulator part from the architecture under study seems to be a major bottleneck. But, after careful analysis, one can observe that, in fact, this intuitive observation may be incorrect. In the conventional case, the typical data flow associated with the management of the memory primitives implies that the minimum global latency is imposed by $L_{CONV} \simeq L_{DEINT} + L_{ENC} + L_{INT} \simeq K_1\,N + 0 + K_1\,N \simeq 2\,K_1\,N$. This value enables a normal operation with no underflow/overflow issues. Similarly, in the case under study, the expression results in $L_{PROP} \simeq L_{DEINT} + L_{ENC} + L_{INT} \simeq K_2\,N + K_2\,N + K_2\,N \simeq 3\,K_2\,N$. Considering the number of errors, to maintain the same performance, the block size of the architecture under study can be intuitively chosen as the block size of the conventional one divided by the number of phases. By doing this, the following expression can be derived: $L_{PROP} \simeq 3\,\frac{K_1}{N}\,N \simeq 3\,K_1$. Thus, as long as the number of parallel phases exceeds 2, the latency of the architecture under study will always be inferior to that of the conventional one.

### 5.4.1.3 Delta-Sigma Topology

To validate the architecture under study, $N$ $\Delta\Sigma$M were designed and implemented. A second-order topology based on an Error-feedback structure (whose block diagram is shown in Fig. 5.20) was chosen due to its short critical path, leading to higher sampling rates. In addition to this, by having just two delays in the feedback loop, only two state registers need to be propagated, reducing the complexity of the module "Synchronous Start-up Control Logic". The modulator's transfer function can be computed by modeling the quantizer by a noise source ($E_q(z)$):

$$Y(z) = STF(z)\,X(z) - NTF(z)\,E_q(z) \tag{5.7}$$

where $STF(z) = 1$ and $NTF(z) = 1 - 2\,z^{-1} + z^{-2}$. This NTF is a high-pass filter, resulting in a $\Delta\Sigma$M with a low-pass response.

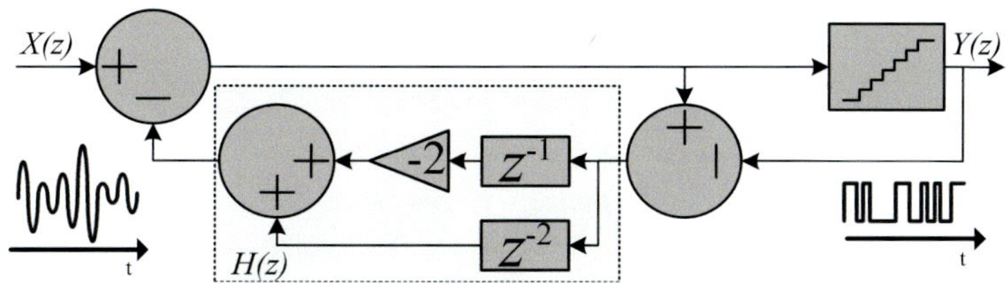


**FIGURE 5.20:** Representation in z-domain of the ΔΣM.

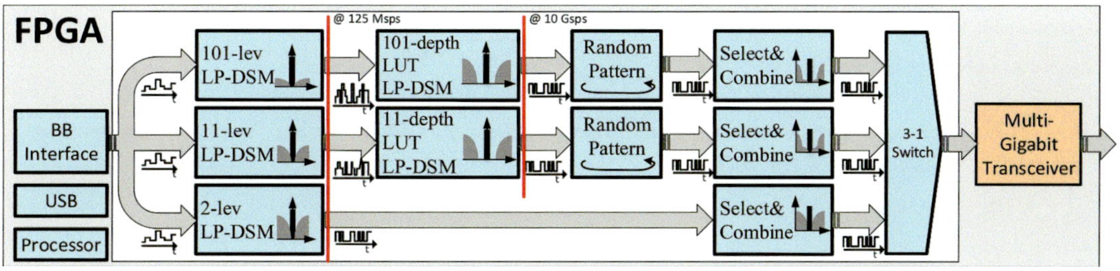


**FIGURE 5.21:** Block diagram of the implemented All-Digital Transmitter.

### 5.4.2 Proposed Architecture

In this section, the FPGA-based ΔΣM transmitter architecture used to demonstrate the potential and flexibility of the studied design methodology is presented and discussed.

The block diagram of the implemented ADT is illustrated in Fig. 5.21. The BB interface is continuously transmitting the samples to the modulator. As previously introduced, the modulator is based on 16 different ΔΣMs according to the architecture presented in Fig. 5.19b. Thereafter, the modulated samples are combined and replicated by the "Digital Upconversion Stage" to generate a parallel output word containing the four components $[I, \overline{Q}, \overline{I}, Q]$ of the desired signal. This output word is serialized by the MGT.

### 5.4.3 Prototype Validation

In this section, the simulated and measured results from the studied architecture are presented and discussed.

To validate the idea, an ADT was fully implemented into a Virtex Ultrascale XCVU095 FPGA using the Xilinx VCU1287 Characterization Kit. This FPGA was used as a proof of concept; any other could be used depending on the requirements. The modulator was optimized to enable a BB sampling rate of 200 MHz, resulting in an MGT sampling rate of 12.8 Gbps. In this way, the obtained carrier frequency will be approximately 3.2 GHz. The MGT was then connected to the Vector Signal Analyzer R&S FSW 8 for signal analysis and demodulation.

To assess the difference between the conventional and the architecture under study, the different techniques were simulated and compared to a single-rate and ideal modulator. The comparison in terms of spectrum is demonstrated in Fig. 5.22, where a 5 Msps 16-QAM signal is transmitted with the different approaches with a $K$ of 16 contiguous samples. It is clear that the architecture under study shows better performance and approaches the ideal spectrum. Thereafter, the in-band signal was filtered, and the error signal between the ideal and the two different approaches was computed (Fig. 5.23). All the signals' magnitudes are in the range of [-0.8, 0.8]. As can be seen, while 16 contiguous samples are insufficient for the conventional architecture to recover from the discontinuities, the architecture under

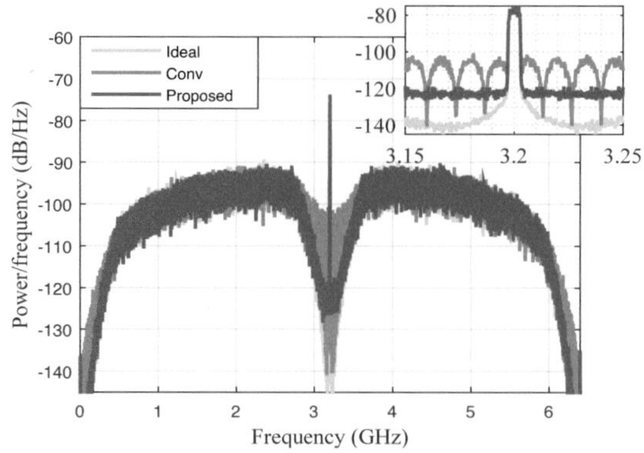


**FIGURE 5.22:** Comparison of the different simulated spectra with $N$ of 16 phases and $K$ of 16 contiguous samples for a 5 Msps, 16-QAM signal.

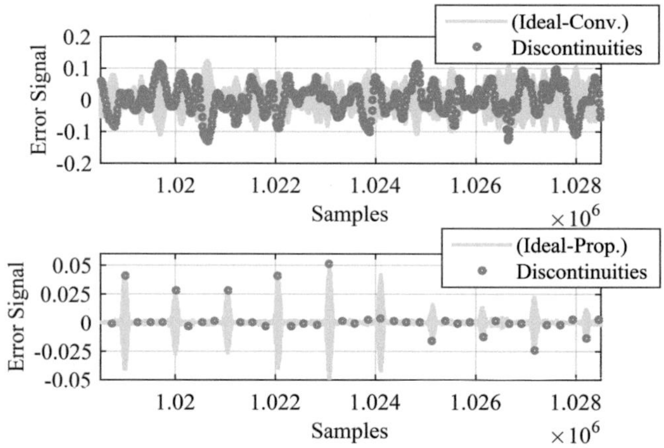


**FIGURE 5.23:** Simulated temporal evolution of the error signals for the different architectures with a $K$ of 16 samples, together with the discontinuities' instants, for a 5 Msps 16-QAM signal.

study is quite superior, maintaining a small and limited error signal. It is also demonstrated that the periodicity of the discontinuities was considerably increased. In practice, it changed from $K$ ticks in the conventional one to $K\,N$ clock cycles in the architecture under study.

The signal's in-band performance as well as the adjacent noise power were also simulated and quantified in terms of Normalized Mean Square Error (NMSE) and ACPR, respectively. The results are depicted in Fig. 5.24. It is clear that the performance of the architecture under study is quite superior to the conventional one: NMSE always inferior to -30 dB, and ACPR always higher than 30 dBc. In particular, this superior performance is higher for a small number of contiguous samples, which was the main objective of this work.

In addition to the simulated results, experimental results were also obtained, and in particular, spectrum and EVM measurements are reported. Different obtained spectra and respective constellations for BWs of 6 MHz, 60 MHz, and 120 MHz are presented in Fig. 5.25 for a small span.

A comparison between different spans (8 GHz and 600 MHz) for a BW of 120 MHz is also depicted in Fig. 5.26. The spurs around the carrier frequency are caused by an inefficient interpolation FIR filter based on ZOH that was implemented as a proof-of-concept.

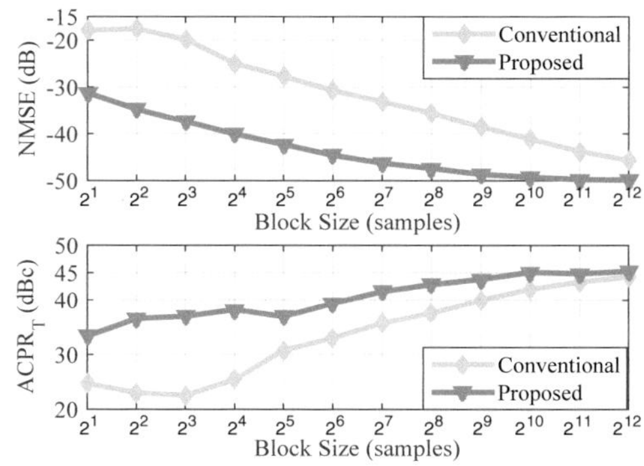


**FIGURE 5.24:** Simulated NMSE and ACPR figures of merit of the different architectures with different block sizes. The input signal was a 5 Msps 16-QAM modulated signal.

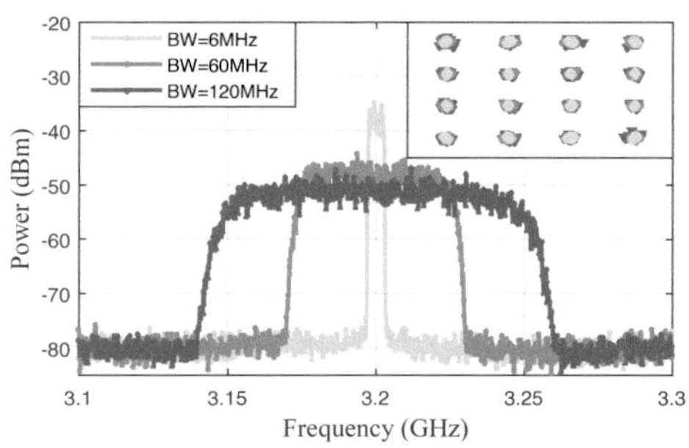


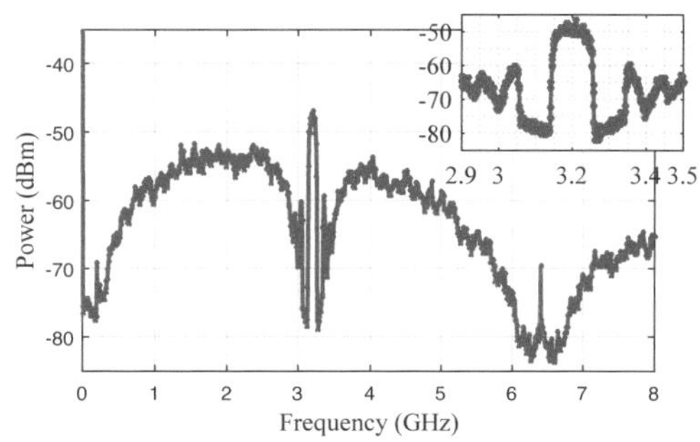


**FIGURE 5.25:** Different obtained spectra and respective constellations of a 16-QAM signal with BWs of 6 MHz, 60 MHz, and 120 MHz (RBW = 100 kHz, VBW = 2 kHz).

**FIGURE 5.26:** Obtained spectrum with different spans of a 16-QAM signal with 120 MHz of BW (RBW = 100 kHz, VBW = 2 kHz).

Table 5.6(a) summarizes the obtained EVM-SNR pairs obtained for the different BWs. All the experimental results present EVM values below 3%, even for the case of 120 MHz of BW, which confirms the validity of the architecture under study. At last, the occupied resources are depicted in Table 5.6 b). The FPGA presents a resource usage that is quite inferior to the conventional approach, with all major primitives below 6%, fulfilling the typical specifications of low- to mid-range FPGAs available on the market.

All the experiments validated the architecture under study. Moreover, it has a high level of scalability and is almost independent of the type of ΔΣM used.

### 5.4.4 Key Remarks

In this section, a new architecture based on propagating the initial conditions of multiple ΔΣMs in a fully parallel and pipelined Delta-Sigma Architecture is studied. The technique was implemented and validated in a FPGA-based ADT. In addition to achieving superior performance in terms of wideband capability with minimal resource usage, the technique is almost independent of the modulator's topology.

Despite the improvement in performance and the reduction in overall resources used, this implementation has still not improved the transmitted bandwidth when compared

**TABLE 5.6:** (a) SNR/EVM pairs for a 16-QAM signal. BW is in MHz, EVM in % and SNR in dB. (b) Comparison of the resources usage of the under study ($K = 16$) and the conventional ($K = 256$) ADT ($Ks$ were chosen according to Fig. 5.24 to ensure the same performance).

**(a)**

| BW | SNR | EVM |
|---|---|---|
| 6 | 36.6 | 1.12 |
| 12 | 35.8 | 1.20 |
| 24 | 33.5 | 1.59 |
| 48 | 31.3 | 2.04 |
| 60 | 30.9 | 2.13 |
| 120 | 28.0 | 2.97 |

**(b)**

| | | Under Study | | Conventional | |
|---|---|---|---|---|---|
| Resources | Total | Used | % | Used | % |
| Flip Flops | 1075200 | 7084 | 0.66 | 15423 | 1.43 |
| LUTs | 537600 | 8558 | 1.59 | 36347 | 6.76 |
| Mem. LUT | 76800 | 2311 | 3.01 | 29641 | 38.60 |
| BRAM | 1728 | 96 | 5.56 | 608 | 35.19 |
| GTHs | 28 | 1 | 3.57 | 1 | 3.57 |

to the one presented in the previous section. However, this architecture has provided the foundation for future optimizations to achieve larger bandwidths. In the next section, the presented architecture will do just that, where it takes advantage of the great performance of this architecture and improves it by changing to an RF-stage modulator with an optimized feedback filter.

## 5.5 RF-stage Architecture with State Propagation

Up to this point, a fully parallel and pipelined Delta-Sigma architecture has been presented that achieves superior performance in terms of wideband capability with minimal resource usage, while also being independent of the modulator's topology. However, the relatively small BW presented by this implementation still needs to be improved to meet the requirements of modern telecommunication protocols, such as 5G [1]. Other wideband implementations include an optimized second-order Time-Interleaved $\Delta\Sigma$M with 488 MHz of BW was also reported in [200]. However, in all the typical SoA works, the application of the DUC after the pulse encoder requires challenging serializer sampling rates (e.g., 20 and 60 Gbps in [126,200], respectively). Nonetheless, in addition to the increased complexity and cost, the design of analog front-ends becomes quite challenging. Thus, the remaining part of this section analyzes a novel architecture capable of transmitting BWs up to 1.25 GHz, embedded into a FPGA, with a serializer bitrate of 10 Gbps. The carrier frequency is 2.5 GHz, but the same architecture can be synthesized to operate at any other carrier frequency ($< 5$ GHz), with minor updates to the implemented hardware.

### 5.5.1 Delta-Sigma Modulator Design

For the ADT's design, an RF-stage modulator was chosen due to the larger available bandwidth provided by this configuration (to refresh the differences between RF- and BB-stage modulators, please consider Section 4.3). Furthermore, the design of the $\Delta\Sigma$ modulator must be done according to the required notch BW. However, at the same time, the critical path must be as short as possible, which is why architectures such as Cascade-of-Resonators, FeedBack/FeedForward (CRFB/F) or Cascaded-of-Integrators, FeedBack/FeedForward (CIFB/F) are not suitable. Thus, the generic Error-Feedback (EF) $\Delta\Sigma$ architecture with a FIR feedback loop (depicted in Fig. 5.27) was selected as a start-

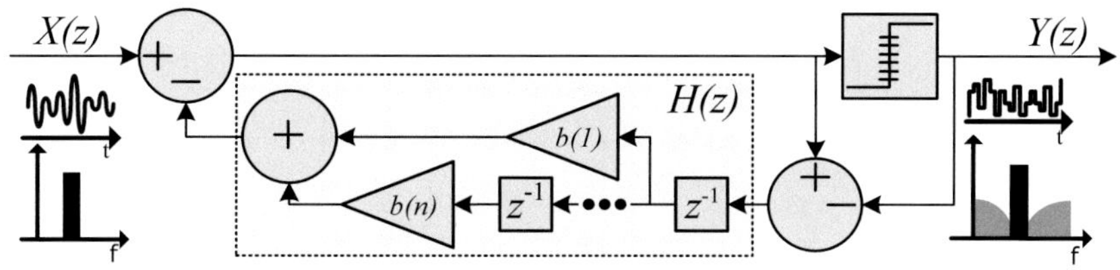


**FIGURE 5.27:** Representation in z-domain of an Error-Feedback ΔΣM with FIR feedback loop.

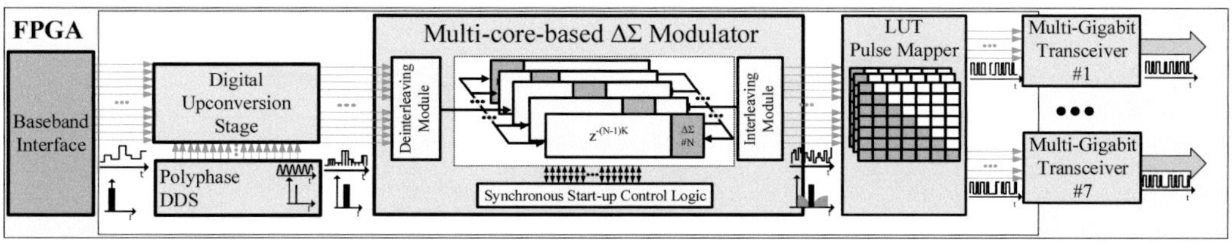


**FIGURE 5.28:** General block diagram of the implemented All-Digital RF Transmitter, embedded into an FPGA.

ing point. This architecture was derived from the conventional second-order EF-ΔΣM, by adding two complex zeros per required notch. For the sake of simplicity, let us restrain the number of notches to 3. By modeling the quantization noise by a random noise signal:

$$\begin{aligned} NTF(z) &= \left(1+\alpha z^{-1}+z^{-2}\right).\left(1+\beta z^{-1}+z^{-2}\right). \\ &\quad .\left(1+\gamma z^{-1}+z^{-2}\right) \end{aligned} \tag{5.8}$$

where $\alpha$, $\beta$ and $\gamma$ are defined as $-2\cos\left(2\pi Fc/Fs\right)$, $Fs$ is the sampling frequency, and $Fc$ is the required notch frequency. As $NTF(z) = 1 + H(z)$, where $H(z)$ is the feedback loop transfer function, $H(z)$ can be computed as:

$$\begin{aligned} H(z) &= b(1)z^{-1} + b(2)z^{-2} + b(3)z^{-3} + b(4)z^{-4} + \\ &\quad + b(5)z^{-5} + b(6)z^{-6} \end{aligned} \tag{5.9}$$

where $b(1) = (\alpha+\beta+\gamma)$, $b(2) = (3+\alpha\beta+\gamma(\alpha+\beta))$, $b(3) = (2(\alpha+\beta)+\gamma(2+\alpha\beta))$, $b(4) = (3+\alpha\beta+\gamma(\alpha+\beta))$, $b(5) = (\alpha+\beta+\gamma)$, and $b(6) = 1$. After deriving the coefficients, simulations were carried out in Matlab to choose the order of the feedback loop suitable to achieve around 1.25 GHz of usable bandwidth (Fig. 5.29). The notches were uniformly distributed into the desired bandwidth. While comparing the second and third FIR curves, it can be realized that the latter is the only one that fulfills the specification with better performance. However, the typical wireless modulated signals (briefly specified in terms of PAPR, modulation order, and bandwidth) preclude the implementation of this architecture with a single-bit output due to stability issues. While the SoA proposes the inclusion of poles to ensure the stability, it is also depicted in Fig. 5.29, that their inclusion reduces the achievable bandwidth. Thus, our solution passes through the use of more bits in the output (in particular, 8 levels were used for the sake of proof-of-concept).

### 5.5.2 Proposed Architecture

The block diagram of the implemented ADT is depicted in Fig. 5.28. The baseband data is interpolated to the serializer's sampling rate, divided into 64 phases, and forwarded to the

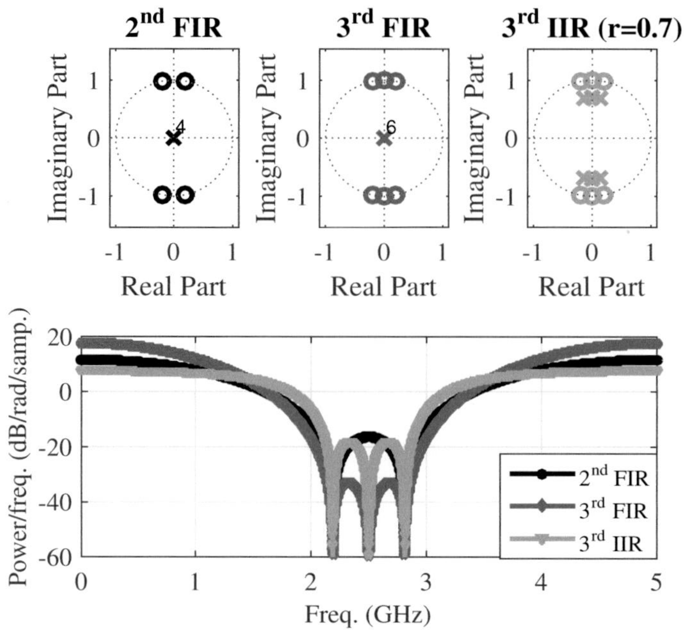


**FIGURE 5.29:** NTF's frequency response and zero-poles placement according to the different feedback-loop transfer functions.

polyphase DUC. This division into $N$ phases is always required due to the limited sampling rate from the FPGA programmable logic subsystems, which is far below the bandwidth of the input signal (typically less than 300 MHz). Each phase is clocked at $Fs_{BB}$, providing an equivalent sampling rate of $Fs_{MGT} = NFs_{BB}$. The DUC is composed of a polyphase DDS, working as LO. The carrier frequency was chosen to be 2.5 GHz. The resultant data is forwarded to a De-interleaving module, which temporarily stores and rearranges the data into $N$ blocks of size $K$. These blocks feed the $N$ phases to the Multi-core 8-level $\Delta\Sigma$ architecture. The computed encoded equivalents are temporarily stored and rearranged again by the Interleaving Module. It must be pointed out that the "Synchronous Start-up Control Logic" was included to ensure the propagation of the state-registers as presented in [46]. This enables a reduction in $K$ while maintaining a given level of performance. Ultimately, latency is improved, as well as memory resource usage. Moreover, in this work, we forced $K$ to be equal to $N$, in order to replace the FIFOs (proposed by the authors to be used in the Deinterleaving/Interleaving modules [45,46]) by just registers and multiplexers. Afterwards, a LUT-based mapper encodes the 3-bit samples into seven single-bit streams, which will be serialized by 7 MGTs.

### 5.5.3 Prototype Validation

To validate the approach under study, a ADT was fully implemented into a Virtex Ultrascale XCVU095 FPGA using the Xilinx VCU1287 Characterization Kit; any other FPGA could be used depending on the requirements. The MGTs sampling rate was configured to 10 Gbps, and lane-to-lane deskew techniques were implemented. The experimental setup is depicted in Fig. 5.30. Due to the challenges of measuring the 7 MGT channels at the same time, an extra MGT was included as an embedded trigger in the acquisition of all the channels with a 4-port Real-time Oscilloscope (Keysight DSA-X 92504Q). The Tektronix AWG610

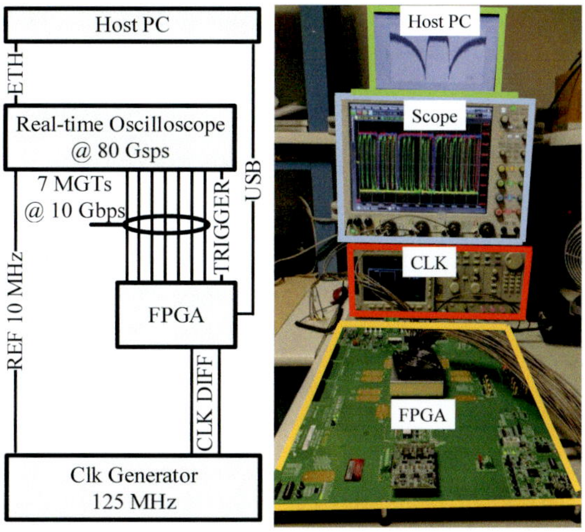


**FIGURE 5.30:** Block diagram and photo of the experimental setup.

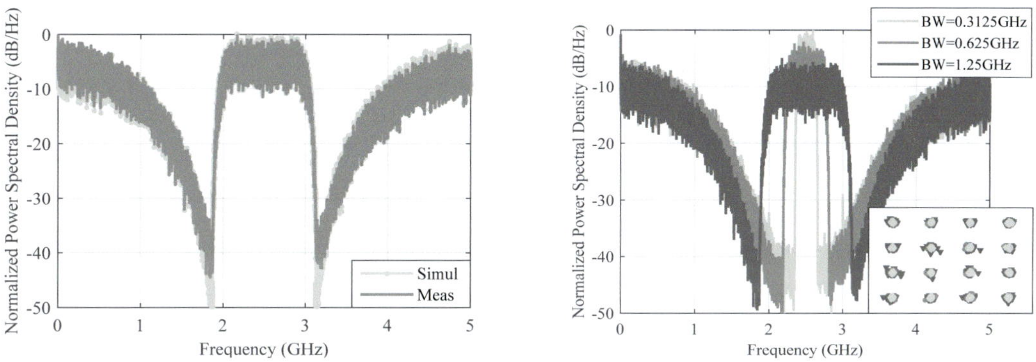


**FIGURE 5.31:** (a) Comparison of the simulated and measured spectra with a 16-QAM modulated signal with 1.25 GHz; (b) Different measured spectra and respective constellations with BWs of 312.5 MHz, 625 MHz and 1.25 GHz.

generates the FPGA reference clock and forwards a 10 MHz reference to the oscilloscope. After having all the channels acquired, the 7 streams are simply combined (note that no post-processing, such as timing/amplitude corrections, is applied), and the signal analysis and demodulation are done through the LabView VSA Software from National Instruments. Experimental results were obtained and, in particular, spectrum and EVM measurements are reported.

The comparison between the simulated and the measured 16-QAM modulated signal with 1.25 GHz (symbol rate of 1 Gsps and RRC of 0.25) is demonstrated in Fig. 5.31a). It can be observed that there is good agreement between the two datasets. The performance drop in terms of SNR (also observed in the same figure) is a result of the amplitude mismatch between the different MGTs. Different measured spectra and respective constellations for bandwidths of 312.5 MHz, 625 MHz, and 1.25 GHz are presented in Fig. 5.31b). These measurements were obtained from three different hardware implementations, with a uniform placement of the three notches to cover the required bandwidth.

**TABLE 5.7:** (a) SNR/EVM pairs for a 16-QAM signal. BW is in GHz, EVM in % and SNR in dB. (b) Resources usage of the architecture under study

(a)

| BW | SNR | EVM |
|---|---|---|
| 0.3125 | 36.90 | 1.05 |
| 0.625 | 36.58 | 1.10 |
| 1.25 | 30.51 | 2.23 |

(b)

| Resources | Used | Total | % |
|---|---|---|---|
| Flip Flops | 35411 | 1075200 | 3.3 |
| LUTs | 94627 | 537600 | 17.6 |
| Mem. LUT | 57882 | 76800 | 75.3 |
| BRAM | 0 | 1728 | 0 |
| DSP48 | 128 | 768 | 16.6 |
| GTHs | 8 | 28 | 28.5 |

Table 5.7 (a) summarizes the obtained EVM-SNR pairs obtained for the different BWs. All the experimental results present EVM values below 2.5%, even for the case of 1.25 GHz of bandwidth, attesting to the validity of the architecture under study. Moreover, to the best of the author's knowledge, this is the maximum bandwidth ever reached with an FPGA-based ADT architecture (the maximum value of 488 MHz was previously reported in [200]). The occupied resources are depicted in Table 5.7 (b). It should be mentioned that the BRAM primitive could be used to balance the high utilization of the Memory LUT.

All the experimental results attest to the validity of the architecture under study. For the sake of this proof of concept, the assessment of the outputs was performed using a high-speed oscilloscope. However, the promising results in terms of signal integrity anticipate that analog combining networks can be incorporated. The envisioned combination network is based on the asymmetric extended H-bridge [39,229], and its design should be optimized to maximize performance in the combination of multiple SMPAs. Following our architecture, analog impairments (such as phase imbalances or magnitude mismatches) can be easily compensated in the digital domain. Thus, the most stringent requirement, from a designer's point of view, is the synchronization between multiple lanes that must be ensured.

### 5.5.4 Key Remarks

The architecture presented in this section is capable of achieving a bandwidth of 1.25 GHz. The technique was implemented and validated in a FPGA-based ADT. In addition to achieving superior performance in terms of wideband capability, the fully digital behavior of this architecture demonstrates strong potential for synthesis in custom application-specific integrated circuits.

---

## 5.6 Wide-band high carrier flexibility ADT

To better understand the existing trade-off in the SoA between carrier agility and wideband transmitters, one needs to know the challenges of creating a $\Delta\Sigma$M-based ADT. The first point to consider in the design of an ADT is whether it is baseband or RF-stage. This choice will primarily affect the carrier agility of the transmitter, as baseband ADTs typically have the upconverter after the modulator, which limits the type of upconverter that can be used. All this, in turn, limits the frequencies to which the upconversion can be made; typically, the only carrier frequency available is equal to one-fourth of the serializer sampling

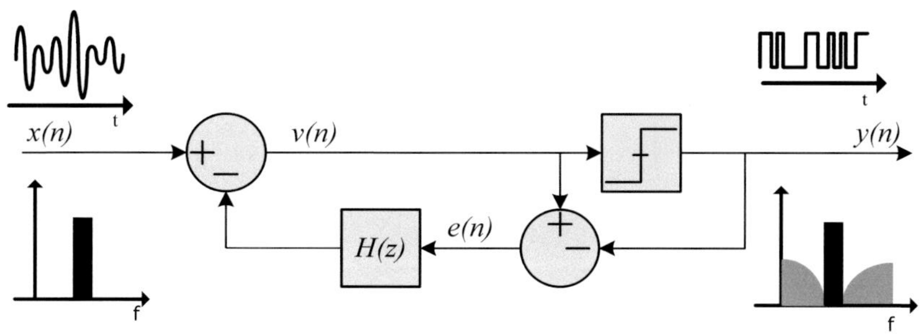


**FIGURE 5.32:** Block diagram of a delta-sigma modulator using the error feedback topology. After this modulator, the signal only needs filtering and amplification to be radiated.

frequency [45]. This leaves the RF-stage modulator as the only option for an agile, i.e., runtime carrier frequency varying, ΔΣM-based ADT.

In an RF-stage ADT, the upconversion is placed before the modulator, which means that the signal that enters the modulator needs to have a high sampling rate. This requires a high operating frequency, i.e., a high clock rate in the FPGA, which is particularly challenging when working with sampling frequencies in the 20-30 GHz range. In [46], a solution to this problem was introduced in the form of a ΔΣM core-based approach with error propagation. Each core is an independent ΔΣM, just as the one presented in Fig. 5.32, and each ΔΣM core processes $k$ consecutive samples. The ΔΣM depicted in Fig. 5.32 is an error feedback modulator, which was chosen not only for its short critical path, but also for its unitary signal transfer function. This block is responsible for transforming the multi-bit samples into single-bit samples. Despite having a discontinuity every $k \times M$ samples, where $M$ is the number of parallel cores, the final results are close to a perfect ΔΣM without the need for the digital logic to operate at GHz frequencies. This ΔΣM core approach, when implemented in an RF-stage architecture, could allow for the creation of a transmitter that is both flexible and wideband.

In this work, a ΔΣM-based ADT was implemented in an FPGA using a combination of filter and architecture optimization. This provides not only a better in-band noise rejection than the state of the art of ADTs, but also allows the implementation of a large number of parallel cores in order to achieve high sampling rates [46,200]. With such optimizations, this work contributes to improving the state of the art of ADTs' carrier agility, with possible frequencies from 1 up to 12 GHz, and bandwidth using only one serializer with carrier frequency agility, with 1 GHz.

### 5.6.1 All-Digital Transmitter Architecture

The general architecture of the transmitter can be seen in Fig. 5.33, where in the figure, it is depicted everything that is implemented in the FPGA, lacking in the figure only the depiction of the random access memory, which stores the baseband signal to be sent to the interpolator, and the serializer, which is placed after the polyphase ΔΣM. The first block in Fig. 5.33 is the interpolator, which uses a polyphase topology in order to increase the sampling rate of the signal without changing the FPGA's clock rate. Then the interpolated complex signal goes to the upconverter, where a polyphase DDS is used to create a complex sine wave at the desired frequency to which the upconversion will be, in this case, this

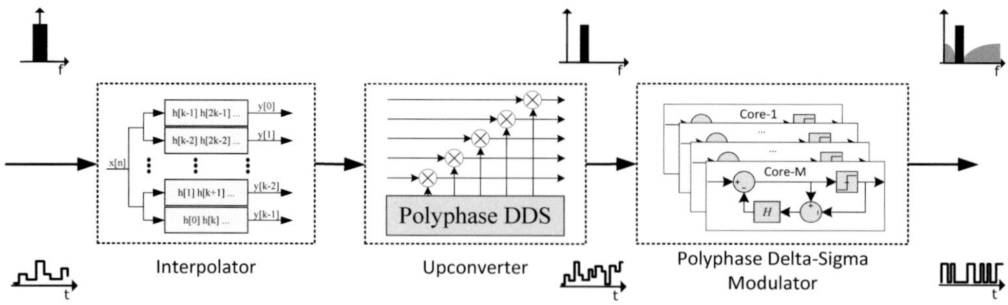


**FIGURE 5.33:** Block diagram of the top level architecture of the ADT, where DDS stands for digital direct synthesis.

frequency will be the carrier frequency, so it varies from 1 GHz up to 12 GHz. After this point, the signal is ready to enter the ΔΣM, depicted in Fig. 5.32.

This work will follow the same ΔΣM core-based approach with error propagation presented in [46]. This approach was chosen because it allows for a large degree of parallel phases, enabling the system to reach higher equivalent sampling rates without having to increase the FPGA's clock rate. Each core features the architecture presented in Fig. 5.32, where the filter determines the NTF. It is known that the higher the order of the filter, the better noise shaping it can produce, and so, a better NTF is presented by the modulator [179]. In [47] and [48], a straightforward way to increase the filter's order in the modulator while maintaining a simple way of controlling the notch frequency was presented. The technique starts by using a simple NTF function and raising it to the power $p$:

$$NTF(z) = \left(1 + \alpha z^{-1} + z^{-2}\right)^p, \tag{5.10}$$

since $\alpha$ is equal to $-2\cos(2\pi F_c/F_s)$, with $F_c$ being the desired carrier frequency and $F_s$ the sampling rate of the signal, it is possible to move the notch in frequency by changing the value of $\alpha$. From this, it is also possible to determine the filter using:

$$H(z) = NTF(z) - 1. \tag{5.11}$$

The method described in the previous paragraph yields FIR filters, which may lead to stability issues for values of $p$ greater than 1. In order to make the modulator stable, the NTF needs to be updated to [47]:

$$NTF(z) = \frac{\left(1 + \alpha z^{-1} + z^{-2}\right)^p}{(1 + R\alpha z^{-1} + Rz^{-2})^p}. \tag{5.12}$$

Using Eq. (5.11) on Eq. (5.12) results in an IIR filter, which, despite the added complexity, is able to generate a stable ΔΣM by tuning the $R$ value. This means that not all values of $R$ will result in a stable ΔΣM. To know the range of possible values for $R$, a simulation of the ΔΣM was conducted using the NTF presented in Eq. (5.12) with $p = 2$. The sampling rate of the ΔΣM was set to 25.6 Gsps, the carrier frequency varied from 1 to 12 GHz, and the $R$ value ranged from 0.3 to 0.95. Using an 800 Mbaud 16-QAM signal, the EVM results of this simulation are presented in Fig. 5.34. Before analyzing the simulation results, several expectations are anticipated. The first is that for low values of $R$, the resulting $H$ filter will have performance close to that of a FIR filter. Therefore, it is expected that the resulting ΔΣM will have poor stability, which will generate a high EVM. Furthermore, from prior

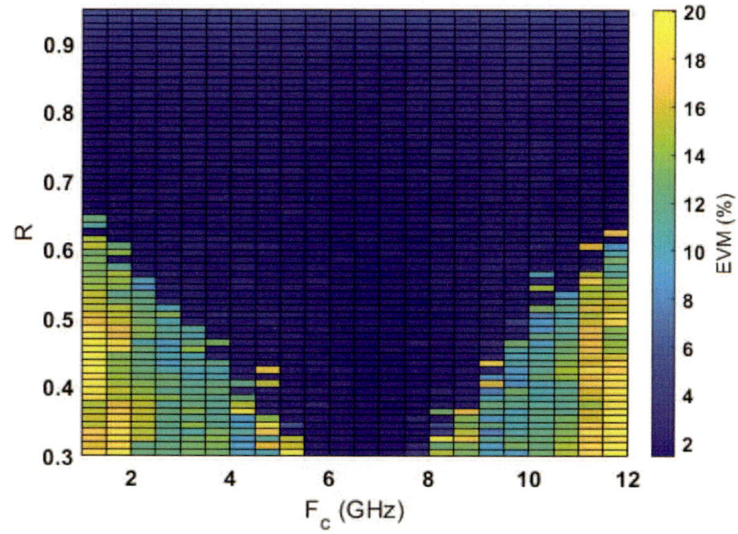


**FIGURE 5.34:** Simulation of the EVM with $R$ varying from 0.3 up to 0.95, for frequencies from 1 to 12 GHz and $p$ equals to 2.

experiments, we know that for a carrier frequency equal to the sampling rate over four, which in this case is 6.4 GHz, a FIR filter is stable. From this, it is expected that a larger range of $R$ values, especially low values, will result in a stable $\Delta\Sigma$M for frequencies around 6.4 GHz.

From Fig. 5.34, it is possible to confirm that not all values of $R$ will result in a stable $\Delta\Sigma$M. In the same figure, stable values of $R$ will result in an EVM represented in blue or dark blue colors, with the darker the color, the lower the EVM. Using values for $R$ between 0.7 and 0.8 is a safe choice; however, it is not the optimal solution for every carrier frequency. The optimal value can vary significantly, with, for example, the best value for 5.5 GHz being 0.32 and for 10 GHz being 0.71. Therefore, for each new frequency, a simulation may be required to determine the optimal $R$ value. It is important to mention that, while this simulation is done offline, after the coefficients of the filter are calculated, they can be stored as LUT in the FPGA programmable logic, allowing the $\Delta\Sigma$M to run in real-time in the FPGA.

### 5.6.2 Hardware Implementation

A proof of concept of the transmitter depicted in Fig. 5.33 was implemented using the Virtex UltraScale+ 56G PAM4 VCU129 FPGA Evaluation Kit from Xilinx. This kit was chosen due to its high sampling rate MGTs, which are capable of transmitting at 28 Gbps.

In order to achieve such a high sampling rate, a high degree of parallelization was required. For that, $M$, which is the number of parallel $\Delta\Sigma$M cores implemented, was set to 128. With this value, it was possible to achieve 25.6 GHz of sampling rate, while maintaining the FPGA's clock at 200 MHz.

Typical implementations of the $\Delta\Sigma$M's filter use the direct filter form; however, due to the use of an IIR filter, this would result in the implementation of multiple adders. Such implementation has a critical path that is too long for the FPGA's specified clock rate.

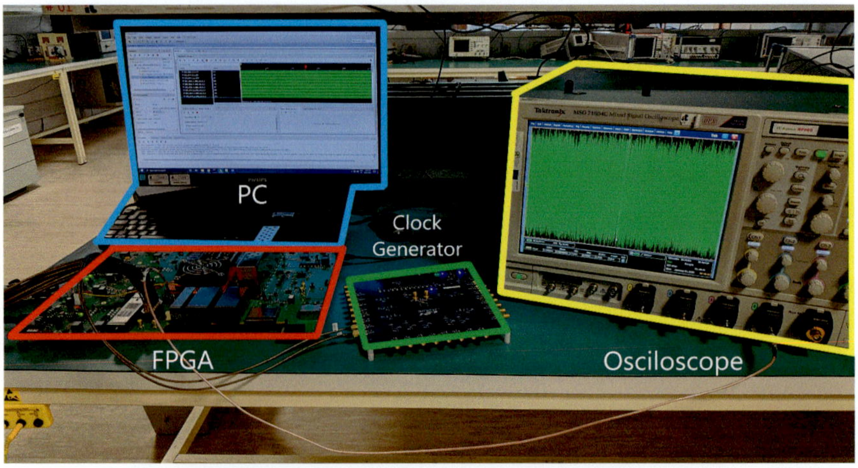


**FIGURE 5.35:** Experimental setup, with the personal computer, clock generator, FPGA, and oscilloscope.

**TABLE 5.8:** Resource usage of the implemented all-digital transmitter.

| Resources | Used | Total | % |
|---|---|---|---|
| Flip Flops | 487882 | 3456000 | 14.12 |
| LUTs | 482427 | 1728000 | 27.92 |
| BRAM | 258 | 2688 | 9.60 |
| DSP48 | 1461 | 12288 | 11.89 |
| GTM | 1 | 24 | 4.17 |

To resolve this problem, the transposed filter form was used. In this form, there is a delay between the consecutive adders, significantly reducing the critical path.

### 5.6.3 Results

In this subsection, the experimental setup and the results obtained will be presented and discussed. Fig. 5.35 depicts the experimental setup, with the FPGA framed in red, in green the clock generator, in yellow the Tektronix MSO71604C oscilloscope, and in blue the personal computer used to program the FPGA and to retrieve the data from the oscilloscope. The resources used by the transmitter implemented in the FPGA are listed in Table 5.8. There, it is possible to see that this implementation is most intensive in the usage of LUTs, flip-flops, and DSPs. Due to this implementation being a proof of concept, there is still further room to optimize the digital signal processing chain resources.

The output spectrum of the all-digital transmitter using an 800 Mbaud, 16-QAM signal at 6.4 GHz can be seen in Fig. 5.36. In the figure, it can be seen that the signal, with approximately 1 GHz of bandwidth, has roughly 30 dB of SNR.

Using the same signal as the one displayed in Fig. 5.36, the carrier frequency was swept from 1 GHz up to 12 GHz. The signal to be transmitted was inspected using the oscilloscope. It is worth saying that a vector signal analyzer was not used because our lab did not have one capable of receiving and demodulating an 800 Mbaud signal. Then, the inspected signal was processed in MATLAB to calculate the EVM of the signal to be transmitted. To corroborate the simulation presented in Fig. 5.34, a measurement sweep is presented in Fig. 5.37 where both the $R$ and $F_c$ were varied, from 0.3 to 0.95 and from 1 to 12 GHz for $R$ and $F_c$,

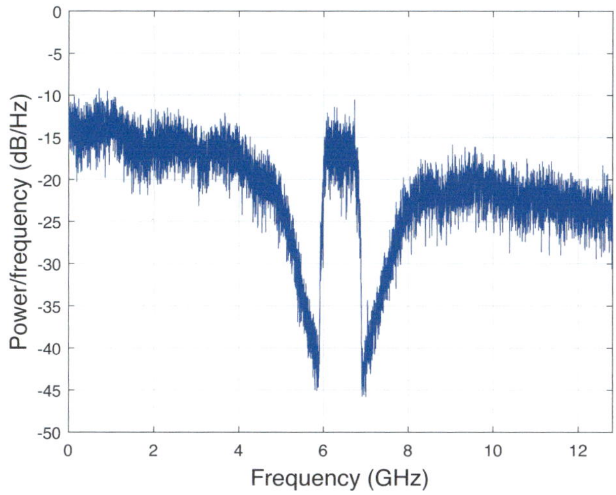


**FIGURE 5.36:** Output spectrum of the all-digital transmitter at 6.4 GHz using an 800 Msps, 16-QAM signal.

respectively. Comparing Fig. 5.34 with Fig. 5.37, it is possible to see that both present similar shapes, which means that both present similar EVMs for the same $[R; F_c]$ pair. However, there are a few regions where the EVM is different. The first one is around 1 GHz, where the measured EVM is higher than the simulated one. The proximity to DC can be one of the reasons for this increase in EVM. The second one is around 12 GHz, where interference from the image present in the second Nyquist zone may be a potential dominant factor. Despite these differences, the general behavior is the same, and the predictions for the best range of $R$ are corroborated.

Using the optimal values of R, Fig. 5.38 was constructed. This figure allows for a clearer representation of the final performance of the ADT. From the figure, it is possible to see that the EVM never goes above 8%, which means that this transmitter would be capable of transmitting a 64-QAM signal, while satisfying the 5G standard for EVM [2]. Furthermore, from 2 to 9 GHz, the EVM never exceeds 5%. This represents a 7 GHz band where it is possible to transmit a 1 GHz bandwidth signal with the EVM around 4%.

Finally, Fig. 5.39 presents the signal after it has been filtered by a commercial off-the-shelf BFHK-6751+ filter from Mini-Circuits. This RF filter is a bandpass one, with its bandpass from 5.9 GHz up to 6.9 GHz and was able to suppress the out-of-band noise generated by the ADT with a high degree of rejection, making it compatible with spectrum masks from protocols such as 5G [2]. Moreover, this test has shown that it is possible to eliminate this noise with a commercial-off-the-shelf RF filter, allowing for a more straightforward implementation of this work into a viable commercial telecommunication transmitter.

#### 5.6.3.1 Comparison with the state of the art

Table 5.9 presents a comparison of this work against the state of the art. In it, it is clearly visible the division where an ADT either has carrier agility or presents wide bandwidth. From the table, it is evident that the maximum bandwidth was 1.25 GHz, as presented by [47]. However, the authors utilized 7 MGTs to transmit the multi-level signal, plus an additional one for synchronization, all operating at 10 Gbps, to achieve these results. When compared to the work presented here, the lower overall bandwidth, of around 1 GHz bandwidth, is achieved using only one MGT, meaning that, in principle, it uses less energy

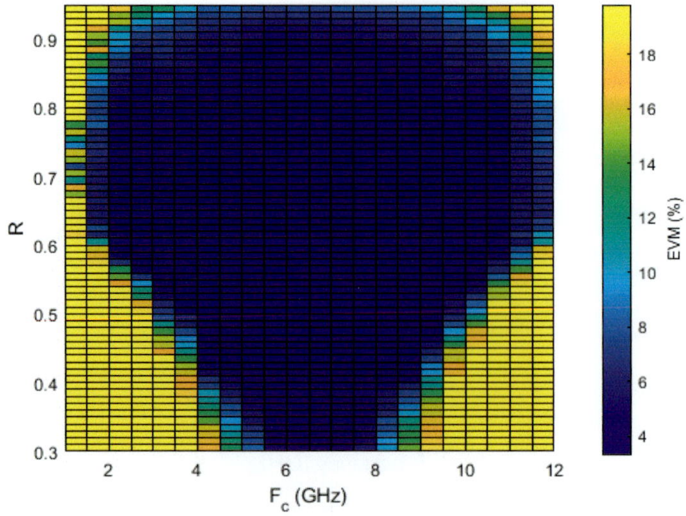


**FIGURE 5.37:** Measurement of the ΔΣM with $R$ varying from 0.3 up to 0.95 for frequencies from 1 to 12 GHz.

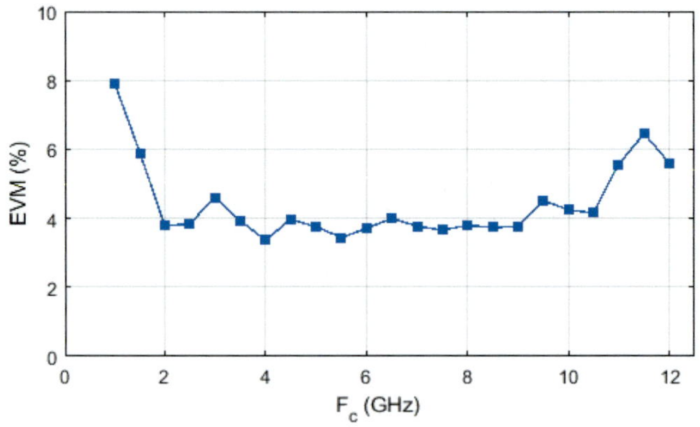


**FIGURE 5.38:** Frequency sweep from 1 GHz up to 12 GHz using an 800 Mbaud, 16-QAM signal.

and space. Moreover, the lack of any carrier frequency agility is a great disadvantage that the work presented here does not have.

For the highest carrier agility presented in an all-digital transceiver, [45] presents the two most agile transmitters, one PWM-based that can transmit from 0.1 up to 6.2 GHz and a ΔΣM-based one that can transmit from 0.1 up to 5.12 GHz. Since this last one is the only one based on ΔΣM, these are the values added to Table 5.9. The transmitter presented here almost doubles this range by being capable of transmitting from 1 up to 12 GHz, all while transmitting a 1 GHz bandwidth signal of up to 64-QAM modulation.

A comparison regarding figures of merit, such as power and area, is more difficult to make. This is due to most authors not revealing these numbers. Nevertheless, one can estimate the overall power consumption based on the resource usage. Having this in mind, the work presented here has an overall resource usage higher than that presented in [47], which would lead to higher power consumption. However, MGTs tend to use more power

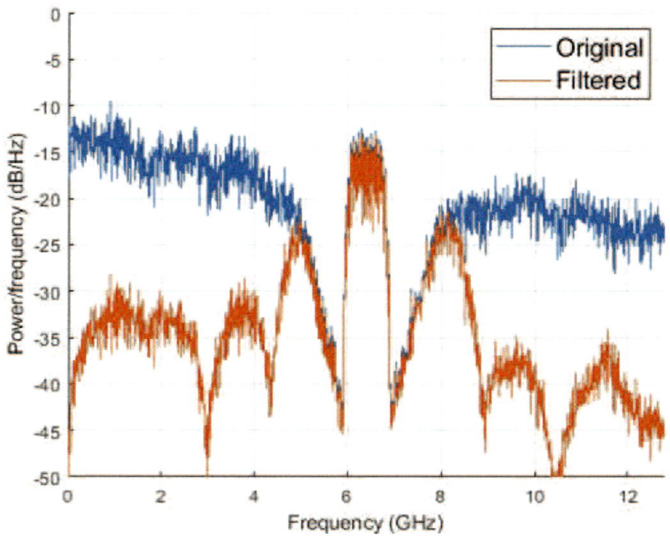


**FIGURE 5.39:** Output spectrum of the ADT with and without filtering.

**TABLE 5.9:** Comparison of carrier agility and transmitted bandwidth with the SoA[1]

| Reference | Min $F_c$ (GHz) | Max $F_c$ (GHz) | Max BW (MHZ) | MGTs |
|---|---|---|---|---|
| [47] | - | 2.5 | 1250 | 8 |
| [200] | - | 4.8 | 500 | 1 |
| [45] | 0.1 | 5.12 | 48 | 1 |
| [43] | 2.375 | 2.625 | 3.75 | 1 |
| This work | 1 | 12 | 1000 | 1 |

[1]In the table "-" stands for not applicable.

when compared to the rest of the digital logic, and [47] uses more of those than the work presented here. Ultimately, it is possible that both would have a similar level of power consumption. Regarding [45], which is the previous state of the art in carrier frequency agility, one can see that its power consumption would be lower; however, here we present almost double the carrier frequency range, when compared to their PWM implementation, which has only 12 MHz of bandwidth. In the end, it is possible to say that the work presented here is resource-intensive; nevertheless, the figures of merit that are achieved more than justify it.

### 5.6.4 Key Remarks

In this section, a highly flexible ΔΣM-based all-digital transmitter is presented. This transmitter enhances techniques used in parallel implementations to create a transmitter capable of operating at 25.6 Gbps. An increase in filter order, as well as the use of an IIR filter with optimized parameters, allows for a 1 GHz bandwidth signal to be transmitted from 1 up to 12 GHz carrier frequency with an EVM of around 4%, from 2 to 9 GHz, and lower than 8%, from 1 to 2 GHz and also from 9 to 12 GHz. With these figures of merit, this transmitter is state-of-the-art in both carrier frequency agility and bandwidth, utilizing only one MGT per transmitter for carrier frequency agility.

## 5.7 Conclusion

In the quest to increase the BW of an ADT, it was clear that the limiting factor was the final sampling rate, which was being limited by the relatively low clock frequency delivered by FPGAs. To overcome this problem, a parallel implementation was needed that could maintain the same low FPGA clock while also providing a higher sampling rate. The time-interleaved architecture, studied in Section 5.2, was proposed to achieve this. The increased performance presented by this implementation enabled the transmission of signals with bandwidths up to 12.5 MHz, accompanied by EVMs lower than 1.52%. However, the critical path presented by this architecture remains proportional to the number of parallel paths used, which lowers the FPGA's clock frequency, ultimately canceling out any benefits of the polyphase equivalent representation. This creates a limit to the number of parallel paths that can be used.

With the problem already identified, which was the increase in the critical path with the increase in the number of parallel paths, a second architecture was proposed, named the Multicore-based Architecture, as seen in Section 5.3. This architecture introduces a new concept where $N$ parallel ΔΣM cores process $k$ samples. Because each core is independent, this architecture enables an increase in the number of parallel paths without extending the critical path. This meant that the final sampling frequency could now be increased to a point where signals with up to 120 MHz BW could be transmitted. Even though this architecture presented a major leap in terms of BW capacity, a closer inspection also revealed that impulsive noise was generated during the transitions between cores. This was exactly the problem addressed in Section 5.4. The solution was the introduction of a module called "Synchronous Start-up Control Logic", which was responsible for propagating the internal state of one core to the next, resulting in delaying the noise spike from every $k$ samples to every $k \times N$ samples.

Section 5.5 demonstrated that an RF-stage modulator combined with an improved modulator design enabled the transmission of a 1.25 GHz signal with an EVM of 2.23%. However, this required the use of eight MGTs. In contrast, Section 5.6 reduced the number of MGTs to one by employing an optimized IIR filter, allowing the transmitter to achieve 1 GHz bandwidth across carrier frequencies from 1 to 12 GHz, with EVM values around 4% throughout most of the band and consistently below 8%.

This result establishes a new state-of-the-art benchmark by combining wideband capability with unprecedented carrier agility using only a single MGT, underscoring the potential of ADTs as practical solutions for future 5G and 6G systems. Although further improvements are possible, these findings confirm that ADTs already represent a viable approach for modern telecommunications deployments.

# *Chapter 6*

# *From Single- to Concurrent Multi-Band Transmission*

## 6.1 Introduction

Despite the apparent ideal and native support for the multi-band capability, design challenges associated with the non-contiguous CA transmission in ADT have hampered the proposal of multi-band solutions. Thus, the reported literature on non-contiguous CA ADTs is quite scarce. Generally speaking, the literature has demonstrated that multi-band transmission could only be achieved by clustering narrowband bands in a limited bandwidth (as briefly introduced in Chapter 4), or with integer multiples of the modulators' sampling frequencies [185], or with reduced sampling rate topologies [109, 121]. Others employ bulky and inefficient power combiners to join different bands before transmission [18, 109]. All these difficulties in designing multi-band transmission arise from the placement of the DUC after the pulse encoding.

This chapter will delve into the specifics of concurrent multi-band RF transmission, supported by the conference paper [44] and the journal articles [43, 47]. The chapter starts by proposing an architecture for a BB-stage ADT that can be real-time reconfigured to transmit one or dual bands of modulated data, with a maximum distance between bands of almost 300 MHz. Then, the proposed architecture has evolved to include a linear pre-compensation mechanism that can improve the dual-band FoM of the dual-band scenario. The chapter is concluded with the proposal of an RF-stage architecture designed for one to three bands that can be digitally generated in the range of 100 MHz to 2.5 GHz.

## 6.2 Dual-band BB-stage Modulator

In this section, we address the frequency agility limitations presented in old implementations, particularly in terms of multi-band transmission, by proposing a new transmitter architecture that enables fine frequency tuning for two non-contiguous frequency bands.

### 6.2.1 Proposed Architecture

The proposed idea is based on the use of a low-complex and tunable $\Delta\Sigma$ architecture to perform the real-time encoding of different frequency bands with negligible degradation of the in-band signals. In particular, the underlying idea explores the real-time displacement of the spectral notches of a $\Delta\Sigma$M to configure the system as a LP-$\Delta\Sigma$M, a Band-Pass Delta-Sigma Modulator (BP-$\Delta\Sigma$M) with tunable center frequency, or a High-Pass Delta-Sigma Modulator (HP-$\Delta\Sigma$M). On the one hand, the use of low- and high-pass topologies

DOI: 10.1201/9781003225713-6

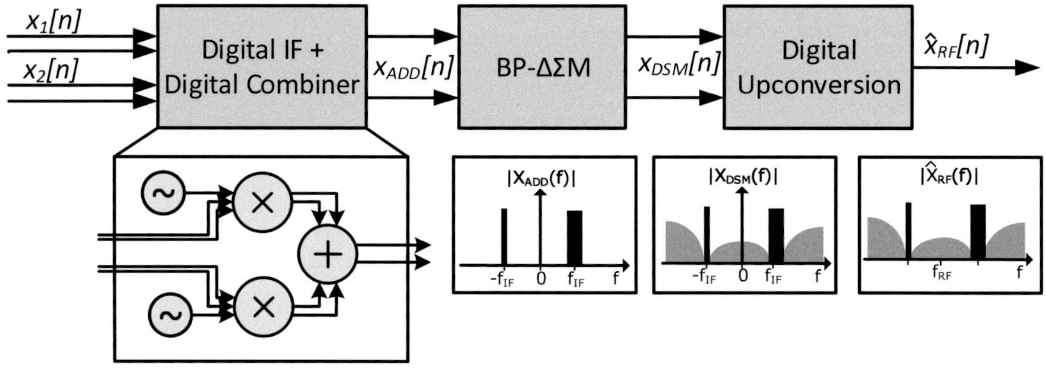


**FIGURE 6.1:** General block diagram with the transmitter's architecture demonstrating the use of this modulator into single- or/and dual-band scenarios.

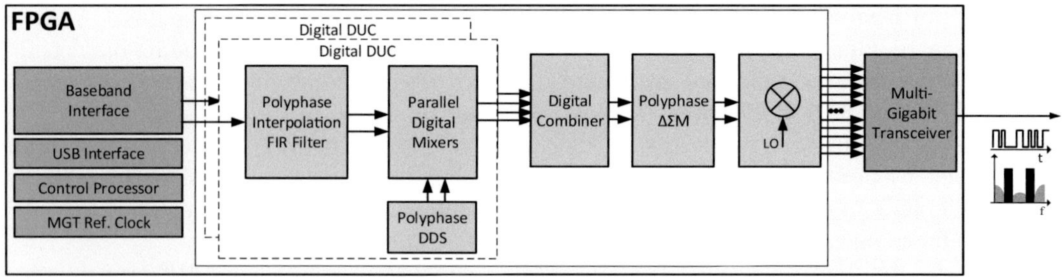


**FIGURE 6.2:** Block diagram of proposed tunable all-digital transmitter implemented in the FPGA, adapted from [44].

enables the development of single-band transmitters. On the other hand, by configuring the system as a bandpass modulator, the two notches in the spectrum can be exploited to develop concurrent dual-band digital transmitters. Focusing on the BP-$\Delta\Sigma$M without loss of generality, the fact that one is dealing with real coefficients means that the frequency response is symmetric, which gives rise to symmetrical notches around the DC component. Thus, in this sense, taking into account that these two bands will preserve their integrity due to the noise-shaping property, they can be used to place up to two different bands (one at $+f_{if}$, one at $-f_{if}$ or two at $\pm f_{if}$) before the digital upconversion stage to RF, such as depicted in Fig. 6.1.

The block diagram of the proposed architecture is presented in Fig. 6.2. For the purpose of this proof-of-concept, one or two baseband test signals (for single- or dual-band, respectively) are transmitted to the FPGA's internal BRAM memory through the interface with the baseband processing. The signals are continuously transmitted through a Polyphase Interpolation FIR filter, which increases the sampling rate by a factor of $M$ distributed across $M$ different paths. Thereafter, the two signals are upconverted to symmetric intermediate frequencies around the DC component through a SSB upconversion stage that employs reconfigurable polyphase DDS modules, which define the frequency notches of the tunable $\Delta\Sigma$M. After that, they are combined and fed to the polyphase $M$-path tunable $\Delta\Sigma$M, whose output samples will then be combined and replicated to generate a parallel output word with the four components of the RF signal ($x_I(n)$, $x_Q(n)$, and their inverted versions), and serialized by the FPGA's MGT.

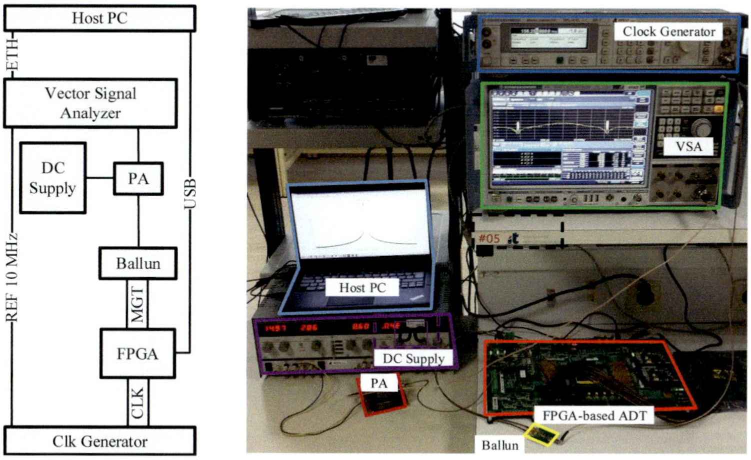


**FIGURE 6.3:** Photo of the experimental setup.

**TABLE 6.1:** Single band scenario: EVM values for different SRs. (a) After ΔΣM (before DUC and serializer). (b) After serializer and PA.

(a)

| | Symbol Rate (Msps) | | | |
|---|---|---|---|---|
| Modulation | 3.125 | 6.250 | 13.021 | 15.625 |
| QPSK | 0.47 | 0.41 | 0.78 | 1.02 |
| 16-QAM | 0.32 | 0.30 | 0.51 | 0.89 |
| 64-QAM | 0.30 | 0.52 | 0.52 | 1.1 |

(b)

| | Symbol Rate (Msps) | | | |
|---|---|---|---|---|
| Modulation | 3.125 | 6.250 | 13.021 | 15.625 |
| QPSK | 1.33 | 1.79 | 2.23 | 2.60 |
| 16-QAM | 1.50 | 1.88 | 2.49 | 3.01 |
| 64-QAM | 1.44 | 1.89 | 2.66 | 2.91 |

### 6.2.2 Prototype Validation

An ADT was fully implemented into a Virtex UltraScale XCVU095 FPGA using the Xilinx VCU1283 Characterization Kit. Focusing on the ΔΣ topology, the design has fallen into the Error-feedback structure, whose illustration was already depicted in Fig. 4.13. As introduced in Chapter 4, this modulator has one direct application: to design RF digital transmitters with a tunable IF. Moreover, taking into account that in the transition from LP-ΔΣM/HP-ΔΣM to BP-ΔΣM, two symmetric NTF's zeros arise, they can be exploited to perform dual-band transmission rather than transmitting just one band. Thus, by varying the $\alpha$ constant, the system's configuration is altered, enabling the use of this modulator in both single- and dual-band applications.

After deriving the $\alpha$ variable, this modulator was optimized, with the aim of increasing the modulator's sampling frequency and, consequently, the maximum distance between bands. To accomplish this, the transfer function $H(z)$ was decomposed into a polyphase equivalent with two different phases and two outputs ($M = 2$) according to the theory discussed in [205] and [36].

The experimental setup of the FPGA-based ADT is depicted in Fig. 6.3. The MGT was connected to a PA (ZHL-1042J), and then to a VSA R&S FSW 8. The MGT's sampling frequency was adjusted to 10 Gbps to provide a center frequency of 2.5 GHz, and the ΔΣM's sampling frequency is 312.6 MHz divided into two different phases.

Experimental results in terms of spectrum and EVM measures were obtained with the VSA for two different scenarios: single-band and dual-band after the PA.

The Table 6.1 shows the results obtained in a single-band scenario with different SRs. Concerning the dual-band scenario, two symmetrical bands with different modulations

**TABLE 6.2:** Dual Band scenario: EVM values for different carrier frequencies. (a) After ΔΣM (before DUC and serializer). (b) After serializer and PA.

(a)

| | Upper Carrier Frequency (GHz) | | | |
|---|---|---|---|---|
| Modulation | 2.506 | 2.538 | 2.568 | 2.600 |
| QPSK | 0.600 | 0.940 | 0.880 | 1.160 |
| | Lower Carrier Frequency (GHz) | | | |
| Modulation | 2.494 | 2.463 | 2.431 | 2.400 |
| 16-QAM | 0.510 | 0.800 | 0.750 | 1.020 |

(b)

| | Upper Carrier Frequency (GHz) | | | |
|---|---|---|---|---|
| Modulation | 2.506 | 2.538 | 2.568 | 2.600 |
| QPSK | 1.67 | 2.52 | 2.660 | 3.47 |
| | Lower Carrier Frequency (GHz) | | | |
| Modulation | 2.494 | 2.463 | 2.431 | 2.400 |
| 16-QAM | 1.77 | 2.46 | 3.15 | 4.34 |

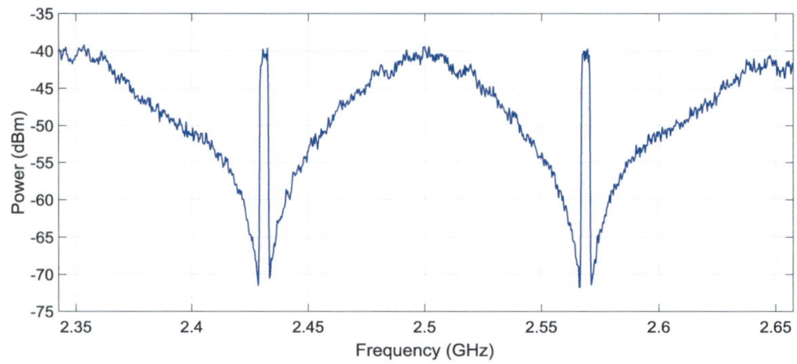


**FIGURE 6.4:** Spectrum of the transmitted signal in a dual-band scenario (RBW = 100 KHz, VBW = 2 KHz, SPAN = 315 MHz).

(QPSK and 16-QAM) were tested for different frequencies, and the results are depicted in Table 6.2. Both tables present EVM values always below 4.5%, resulting in a well-defined constellation.

Additionally, measurements after the polyphase ΔΣM block were carried out, showing that a degradation of approximately 1.5% was introduced by the analog front-end (output buffer and PA). The choice of higher SR values for the single-band scenario is explained by the location of its ΔΣM's zeros (recall Fig. 4.14). In this case, the ΔΣM is tuned to a low-pass configuration, resulting in two overlapped zeros in the NTF, which increases the processing bandwidth and accommodates carriers with higher SR. In the dual-band scenario, any two carrier frequencies may be chosen through the adequate selection of the central frequency and the spacing between the two carriers, provided that the distance between them does not exceed the modulator's sampling frequency. Moreover, if the hardware complexity is kept low, the sampling frequency can be raised at the expense of increasing the number of paths in the polyphase decomposition, resulting in higher frequency separation between the two different carriers.

For the used FPGA, the proposed system enables the design of a single-carrier tunable transmitter at a carrier frequency of 2.5 GHz, or a dual-band configuration with a central frequency of 2.5 GHz and a separation between carriers of up to 300 MHz. The carrier frequency can also be changed. However, as previously introduced, it is entangled with the serializer's sampling rate, and thus, its change is quite limited, leading to a modification of the Pulse Encoder's OSR.

Fig. 6.4 shows the obtained spectrum after the PA for two modulated signals with a SR of 3.125 MHz: a QPSK at 2.5688 GHz and a 16-QAM at 2.431 GHz.

## 6.3 Precompensated Dual-band BB-stage Modulator

The previous section demonstrates that the proposed architecture results in a significant degradation of the FoM in the dual-band experiment. To enhance system performance in the dual-band scenario, it is worthwhile to investigate the DUC in greater detail, considering that it is the only module between the ΔΣM and the serializer that can cause performance degradation.

### 6.3.1 Analysis of the Digital Upconversion Chain

In Section 2.2.5, the DUC used in BB-stage modulators was briefly introduced and, subsequently, it was demonstrated why the replication and combination of the encoded signals in the form of $[x_I[n], -x_Q[n], -x_I[n], x_Q[n]]$ leads to the upconversion from BB to one fourth of the serializer's sampling rate. However, an issue must be pointed out regarding this upconversion chain. Recalling Eq. (2.15), one should note that just the real part of the output signal is considered. This leads to:

$$\begin{aligned} \widehat{x}_{RF}[n] &= Re\left[x_{INT}[n]e^{jw_c n}\right] \\ &= \frac{x_{INT}[n]e^{jw_c n} + x^*_{INT}[n]e^{-jw_c n}}{2} \end{aligned} \tag{6.1}$$

Eq. (6.1) demonstrates that, by using this upconversion scheme, the spectral content of a conjugated signal image will be overlapped with the original spectrum, possibly corrupting the band of interest. This problem is exacerbated in the case of ADTs due to the use of a discrete ZOH filter with accentuated sidelobes extending across the entire spectrum of frequencies. As it is graphically illustrated in Fig. 6.5, when using an IF shifted from the carrier frequency, an aliased image appears in the symmetric IF value. Until now, this problem has remained unknown and disregarded because the aliased image is hidden behind the quantization noise. However, this is not the case when symmetric bands are operated for the transmission/reception of different bands.

To minimize this problem, the first step involves modeling the whole system. In terms of the frequency domain, the first two modules (upsampler and discrete ZOH filter) have the following frequency response: [120]

$$H_{zoh}(k) = e^{-j\left(\frac{\pi k(L-1)}{N}\right)} \left(\frac{\sin(\frac{\pi k L}{N})}{\sin(\frac{\pi k}{N})}\right) \tag{6.2}$$

where $N$ is the length of the interpolated signal and $k$ ranges from $-N/2$ to $N/2-1$. Considering that $f = \frac{k.Fs_{RF}}{N}$, the previous equation can be reinterpreted in order to get a function of $f$:

$$H_{zoh}(f) = e^{-j\left(\frac{\pi f}{Fs_{RF}}\left(\frac{Fs_{RF}}{Fs_{BB}}-1\right)\right)} \left(\frac{\sin(\frac{\pi f}{Fs_{BB}})}{\sin(\frac{\pi f}{Fs_{RF}})}\right) \tag{6.3}$$

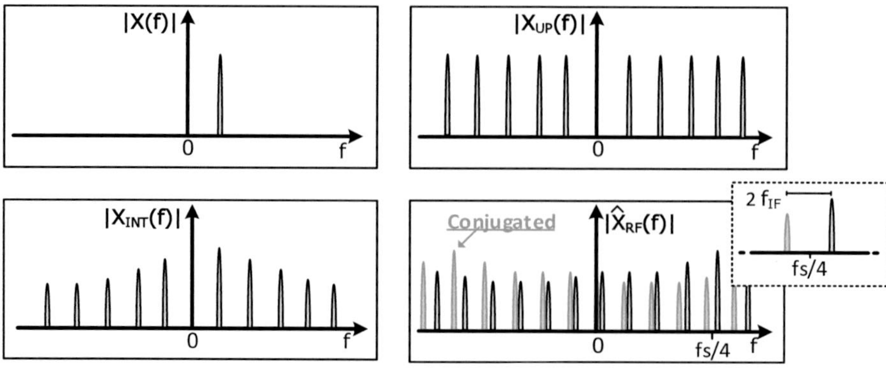


**FIGURE 6.5:** Illustration of a single-tone excitation considering the blocks from Figure 2.16.

To include the final upconversion stage that is required to translate the carriers from BB/IF to RF, this equation can be shifted to the center frequency $w_c$. The Discrete Fourier Transform (DFT) of $e^{jw_c n}$ is $\delta(w - w_c)$, resulting in:

$$H_{zoh}(f - f_c) = e^{-j\left(\frac{\pi(f-f_c)}{Fs_{RF}}\left(\frac{Fs_{RF}}{Fs_{BB}}-1\right)\right)} \left(\frac{\sin(\frac{\pi(f-f_c)}{Fs_{BB}})}{\sin(\frac{\pi(f-f_c)}{Fs_{RF}})}\right) \tag{6.4}$$

Finally, as previously introduced in Eq. (6.1):

$$\begin{aligned} \widehat{X}_{RF}(f) &= \frac{X(f) * H_{zoh}(f - f_c)}{2} + \\ &\quad \frac{X^*(-f) * H^*_{zoh}(-f + f_c)}{2} \end{aligned} \tag{6.5}$$

### 6.3.2 Proposed Architecture

As previously introduced in the last subsection, the poor performance of the previous dual-band architecture is caused by a weak interpolation filter, which introduces crosstalk between the two modulated bands.

In practical terms, focusing on the proposed dual-band scenario with two symmetric notches around a center frequency of $(Fs_{RF}/4)$, one can model this issue assuming that two modulated signals are being injected in two different and symmetrical bands ($X_1$ and $X_2$) around the DC component ($f_i$ and $-f_i$, respectively), which will be propagated along the upconversion chain and will result in two different and symmetrical bands ($Y_1$ and $Y_2$) around the center frequency ($Fs_{RF}/4 + f_i$ and $Fs_{RF}/4 - f_i$). In an ideal situation, no crosstalk would happen, and, because of that, the system could be simply modeled as:

$$\begin{bmatrix} Y_1 \\ Y_2 \end{bmatrix} = I(z) \begin{bmatrix} X_1 \\ X_2 \end{bmatrix} = \begin{bmatrix} 1 & 0 \\ 0 & 1 \end{bmatrix} \begin{bmatrix} X_1 \\ X_2 \end{bmatrix} \tag{6.6}$$

where $Y_1$ and $Y_2$ are the modulated signals from $X_1$ and $X_2$, respectively, merely translated into different carriers ($X_1 \rightarrow Y_1$ and $X_2 \rightarrow Y_2$).

The consideration of only the real-valued modulated signals results in the use of a *widely linear* or *linear-conjugate-linear* transformation which depends linearly on the variable and its conjugate [176]. In particular:

$$\begin{bmatrix} Y_1 \\ Y_2 \end{bmatrix} = H(z) \begin{bmatrix} X_1 \\ X_2 \end{bmatrix} + C(z) \begin{bmatrix} X_1^* \\ X_2^* \end{bmatrix} \tag{6.7}$$

And, in this case:

$$\begin{aligned} H(z) &= \begin{bmatrix} H_{11} & 0 \\ 0 & H_{22} \end{bmatrix} \\ C(z) &= \begin{bmatrix} 0 & C_{12} \\ C_{21} & 0 \end{bmatrix} \end{aligned} \tag{6.8}$$

Both the $H's$ and $C's$ coefficients are linear transfer functions intrinsic to the upconversion chain and, because of that, invariant with the signal excitation. In practice, $H_{11}$ and $H_{22}$ characterize the attenuation/gain suffered by the modulated signals (namely, $X_1 \rightarrow Y_1$ and $X_2 \rightarrow Y_2$, respectively), while $C_{21}$ and $C_{12}$ characterize the amount of the conjugate input signals that is aliased to the opposite and symmetric band (namely, $X_1^* \rightarrow Y_2$ and $X_2^* \rightarrow Y_1$, respectively). According to the given formulation and neglecting an equalization (assuming the output should be $Y_1 = H_{11}X_1$ and $Y_2 = H_{22}X_2$), two input signals can be computed in order to perform a pre-compensation to minimize this cross-signal overlap between the two different bands:

$$W_1 = \frac{H_{11}^* H_{22} X_2 - C_{21} H_{11}^* X_1^*}{H_{11}^* H_{22} - C_{12}^* C_{21}} \tag{6.9}$$

$$W_2 = \frac{H_{11} H_{22}^* X_1 - C_{12} H_{22}^* X_2^*}{H_{11} H_{22}^* - C_{12} C_{21}^*} \tag{6.10}$$

Moreover, the crosstalk effect $(C_{12}, C_{21})$ is substantially lower than the system gain $(H_{11}, H_{22})$. This means that $|H_{11}H_{22}| >> |C_{12}C_{21}|$, and so, $H_{11}^*H_{22} - C_{12}^*C_{21} \simeq H_{11}^*H_{22}$ and $H_{11}H_{22}^* - C_{12}C_{21}^* \simeq H_{11}H_{22}^*$, leading to:

$$W_1 \simeq X_2 + G_{21}X_1^* \tag{6.11}$$

$$W_2 \simeq X_1 + G_{12}X_2^* \tag{6.12}$$

with the pre-compensation filters $G_{21} = -\frac{C_{21}}{H_{22}}$ and $G_{12} = -\frac{C_{12}}{H_{11}}$. From an implementation perspective, the proposed pre-compensation scheme is based on a simple interconnect matrix between the two modulated bands, which must be applied in the baseband processing before the upconversion chain.

To summarize, since one is dealing with linear effects, the proposed solution is to minimize the crosstalk overlap between the two different bands by including a pre-compensation signal in each band. To accomplish this, each signal's transfer function must have the same frequency response as the aliased image that one wants to remove (with an opposite phase) and must be amplified by the same amount that is lost along the upconversion chain.

To proceed with the validation of this pre-compensation method, the first step is to identify the $H's$ and $C's$ transfer functions, and, to do that, the frequency response from the entire DUC chain must be taken into account (Eq. 6.4). In particular, one can verify that the transfer functions $H_{11}$ and $H_{22}$ are the magnitude/phase pair from all the bins of interest evaluated in $H_{zoh}(f - f_c)$, while, on the other hand, $C_{12}$ and $C_{21}$ are evaluated in $H_{zoh}^*(-f + f_c)$ (as can be seen in Fig. 6.6).

#### 6.3.2.1 Frequency-domain Filtering

The most straightforward implementation for assessing the performance of this compensation method is based on frequency-domain filtering, where frequency sampling is performed

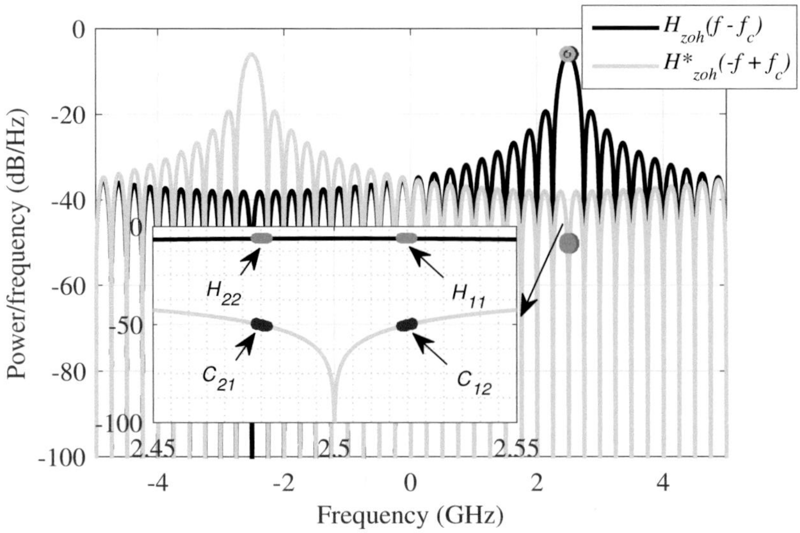


**FIGURE 6.6:** Frequency response of the upconversion chain (focusing in the addition between the complex-part with its conjugated to produce the real-part, as demonstrated in (6.5)) together with the interest bins of each transfer function ($H_{11}$, $C_{12}$, $C_{21}$, $H_{22}$) for an IF of 20 MHz and a SR of 3.125 Msps.

to select the bins of interest, rather than relying on continuous-time solutions. Basically, all the operations between the transfer functions $G_{12}$, $G_{21}$ and the input signals $X_1$ and $X_2$ are algebraically performed in the frequency-domain. To perform this, after selecting the bins of interest of each transfer function, a Fast Fourier Transform (FFT) (with the RF signal's length) is applied to the conjugate of each input signal ($X_1^*$ and $X_2^*$) before upconversion to IF. After having the modulated signals in the frequency-domain, frequency domain filtering is applied, and then, an Inverse Fast Fourier Transform (IFFT) is required to get the two compensation signals ($w_1$ and $w_2$) in the time-domain:

$$w_1[n] = x_1[n] + \mathrm{IFFT}(\mathrm{FFT}(x_2^*[n])G_{12}) \tag{6.13}$$
$$w_2[n] = x_2[n] + \mathrm{IFFT}(\mathrm{FFT}(x_1^*[n])G_{21}) \tag{6.14}$$

An offline implementation of this compensation technique is straightforward, as no additional hardware is required. The pre-compensation signals must be computed in the baseband processing and combined with the input signals. Thus, the overall architecture, as shown in Fig. 6.2, can be maintained. However, considering real-time constraints, an implementation of this technique is not particularly attractive in terms of required hardware. In this case, FFT/IFFT modules with a large number of points included within the FPGA may be required.

#### 6.3.2.2 Time-domain Filtering

Rather than performing all the operations in the frequency-domain, the filters can be transferred to the time-domain and applied directly to the conjugate of the input signals:

$$w_1[n] = x_1[n] + x_2^*[n] * \mathrm{IFFT}(G_{12}) \tag{6.15}$$
$$w_2[n] = x_2[n] + x_1^*[n] * \mathrm{IFFT}(G_{21}) \tag{6.16}$$

A major advantage of this implementation is that, as previously introduced, $G_{12}$ and $G_{21}$ are intrinsic transfer functions of the upconversion, which confers to the system the capability of performing online pre-compensation schemes. The respective filtering topologies

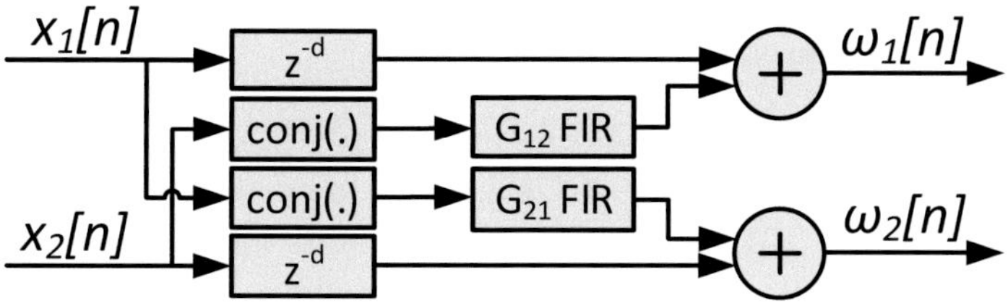


**FIGURE 6.7:** Block diagram of the pre-compensation mechanism that was included in the FPGA's hardware between the FPGA's internal memory (storing a baseband test signal) and the polyphase interpolation filters. The pre-compensation filters were designed with a fourth-order Transposed FIR filter topology, which presents a short critical path.

were optimized with the Parks-McClellan algorithm to just four complex coefficients for the filter ($g_{12,x}$ and $g_{21,x}$). More information related to the computation of these taps can be found in [43].

In conclusion, a pre-compensation method which requires computing and reloading four different complex coefficients for each filter ($g_{12,x}$ and $g_{21,x}$) for each IF value was proposed as an architectural solution to enhance the dual-band FoM. Thus, the module responsible for the pre-compensation mechanism (depicted in Fig. 6.7) must be included between the FPGA's internal memory (that is emulating the BB interface) and the polyphase interpolation FIR filters (recall Fig. 6.2). The main block of this module comprises two complex-coefficient fourth-order filters, which, from the implementation point of view, are almost negligible.

### 6.3.3 Prototype Validation

To validate both the single- and dual-band agile transmitters, as well as to assess the performance of the proposed pre-compensation method, the overall architecture was fully implemented in a Virtex UltraScale XCVU095 FPGA using the Xilinx VCU1287 Characterization Kit. The MGT's output was connected to a PA (ZHL-1042J), and then to a VSA R&S FSW 8, as previously depicted in the photo of the experimental setup in Fig. 6.3.

Spectrum and EVM measurements were obtained with the VSA for three different scenarios:

- Single-band with a LP-ΔΣM with no IF for different SR values (Table 6.3);
- Single-band with a BP-ΔΣM for different IF values (Fig. 6.8);
- Dual-band with a BP-ΔΣM for different IF values (Fig. 6.9). Measurements with the pre-compensation mechanism are also included in the same figure.

Changing the IF value is done in a straightforward way and in real-time by updating the set of variables that control the DDS, compensation filters and ΔΣM. In all the experiments, the MGT's sampling frequency was adjusted to 10 Gbps to provide a center frequency of 2.5 GHz, and the ΔΣM's global sampling rate is 250 MHz (2 phases running at 125 MHz). In the dual-band scenario, two symmetrical bands were tested with different modulations (QPSK and 16-QAM) and for different frequencies.

All the experimental results present EVM values below 2.6% resulting in a well-defined constellation (Table 6.3, and Figs. 6.8 and 6.9). In particular, the single-band with a LP-ΔΣM shows the best results with EVMs always below 2%, even for an SR of 15.625 Msps. Concerning the dual-band scenario, two symmetrical bands were transmitted in different

**TABLE 6.3:** Single-band performance with a LP-ΔΣM at 2.5 GHz. The results demonstrate the EVM (in %) after serializer and PA.

| | Symbol Rate (Msps) | | | |
|---|---|---|---|---|
| Modulation | 3.125 | 6.250 | 12.5 | 15.625 |
| QPSK | 0.76 | 0.90 | 1.36 | 1.70 |
| 16-QAM | 0.78 | 0.92 | 1.33 | 1.82 |
| 64-QAM | 0.78 | 0.93 | 1.37 | 1.83 |

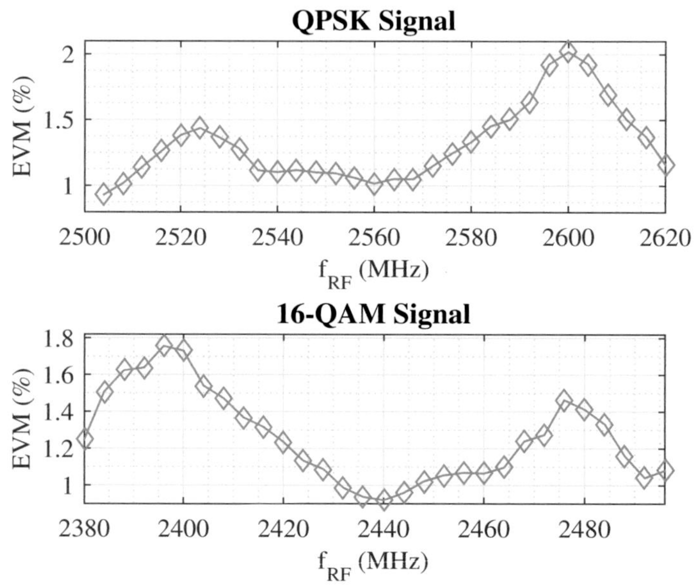


**FIGURE 6.8:** Single-band performance in a BP-ΔΣM scenario with a center frequency of 2.5 GHz and an IF ranging from ±2 MHz to ±120 MHz.

IF values. In Fig. 6.9, the effectiveness of this pre-compensation mechanism is verified with both implementation strategies (frequency-domain and time-domain). In fact, it was possible to decrease the EVM values down to the single-band ones, used as a baseline for proper comparison. It must be noted that to perform a fair comparison between the dual-band scenario with the single-band one, the input range of the latter was reduced to half, explaining the EVM's increase compared to the Fig. 6.8. Furthermore, the EVM's periodical pattern is also expected, and it can be explained through the illustration in Fig. 6.6. The amount of noise from the transfer function $H_{zoh}(-f + f_c)^*$ will interfere more or less with the original band, depending on the IF value. Thus, the periodical pattern in the transfer function's magnitude response will directly affect the signal's quality. In particular, the signal's quality will be higher (lower EVM's value), every time the frequency value matches one of the sinc zeros.

To gain a clearer understanding of the pre-compensation scheme's operation, a final measurement was performed with pre-compensation applied in a single-band scenario using a BP-ΔΣM, whose results are shown in Fig. 6.10. In this experiment, the aliased image from the complex upconverted carrier in the symmetric band is almost completely eliminated when pre-compensation is turned on, lowering the noise floor by approximately 15 dB in this specific band.

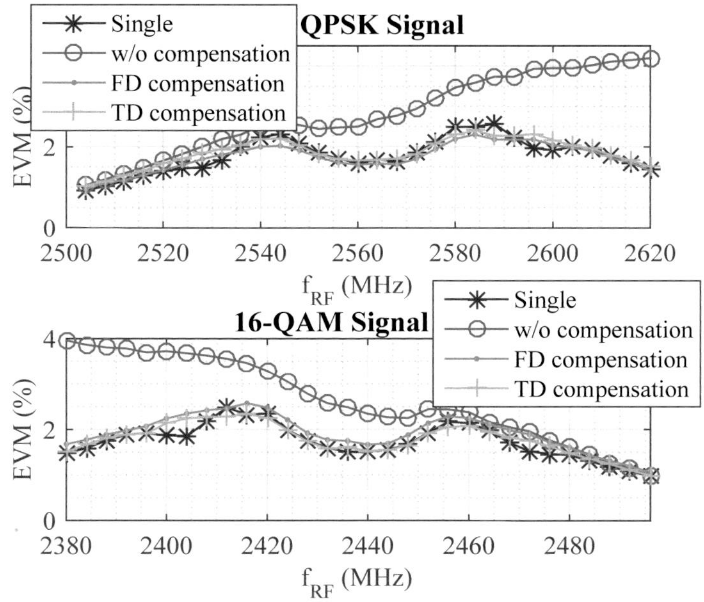


**FIGURE 6.9:** Dual-band performance in a BP-ΔΣM scenario with a center frequency of 2.5 GHz and an IF ranging from ±2 MHz to ±120 MHz. *FD compensation* and *TD compensation* represent the EVM results with the two compensation implementations (frequency- and time-domain, respectively).

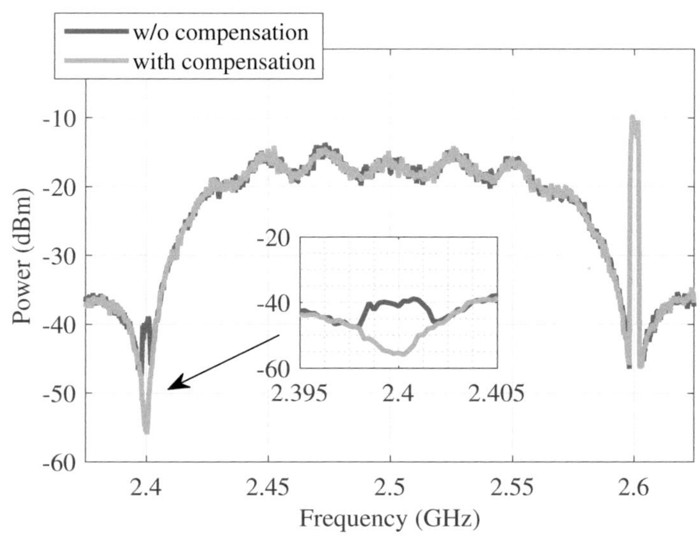


**FIGURE 6.10:** Pre-compensation mechanism applied into a single-band with a BP-ΔΣM. The transmitted signal is a QPSK with an SR of 3.125 Msps at 2.6 GHz (Resolution Bandwidth (RBW) = 100 kHz, Video Bandwidth (VBW) = 2 kHz and SPAN = 250 MHz).

Figs. 6.11 and 6.12 show two obtained spectra and constellations of a single-band with LP-ΔΣM and of a dual-band with a BP-ΔΣM, respectively.

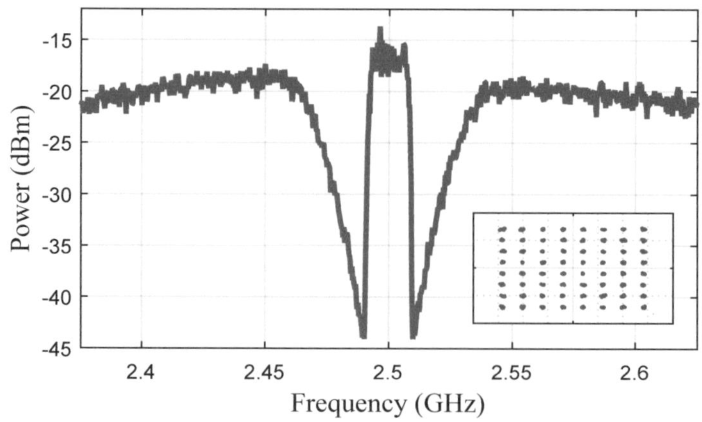


**FIGURE 6.11:** Spectrum of the transmitted signal (64-QAM with an SR of 15.625 Msps at 2.5 GHz) in a single-band scenario with a LP-ΔΣM (RBW = 100 kHz, VBW = 2 kHz and SPAN = 250 MHz).

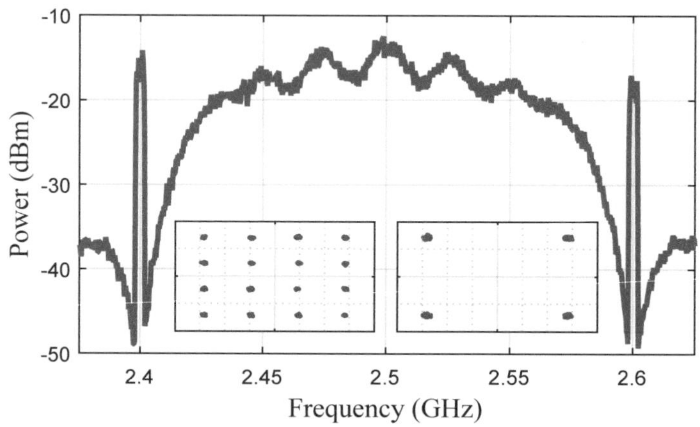


**FIGURE 6.12:** Spectrum of the transmitted signal in a dual-band scenario (QPSK at 2.6 GHz and 16-QAM at 2.4 GHz, both with an SR of 3.125 Msps) with a BP-ΔΣM (RBW = 100 kHz, VBW = 2 kHz and SPAN = 250 MHz).

## 6.4 Triple-band RF-stage Modulator

In order to design a triple-band modulator, the use of a BB-stage ADTs with a very limited NZ starts to become problematic. The maximum span between bands is always limited to the global sampling rate from the Pulse Encoder. Namely, the spans of 300 MHz and 250 MHz were achieved in the two previous sections. If three bands with reasonable bandwidth must be included, one can conclude that the span is too limited for a non-contiguous transmission. Thus, novel strategies must be explored to achieve a higher NZ, while taking into account the physical limitations imposed by the FPGA. For this reason, the architectures presented in the two previous sections cannot be considered, and a different one must be pursued.

### 6.4.1 Proposed Architecture

The proposed idea is to extend the single-band all-digital RF-stage architecture (proposed in [45] and introduced in Chapter 4) to enable the synthesis of triple-band RF transmitters. By taking advantage of the direct synthesis of RF waveforms in the digital domain before the Pulse Encoder, unprecedented frequency agility is achieved. To meet the specifications, highly parallelized architectures that allow for high equivalent sampling rates must be explored.

At the same time, as the RF-stage architecture intensively utilizes logical resources, replicating fully digital RF chains to enable multi-band capability is only possible with novel design methodologies and optimization techniques. Thus, two major modifications are proposed: the integration of the propagation of state-registers in the $\Delta\Sigma$ architecture (proposed in [46] and briefly reported in Chapter 5) as a way to relax the resources usage; and the replacement of the FIFOs by a *Corner-Bender Matrix Transposer* from the Deinterleaving/Interleaving modules. The following subsections provide a brief overview of the proposed architecture.

#### 6.4.1.1 Digital Up-conversion Stage and Bands Combination

To enable the transmission of three different bands, the DUC stage must be replicated three times. The block diagram of each DUC was already introduced in Chapter 4. Similarly to the single-band architecture, the final bitrate $Fs_{RF}$ defines the modulator's first Nyquist Zone (from 0 to $Fs_{RF}/2$). This means that the carrier frequency from each band can be located in any bin from the frequency grid given by the DDS' frequency resolution Eq. (4.2) within this Nyquist Zone.

Subsequently, the three interpolated and upconverted bands are combined to generate an equivalent representation with $N$ different phases, which contains the three distinct bands. At the same time, the resultant signal must be scaled down to a suitable dynamic range of the Pulse Encoder.

#### 6.4.1.2 RF-stage Enhanced Multicore-based $\Delta\Sigma$M

The Pulse Encoder must be designed to reduce the number of output levels while ensuring minimal distortion in the three different bands. Moreover, the three different bands must be independently controlled and tunable, which typically implies the use of feedback loops to ensure some form of noise shaping. To accomplish these requirements, $\Delta\Sigma$ techniques were selected due to their inherent noise-shaping property that minimizes the amount of in-band distortion.

The propagation of state registers is included in the Pulse Encoder to relax the resource usage. By adopting this technique, the periodicity of the impulsive noise can be reduced by a factor of $N$. Additionally, this architecture enables the matching of the number of phases, $N$, with the number of contiguous samples, $K$. This is of utmost importance because the Deinterleaving/Interleaving modules can be reduced to simply matrix transposers. When $N$ is different from $K$, the design of Deinterleaving/Interleaving modules is typically done with resort to FIFOs. However, the allocation of distributed/block RAM for FIFOs typically leads to a low-occupancy rate. On the other hand, the design of matrix transposers is quite simple. The use of a *Corner-Bender Matrix Transposer* [6] is proposed to maintain the desired functionality by just using registers and multiplexers.

The Corner-Bender Matrix Transposer is depicted in Fig. 6.13. This design does not utilize any FIFO memories; it is fully synchronous, does not require any control signals, and only employs fixed-size shift register delays. The latency is the minimal possible and equal to the number of contiguous samples ($K$). Moreover, the same module can be used for

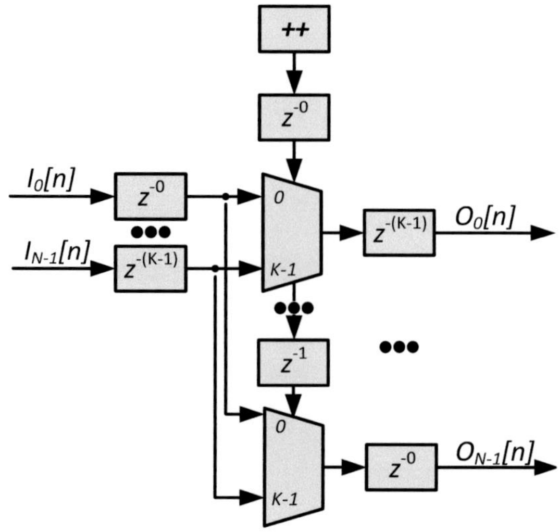


**FIGURE 6.13:** Illustration of the block diagram of a Corner-Bender Matrix Transposer. The designed Interleaving and Deinterleaving modules are based on this transposer.

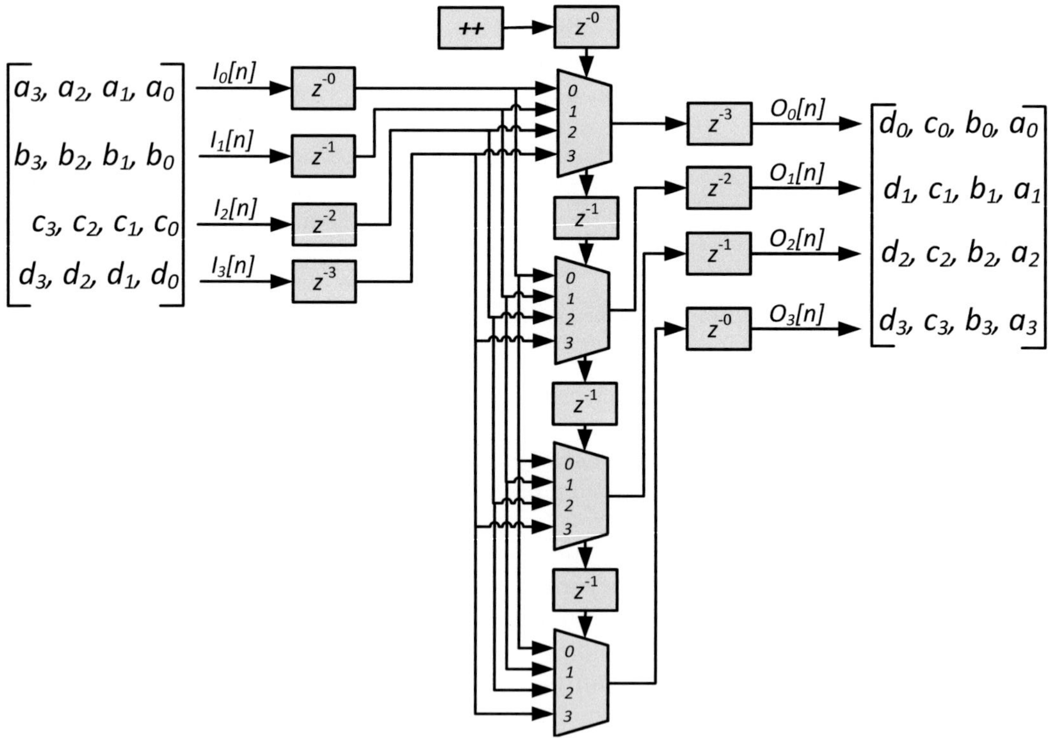


**FIGURE 6.14:** Illustration of a Corner-Bender with four phases and four contiguous samples ($N = 4$ and $K = 4$).

the Deinterleaving and Interleaving operations. In short, it will be seen that the inclusion of this module enabled a drastic reduction in BRAM primitive resource usage. To provide a clearer understanding, an example with four phases and four contiguous samples ($N = 4$ and $K = 4$) is also illustrated in Fig. 6.14.

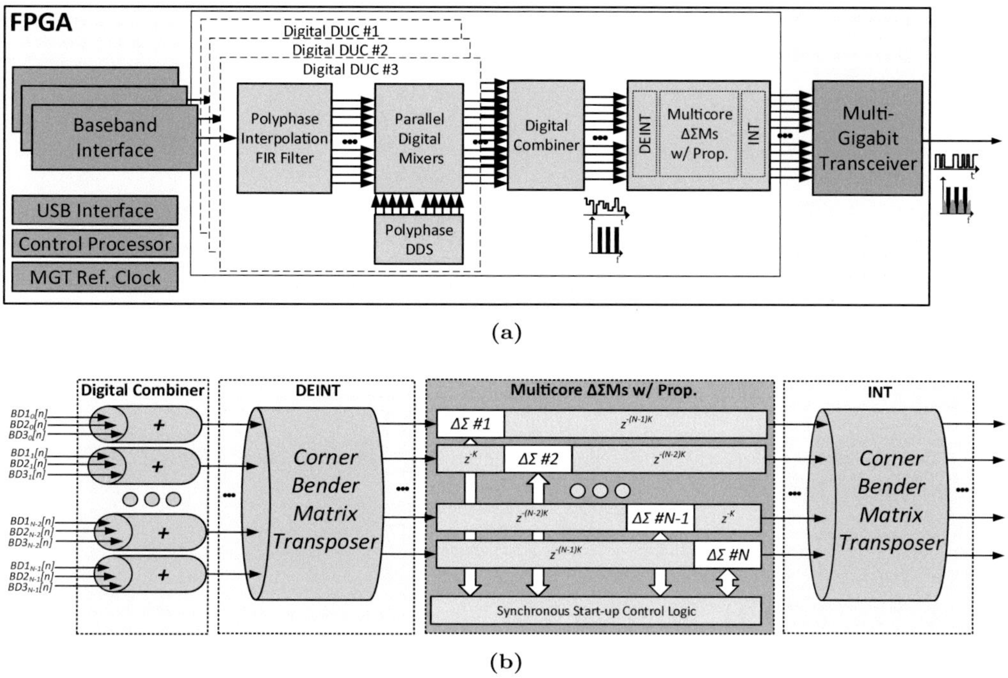


**FIGURE 6.15:** (a) General block diagram of the implemented All-Digital RF transmitter, embedded into an FPGA; (b) Detailed illustration of the Digital Combiner and the ΔΣM with $N$ parallel stages, presented in (a).

#### 6.4.1.3 General Block Diagram

The general block diagram of the proposed architecture is depicted in Fig. 6.15a. The detailed illustration of the Pulse Encoder is depicted in Fig. 6.15b. One can clearly visualize the utilization of the Corner Bender Matrix Transposers as Deinterleaving and Interleaving Modules (referred to in the illustration as *DEINT/INT*). In addition to this, the displacement of the ΔΣMs, together with the *Synchronous Start-up Control Logic* module, to enable the propagation of state registers [46] is also illustrated.

A ΔΣ modulator must be designed and replicated $N$ times, according to the architecture introduced in the last subsection. The choice of it must be made according to two different requirements. First, the architecture must present an NTF with three well-defined notches, to ensure the integrity of the three different bands. Second, the notches must also be tunable and independently controlled to provide the desired real-time agility to the overall system. At last, the modulator's critical path must be as short as possible, which is why architectures such as CRFB/F or CIFB/F cannot be chosen.

Thus, the Error-Feedback ΔΣ architecture (depicted in Fig. 4.13) was selected as a starting point for this work. By modeling the quantization noise as a random noise signal ($E_q(z)$), the following transfer function can be derived:

$$Y(z) = STF(z)X(z) - NTF(z)Eq(z) \tag{6.17}$$

where $STF(z) = 1$, $NTF(z) = (1 + H(z))$ and $H(z) = \alpha z^{-1} + z^{-2}$. The modulator's behavior depends on the variable $\alpha$, defined as $\alpha = -2\cos(2\pi F_{c1}/Fs)$, where $F_{c1}$ is the notch center frequency and $Fs$ is the modulator's sampling rate. In particular, it has two complex zeros that allow the adjustment of the notch carrier as desired. An intuitive approach to achieve a triple-band modulator could involve the inclusion of 2 pairs of other complex zeros, defined by $\beta$ and $\gamma$:

$$NTF(z) = \left(1 + \alpha z^{-1} + z^{-2}\right) . \left(1 + \beta z^{-1} + z^{-2}\right) . \left(1 + \gamma z^{-1} + z^{-2}\right) \tag{6.18}$$

where $\beta = -2\cos(2\pi F_{c2}/Fs)$, $\gamma = -2\cos(2\pi F_{c3}/Fs)$ with $F_{c2}$ and $F_{c3}$ being the notches center frequencies.

However, for the used wireless modulated signals, this architecture may lead to stability issues, specifically depending on the placement of the notches. Thus, to control the NTF's gain, 6 complex poles were also included in the architecture. Each pair of poles closely follows each pair of zeros, to ensure the stability of the system, according to [170]. This leads to:

$$NTF(z) = \frac{1 + \alpha z^{-1} + z^{-2}}{1 + r\alpha z^{-1} + r^2 z^{-2}} . \frac{1 + \beta z^{-1} + z^{-2}}{1 + r\beta z^{-1} + r^2 z^{-2}} . \frac{1 + \gamma z^{-1} + z^{-2}}{1 + r\gamma z^{-1} + r^2 z^{-2}} \tag{6.19}$$

where $r$ ranges from 0 to 1, and controls the closeness between poles and zeros. This equation can be simplified to:

$$NTF(z) = \frac{1 + A_0 z^{-1} + B_0 z^{-2} + C_0 z^{-3} + D_0 z^{-4} + E_0 z^{-5} + z^{-6}}{1 + F_0 z^{-1} + G_0 z^{-2} + H_0 z^{-3} + I_0 z^{-4} + J_0 z^{-5} + K_0 z^{-6}} \tag{6.20}$$

where $A_0 = (\alpha + \beta + \gamma)$, $B_0 = (3 + \alpha\beta + \gamma(\alpha + \beta))$, $C_0 = (2(\alpha + \beta) + \gamma(2 + \alpha\beta))$, $D_0 = (3 + \alpha\beta + \gamma(\alpha + \beta))$, $E_0 = (\alpha + \beta + \gamma)$, $F_0 = r(\alpha + \beta + \gamma)$, $G_0 = r^2(3 + \alpha\beta + \gamma(\alpha + \beta)$, $H_0 = r^3(2(\alpha + \beta) + \gamma(2 + \alpha\beta)$, $I_0 = r^4(3 + \alpha\beta + \gamma(\alpha + \beta)$, $J_0 = r^5(\alpha + \beta + \gamma)$ and $K_0 = r^6$.

In turn, the feedack loop transfer function ($H(z)$) can also be computed:

$$H(z) = \frac{A_1 z^{-1} + B_1 z^{-2} + C_1 z^{-3} + D_1 z^{-4} + E_1 z^{-5} + F_1 z^{-6}}{1 + F_0 z^{-1} + G_0 z^{-2} + H_0 z^{-3} + I_0 z^{-4} + J_0 z^{-5} + K_0 z^{-6}} \tag{6.21}$$

where $A_1 = (\alpha + \beta + \gamma)(1 - r)$, $B_1 = (3 + \alpha\beta + \gamma(\alpha + \beta))(1 - r^2)$, $C_1 = (2(\alpha + \beta) + \gamma(2 + \alpha\beta))(1 - r^3)$, $D_1 = (3 + \alpha\beta + \gamma(\alpha + \beta))(1 - r^4)$, $E_1 = (\alpha + \beta + \gamma)(1 - r^5)$ and $F_1 = (1 - r^6)$.

The impact of placing poles in $H(z)$ in terms of the magnitude of the NTF is depicted in Fig. 6.16. The magnitude is reduced, reducing the stability issues. However, as a drawback, the bandwidth of the notches becomes lower.

#### 6.4.1.4 Extension to Multi-Level Output

To enhance the system performance (assessed in terms of ACPR, EVM, and SNR), a final stage that performs an extension to a multi-level output can also be included. The synthesis of this multi-level output is based on the combination of pulsed trains. The final architecture is depicted in Fig. 6.18. Three major differences can be pointed out: the single-bit quantizer from each ΔΣM was replaced by a 7-level quantizer; 7 different MGTs were activated instead of just 1; and a LUT Mapper was included before the MGTs. This LUT Mapper is the core of this last stage and converts a multi-level input signal into a combination of 7 different pulsed train sequences. The mapper must be designed according to the analog combination network used.

The envisioned combination network to be applied in this architecture was first proposed in [39], and an extension to 5-level was later reported in [229]. This increase in the number of levels is possible due to the use of an Asymmetric Extended H-bridge Combining Network, whose conceptual diagram is depicted in Fig. 6.17. In particular, the switches from 4 up to 7 must have double-sized transistors when compared to the remaining ones. In the case of this combining network, the mapper is depicted in Table 6.4. In particular, one can realize that this mapper can generate a 3-level output (with a dark gray background), a 5-level output (with both dark gray and light gray backgrounds), or a 7-level output (by utilizing the full table).

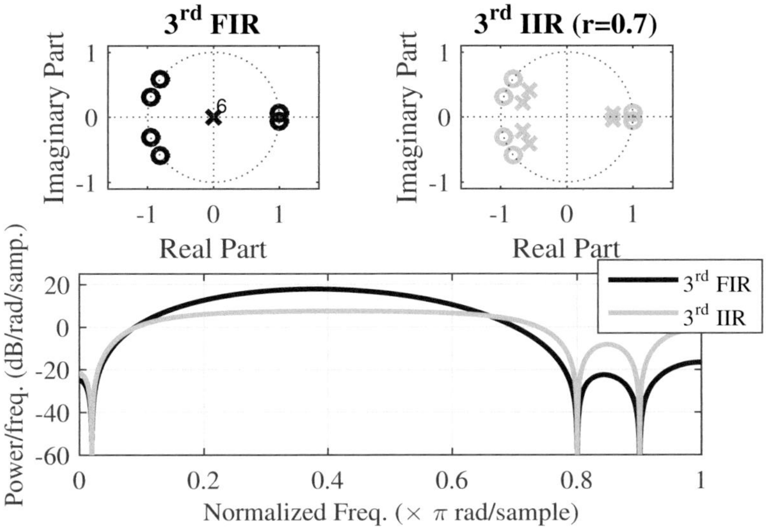


**FIGURE 6.16:** NTF's frequency response and zero-poles placement according to the different feedback-loop transfer functions.

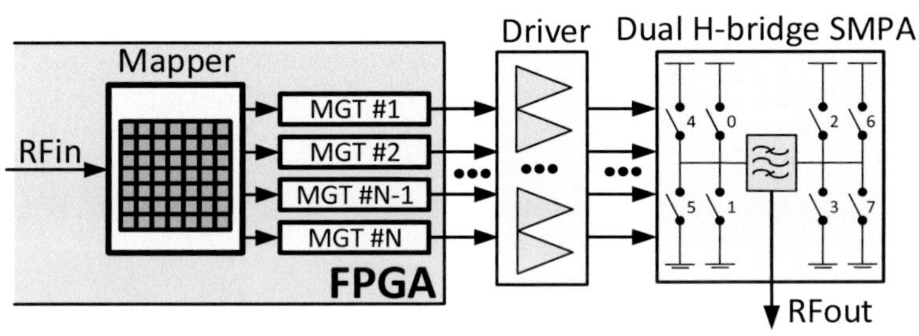


**FIGURE 6.17:** Detailed illustration of the proposed amplification scheme based on a dual H-bridge SMPA.

### 6.4.2 Prototype Validation

To validate the proposed multi-band architecture, the two different ADTs (depicted in Figs. 6.15 and 6.18) were implemented in a Virtex 7XC7VX485T FPGA using the Xilinx FPGA VC707 Evaluation Kit.

The maximum timing constraints in the single-bit project imposed a maximum $Fs_{BB}$ of 78.125 MHz, and consequently, with 64 different phases ($N = 64$), a bitrate of 5 Gbps was chosen. For the multi-level case, implementing a multi-bit quantizer resulted in a reduction of the maximum BB sampling rate to 62.5 MHz, which in turn led to a serializer bitrate of 4 Gbps. The frequency resolutions can be computed by Eq. (4.2). In particular, the different bitrates together with the 10-bit resolution of the DDS' phase accumulator lead to 4.8828 MHz and 3.906 MHz, for the single-bit and 8-level case, respectively.

Due to the lack of a proper combining network based on this proposal, an alternative option had to be chosen for the sake of validating the proof of concept. Thus, a conventional RF Power Combiner was utilized to combine the outputs of the different serializers. In this case, the mapper had to be updated to Table 6.5. This new mapper can be simply seen as a linear combination of the pulse trains to achieve a higher number of output levels. One can also realize from Table 6.5 that the column related to the $MGT_0$ is constant. This means

**TABLE 6.4:** Proposed mapper for the amplification scheme depicted in Fig. 6.17.

| $RF_{in}$ | 0 | 1 | 2 | 3 | 4 | 5 | 6 | 7 |
|---|---|---|---|---|---|---|---|---|
| 3 | 1 | 0 | 0 | 1 | 1 | 0 | 0 | 1 |
| 2 | 0 | 0 | 0 | 0 | 1 | 0 | 0 | 1 |
| 1 | 1 | 0 | 0 | 1 | 0 | 0 | 0 | 0 |
| 0 | 0 | 1 | 0 | 1 | 0 | 0 | 0 | 0 |
| -1 | 0 | 1 | 1 | 0 | 0 | 0 | 0 | 0 |
| -2 | 0 | 0 | 0 | 0 | 0 | 1 | 1 | 0 |
| -3 | 0 | 1 | 1 | 0 | 0 | 1 | 1 | 0 |

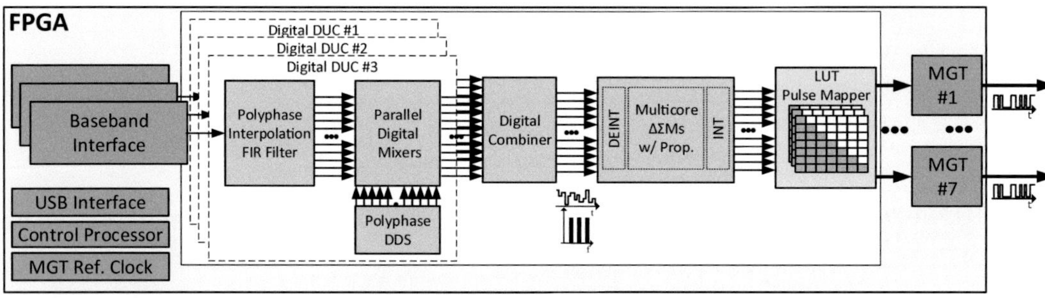


**FIGURE 6.18:** General block diagram of the implemented All-Digital RF transmitter, embedded into an FPGA with an extension to a multi-level output.

that one of the serializers could be removed. However, it was decided to keep it to maintain the coherence with the ideal combining network based on the H-bridge.

The experimental setups are depicted in Figs. 6.19a and 6.19b, for the single- and multi-bit projects, respectively. In the first one, the MGT was directly connected to a VSA Keysight EXA N9010A. In the multi-bit project, lane-to-lane deskew techniques were implemented in the digital hardware, and extra delay lines were attached to some MGTs to ensure lane-to-lane synchronism. Moreover, the amplitude swing of each MGT was slightly changed to ensure the same output power. Afterwards, the 7 MGTs were combined into a Power Combiner ZN8PD1-63W+ from Mini-Circuits. The combined output was connected to the same VSA. The VSA was configured to perform ACPR measurements as well as to record the I/Q data to be further analyzed in the VSA LabView Software running on the computer.

To enable the sweep of different carrier frequencies, modulations, bandwidths, and to record such FoM as spectrum, MER, and EVM, an automated testbed was designed. This testbed has the host computer controlling/managing the FPGA-based ADT through a USB connection, and it is connected to the VSA via an Ethernet connection. The different experimental setups are depicted in Fig. 6.19. More information related to the sweep of carrier frequencies can be found in [47].

#### 6.4.2.1 Single-bit RF Transmission

This section details the experimental results for the single-bit project.

Fig. 6.20a shows the measured ACPR results (lower and upper) in the range of 100 MHz up to 2.5 GHz, with a step of 48.828 MHz for three different bandwidths per band: 4.688 MHz, 9.375 MHz, and 18.75 MHz. It can be seen that all the measurements maintain the same coherency: the increase in bandwidth leads to higher quantization noise

**TABLE 6.5:** Implemented mapper for the use of assessment of the multi-level output with a Power-Combiner.

| $RF_{in}$ | $MGT_0$ | $MGT_1$ | $MGT_2$ | $MGT_3$ | $MGT_4$ | $MGT_5$ | $MGT_6$ | $MGT_7$ |
|---|---|---|---|---|---|---|---|---|
| 3 | 1 | 1 | 1 | 1 | 1 | 1 | 1 | 1 |
| 2 | 1 | 1 | 1 | 1 | 1 | 1 | 1 | 0 |
| 1 | 1 | 1 | 1 | 1 | 1 | 1 | 0 | 0 |
| 0 | 1 | 1 | 1 | 1 | 1 | 0 | 0 | 0 |
| -1 | 1 | 1 | 1 | 1 | 0 | 0 | 0 | 0 |
| -2 | 1 | 1 | 1 | 0 | 0 | 0 | 0 | 0 |
| -3 | 1 | 1 | 0 | 0 | 0 | 0 | 0 | 0 |

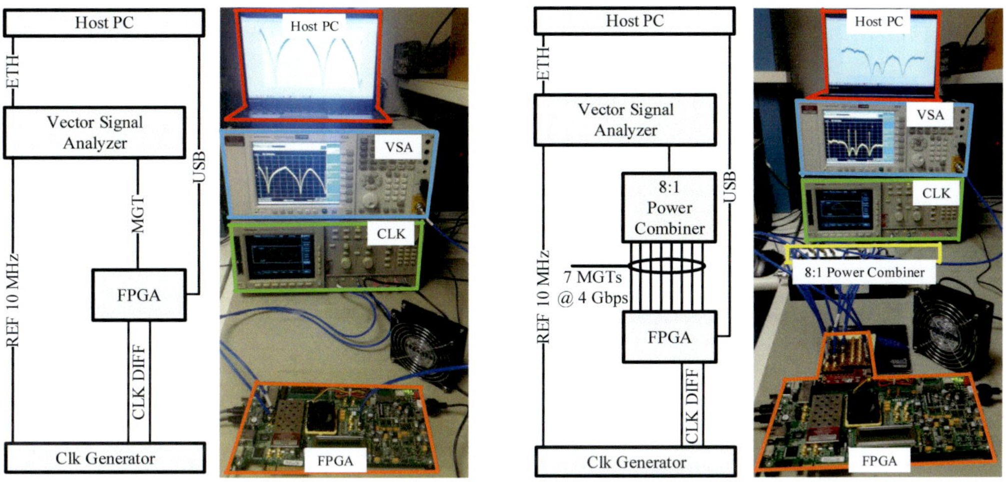


(a) (b)

**FIGURE 6.19:** Photo of the experimental setup in the single-bit architecture (a) and in the 7-level architecture (b).

around the carrier frequencies, which consequently increases the ACPR values. The signals with higher bandwidth present the worst results in terms of ACPR: between -20 and -30 dBc.

MER and EVM obtained pairs are reported in Fig. 6.20b. Again, the same intuitive conclusions can be extracted. Narrow bandwidths provide better FoMs. In particular, the maximum obtained MERs were 38, 35, and 31.5 dB, for the aforementioned bandwidths.

The spectrum in a span of 2.5 GHz, for carrier frequencies of 600.59 MHz, 2.002 and 2.4023 GHz, is shown in Fig. 6.22a.

#### 6.4.2.2 7-Level RF Transmission

This subsection details the experimental results for the concurrent 7-level 3-Band RF Transmission. Fig. 6.21a shows the measured ACPR results (lower and upper) in the range of 100 MHz up to 2.0 GHz, with a step of 39.062 MHz for five different bandwidths per band: 3.75 MHz, 7.5 MHz, 15 MHz, 18.75 MHz and 37.5 MHz. It can be seen that all the measurements maintain the same coherency: the increase of the bandwidth leads to higher quantization noise around the carrier frequencies, with the inherent increase in the ACPR

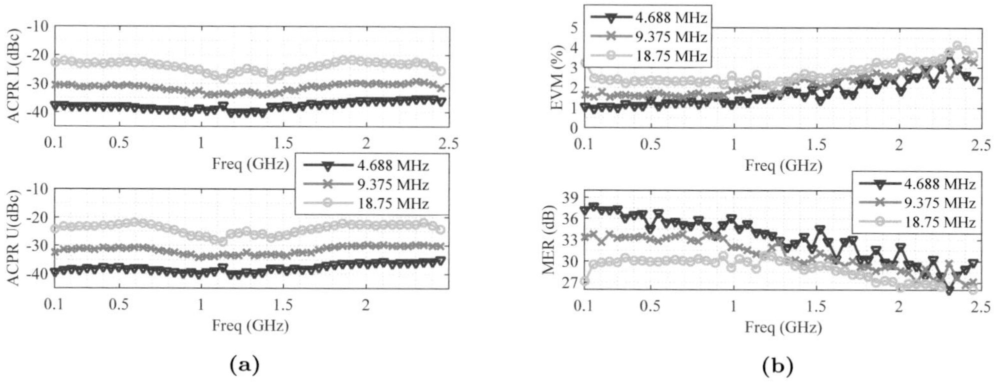


**FIGURE 6.20:** Measured results of a sweep in the carrier frequency with a 16-QAM modulated signal with 4.588, 9.375 and 18.75 MHz in terms of: (a) Adjacent-Channel Power Ratio Lower (ACPR-L) and Adjacent-Channel Power Ratio Upper (ACPR-U); (b)EVM and MER.

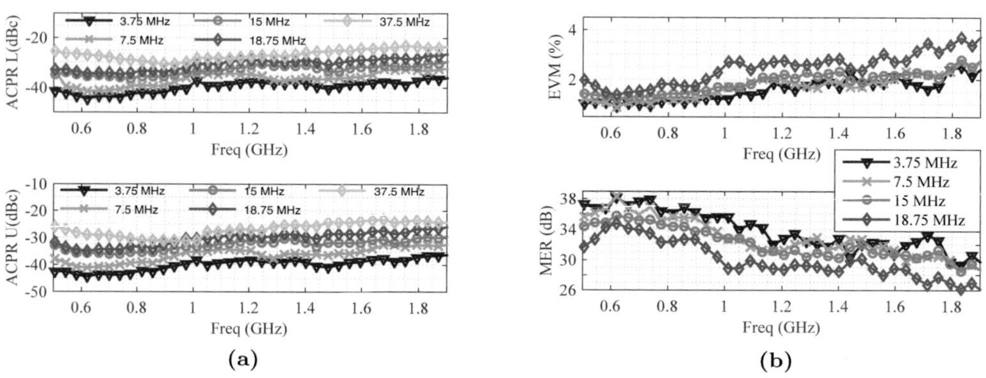


**FIGURE 6.21:** Measured results of a sweep in the carrier frequency with a 16-QAM modulated signal with 3.75, 7.5, 15 and 18.75 MHz in terms of: (a) ACPR-L and ACPR-U; (b)EVM and MER.

values. The signals with higher bandwidth present the worst results in terms of ACPR: between -20 and -30 dBc. By comparing this figure with the single-bit one (Fig. 6.20a), it is possible to understand the impact of using a multi-bit quantizer: the same results were achieved with twice the signal bandwidth.

MER and EVM obtained pairs are demonstrated in Fig. 6.21b. Due to the limited available I/Q bandwidth from the VSA (25 MHz), it was not possible to obtain measurements for the 37.5 MHz case. Again, the same intuitive conclusions can be extracted. Narrow bandwidths provide better FoMs. In particular, the maximum obtained MERs were 38, 37, 36, and 34 dB, for the aforementioned bandwidths.

The spectrum in a span of 2.0 GHz, for carrier frequencies of 500 MHz, 1.6016 and 1.9219 GHz is shown in Fig. 6.22b.

Finally, the occupied resources in the FPGA for the single-bit and multi-level architectures are presented in Table 6.6. A high percentage of DSP slices is used in the parallel

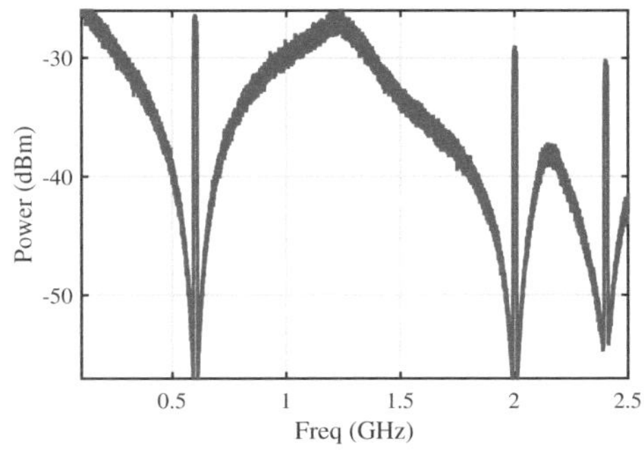


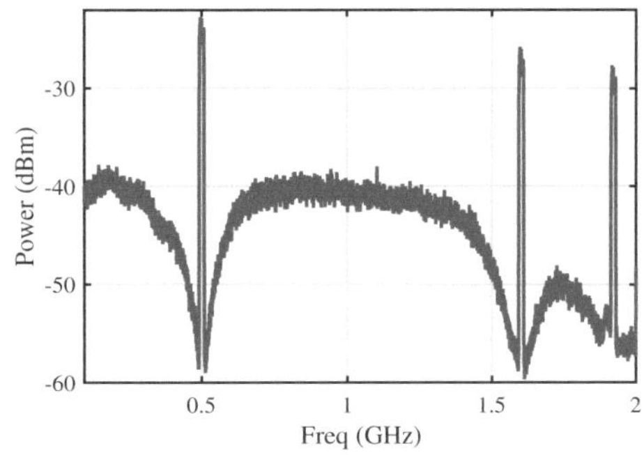


**FIGURE 6.22:** (a) Obtained spectrum in the single-bit architecture, with a full span of 2.5 GHz for carrier frequencies of 600.59 MHz, 2.002 and 2.4023 GHz, with a bandwidth of 18.75 MHz; (b) Obtained spectrum in the 7-level architecture, with a full span of 2 GHz for carrier frequencies of 500 MHz, 1.6016 and 1.9219 GHz, with a bandwidth of 18.75 MHz.

**TABLE 6.6:** Occupied resources for the single-bit and multi-level RF transmitter design

| | | Single-bit | | Multi-level | |
|---|---|---|---|---|---|
| Logic Resources | Total | Used | % | Used | % |
| Flip Flops | 607200 | 164944 | 27.16 | 174533 | 28.58 |
| LUTs | 303600 | 113123 | 37.26 | 140027 | 46.12 |
| Memory LUT | 130800 | 60045 | 45.91 | 74004 | 56.58 |
| BRAM | 1030 | 192 | 18.64 | 192 | 18.64 |
| DSP48 | 2800 | 1876 | 67.00 | 1876 | 67.00 |
| GTXs | 27 | 1 | 3.7 | 7 | 25.9 |

$\Delta\Sigma$M, as well as in the polyphase interpolation FIR filters. These filters were designed with a four-tap transposed direct-form architecture per phase. The inclusion of the Corner-Bender Matrix Transposer enabled considerable savings in the BRAM primitive, as expected. In particular, in this project, the BRAM is just being used to accommodate the DDS's ROMs. Nonetheless, depending on the requirements, there is enough margin to implement an even more efficient memory usage through the exploration of the symmetry of the DDS' sinusoidal waveforms.

#### 6.4.2.3 Comparison to the SoA

Table 6.7 presents an overview of the relevant modulators presented in the literature. The table aimed to focus on the number of transmitted bands, the achieved bandwidth per band, the maximum span between bands, the serializer's sampling rate, the modulator type, the number of levels from the output signal, the number of transmitters, the dependence of the carrier frequency with the serializer bitrate, the hardware implementation and, last but not least, the agility. By agility, one is referring to implementations or techniques that allow the adjustment of the carrier frequency without changing the output's sampling rate.

Clearly, the two proposed implementations surpass the SoA in several aspects. In the first place, they are the first architectures to synthesize three different bands with maximum spans of 2.5 and 2 GHz with an online FPGA implementation. The maximum spans from

**TABLE 6.7:** Comparison of the proposed architectures with some relevant modulators from the SoA[1]

| Reference | No. Bands | BW per Band (MHz) | Max. Span (GHz) | Fout (Gbps) | Mod. Type | No. Lev. | No. Tx | Fc indep. Fout | HW. Impl. | Agility |
|---|---|---|---|---|---|---|---|---|---|---|
| **This (Impl.1)** | **3** | **18.75** | **2.5** | **5** | **3rd EF RF-BPDSM** | **2** | **1** | **Yes** | **FPGA online** | **Yes** |
| **This (Impl.2)** | **3** | **37.5** | **2** | **4** | **3rd EF RF-BPDSM** | **7** | **7** | **Yes** | **FPGA online** | **Yes** |
| [121] (Impl.1) | 2 | 5 | 0.7 | 3.9 | 6th CRFB RF-BPDSM | 2 | 1 | Yes | FPGA offline | No |
| [121] (Impl.2) | 3 | 5 | 1 | 3.9 | 6th CRFB RF-BPDSM | 2 | 1 | Yes | FPGA offline | No |
| [29] | 3 | 10 | 1.8 | 7 | 4th CRFB RF-BPDSM | 32 | 2 | Yes | AWG offline | No |
| [30] | 2 | 20 | 0.60 | 24.5 | 6th CRFB RF-BPDSM | 3 | 2 | Yes | AWG offline | No |
| [43] | 2 | 3.75 | 0.25 | 10 | first EF BB-BPDSM | 2 | 1 | No | FPGA online | Lim. |
| [185] | 2 | 2 | 0.45 | 7.2 | 2nd CIFB BB-LPDSM | 2 | 1 | No | FPGA online | Lim. |
| [186] | 3 | <2 | 0.378 | 9 | 2nd CIFB BB-LPDSM | 2 | 2 | No | FPGA online | Lim. |
| [109] | 2 | 10 | 0.6 | 6.14 | 8th CRFB BB-BPDSM | 4 | 3 | No | Simul. | - |

[1]In the table, "n/a" and "-" stand for "not-available" and "not-applicable", respectively.

the SoA (1 and 1.8 GHz) were achieved with offline FPGA/Arbitrary Waveform Generator (AWG)-based implementations [29, 121].

The proposed architectures also demonstrate the highest bandwidth per band (18.75 and 37.5 MHz), resulting in the highest aggregated bandwidth (56.25 and 112.5 MHz) among all the SoA. The previous maximum aggregated bandwidth (40 MHz) was reported in [30] with an offline AWG-based implementation and with 32 output levels.

In addition to that, it must be pointed out that this work is the first to present no additional OSR between the maximum span and the output sampling rate (generation is possible up to the Nyquist limit). Ultimately, this results in a minimal output sampling rate, which is crucial for minimizing the switching losses from the amplification stage.

Lastly, one can also recognize that the proposed architectures are the first from the SoA to achieve real-time agility. This means that the three bands can be located in any bin from the frequency grid given by the DDS' frequency resolution within the first NZ (2.5 and 2 GHz, in the single-bit and multi-bit implementations, respectively). All the other approaches based on FPGA online implementations do not possess this capability due to the fact that the DUC is applied after the Pulse Encoder (the concept of BB-stage ADTs). It must also be stated that, as proposed in [45], when a fine frequency resolution is required, a single-rate digital-IF together with a single-rate high-resolution DDS should be included before the polyphase interpolation FIR filters.

One final note regarding the scalability of this architecture is worth highlighting. The main focus of this work was the proposal of a fully agile multi-band RF-stage transmitter, suitable for contiguous/non-contiguous CA scenarios, that could be embedded into programmable logic. First, even though a triple-band RF transmitter was implemented, the formulated design methodology is independent of the number of bands. However, as more bands are included, there is an additional increase in the critical path, which forces a reduction in the serializer's sampling rate. Additionally, more resources are also required. Thus, from a system designer's point of view, the inclusion of optimization techniques in terms of timing/area must be adopted. As far as the resource usage is concerned, it must be highlighted that the amount of logic resources that were reported in both implementations should not be taken as the minimum required values. Several different optimization techniques, in the critical subsystems, can always be applied according to different requirements.

---

## 6.5 Conclusion

This chapter presented three different works related to the design of concurrent multi-band ADT.

The first part of the chapter reported a BB-stage architecture that proposed the real-time displacement of the notches from a $\Delta\Sigma$ architecture to perform the single- or dual-band transmission of modulated data. The technique was proposed and validated in an FPGA. The system is able to adjust the distance between bands, provided that their maximum span does not exceed the modulator's sampling rate (300 MHz).

The proposed architecture is evolved in the second part of the chapter. Its major limitations were reported and theoretically formulated, and a precompensation technique was proposed and experimentally validated. The presented precompensation technique achieved a substantial improvement in signal quality compared to the previous dual-band work. Moreover, the precompensation technique was optimized to be implemented with minimal modifications in the proposed hardware.

The chapter was concluded with the proposal of an RF-stage architecture designed to generate up to three bands digitally, spanning the range of 100 MHz to 2.5 GHz. This architecture is particularly suited for designing agile and real-time reconfigurable ADTs for contiguous or non-contiguous CA scenarios. The technique was proposed and validated in an FPGA. The reported results demonstrate superior performance in terms of agility, the maximum span between bands (up to 2.5 GHz), and aggregated bandwidth (up to 56.25 MHz). To enhance the system performance, an extension to a multilevel architecture was also proposed, enabling the achievement of 112.5 MHz of aggregated bandwidth. In addition to the superior performance, as compared to the SoA counterparts, the fully digital behavior of this architecture shows strong potential to be synthesized in custom application-specific integrated circuits.

In all the proposed architectures, the whole system frequency response can be updated in real time, allowing one to choose between single-band scenarios, dual-band, and triple-band scenarios.

# Chapter 7

# Applications of All-Digital Transmitters

## 7.1 Introduction

This chapter will focus on different applications of ADTs that can take advantage of the specific characteristics provided by this type of transmitter, and it is supported by the conference papers [33, 51, 118, 155] and by the journal articles [8, 32]. The first application takes advantage of the binary nature of the ADT's output in order to create a simple RoF solution. In this case, the usage of an ADT not only simplifies the electro-optical conversion but also provides a simpler RRH due to the fact that the signal that reaches the RRH only needs to be filtered and amplified to be sent to the antenna. Then the second application builds upon this concept and introduces optical upconversion, essentially expanding the frequency agility of the ADTs into the mm-wave band. In the third application, a complete transceiver is introduced, demonstrating that it is possible to integrate both transmission and reception into a single, all-digital device. The fourth application regards the creation of an all-digital antenna array. Here, several ADTs are used to drive an antenna array, with the signals having the necessary phase and amplitude to steer a beam. Finally, the fifth application proposes a different approach from the previous application. Here, a single ADT is used to drive the antenna array, but using digital True Time Delay (TTD) to perform phase control to steer the beam.

## 7.2 Designing Radio-over-fiber solutions

In this section, a new RoF approach is shown using a two-level digital RF signal from an ADT and a Gbps optical link to transmit the signal to a very simple RRH, also called RU, that uses only a bandpass filter and an RF amplifier. This solution stands out for its simplicity in the RRH unit, similar to the analog RF-over-fiber solution but utilizing digital modulation in the fiber.

### 7.2.1 All-Digital Transmitters Architecture

All-digital transmitters have been explored over the last few years as a SoC solution for SDR due to their potential flexibility and reconfigurability. While the completely digital generation of RF signals is already a well-known concept [108], the full advantages in flexibility, reconfigurability, and integration are yet to be fully explored.

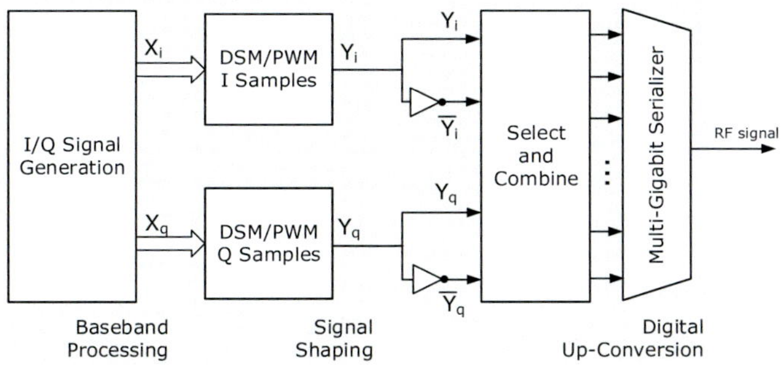


**FIGURE 7.1:** FPGA-based all-digital transmitter architecture.

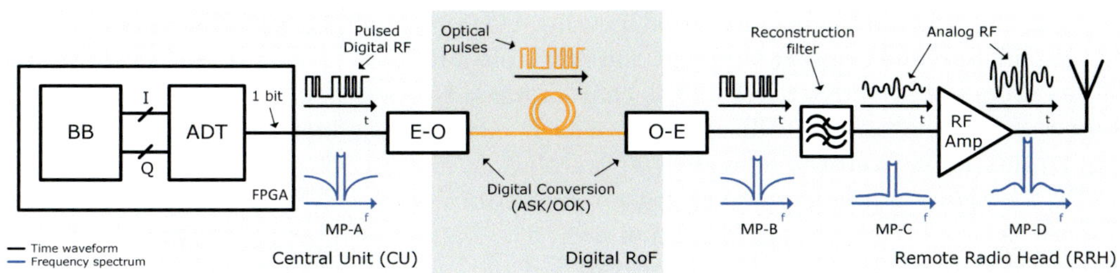


**FIGURE 7.2:** Proposed RoF transmission system with waveforms and frequency spectra along the transmitter chain.

A technology that promises to be an interesting platform for the ADT concept is the FPGA. Its hardware integration, reconfigurability, flexibility, and processing power in modern FPGA's seem to be a convenient solution for a modern fully integrated SDR system.

The top-level architecture of a ADT is shown in Fig. 7.1. The baseband signal is processed using a pulse modulator, with $\Delta\Sigma$M or PWM, and a digital up-conversion stage that modulates the binary signal with a square wave at RF.

## 7.2.2 All-Digital Transmitter with Remote Radio Head

The proposed system architecture is shown in Fig. 7.2. It is composed of a completely digital Central Unit (CU) that receives the digital information to be transmitted. The CU applies the baseband processing and then the RF up-conversion. The signal at the fiber is binary and is generated by the ADT and converted with an Electrical-Optical (E-O) converter in the CU. This signal can be converted back into a digital electric signal at the RRH using an Optical-Electrical (O-E) converter. The RRH unit receives the signal that only needs to be filtered and amplified to be sent to the antenna.

### 7.2.2.1 ADT Central Unit

An ADT is used to implement the signal shaping and up-conversion up to RF frequencies of the IF signal. The output of the ADT is a 2-level electric signal centered at the carrier frequency and with quantization noise in the side bands. The quantization noise may vary depending on the pulse modulator chosen, $\Delta\Sigma$M or PWM. Here, it was used $\Delta\Sigma$M to shape the quantization noise out of the signal's band and improve its signal-to-quantization noise ratio. The transmitter architecture (Fig. 7.1) uses a Time-Interleaved Delta Sigma approach that achieves higher equivalent frequency for the delta-sigma modulator by parallelizing its outputs [54].

The ΔΣM output is then up-converted to RF using a bit combiner block and an FPGA's integrated multi-gigabit serializer. Jointly, these blocks perform the equivalent of a multiplication by a square wave, up-converting the signal to the desired carrier frequency.

The high-frequency pulsed signal is then converted into a series of optical pulses at a certain wavelength $\lambda$ using a E-O converter. This conversion uses a very simple Amplitude-Shift Keying (ASK) modulation; in this particular case, it uses On-Off Shift Keying (OOK). By using such simple 2-level modulations, the system becomes insensitive to the potential non-linearities of the laser-driver and diode.

#### 7.2.2.2 Remote Radio Head

At the RRH, the optical signal is converted back into an electric signal, thereby reconstructing the original signal generated by the ADT. This signal can therefore be filtered and amplified as the original signal at the ADT output.

The proposed system's RRH unit is characterized by its simplicity, similar to the analog RF-over-fiber case, which utilizes only an O-E converter, a bandpass filter to attenuate quantization noise from the delta-sigma modulation, and an RF amplifier.

For the signal amplification, at least two distinct approaches can be used. The most common is to amplify the analog signal after the bandpass filter as depicted in Fig. 7.2. However, a potentially more efficient alternative can be used with a switched class-S amplifier, directly fed by the digital signal from the O-E converter, followed by a reconstruction bandpass filter.

### 7.2.3 Tests and Results

A laboratory prototype implemented as a proof of concept for the proposed system is shown in Fig. 7.3. The ADT CU was implemented using a KC705 board from Xilinx with a Kintex-7 XC7K325T FPGA. The ΔΣM equivalent sampling frequency is 920 MHz (4 parallel paths at 230 MHz). The RF output generated signal is centered at 920 MHz and has a bit rate of 3680 Gbps, which is the rate of the serializer output and therefore the digital rate of the signal in the fiber.

The E-O and O-E conversions were made using Small Form-factor Pluggable (SFP) transceivers, with 850 nm wavelength. At the ADT's output, the SFP was connected to the KC705 board, and at the RRH's input, a Finisar FDB-1027 development kit was used to recover the electrical signal. A saw filter centered at 915 MHz with a 26 MHz bandwidth was used to filter the quantization noise. The amplification stage uses a wideband amplifier from Mini-Circuits (ZHL-1042J).

Several baseband signals were created and pre-processed offline using MATLAB and were loaded into the FPGA's internal memory. The SNR along the transmitter chain was measured to evaluate the whole transmission path using a VSA from Rohde & Schwarz (model FSQ 8). The measure points (MPs) A, B, C, and D are shown in Fig. 7.2 and correspond to:

- MP-A—ADT output.
- MP-B—O-E conversion output.
- MP-C—Reconstruction bandpass filter output.
- MP-D—RF amplifier output.

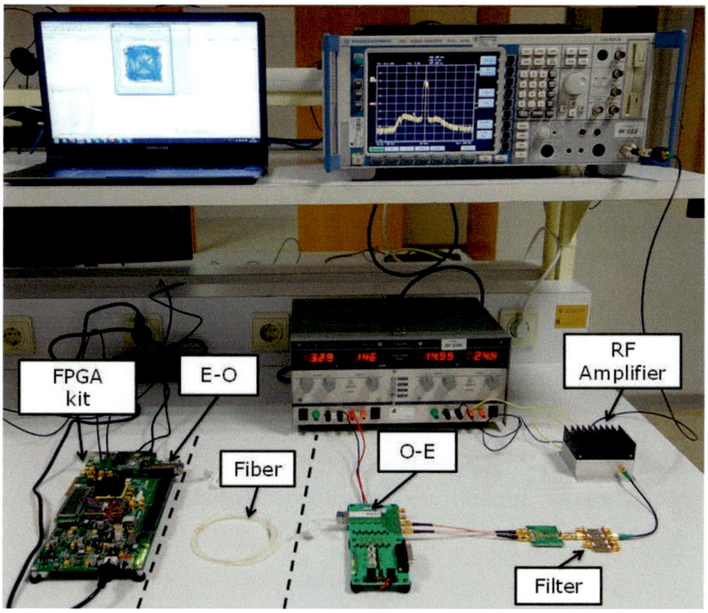


**FIGURE 7.3:** ADT-based Digital RoF system.

The images presented in Figs. 7.4 and 7.5 show the spectra measured at the defined measurement points.

The quantization noise is visible in Fig. 7.4 for the ADT and O-E conversion outputs. This is because both of these signals are 2-level signals resulting from the $\Delta\Sigma$M.

Table 7.1 presents the SNR results for several test signals transmitted using the proposed system and demodulated using the VSA. The SNR measurements along the transmission path were used to evaluate the system's behavior. The overall SNR degradation for this system was computed for the test signals as a figure of merit for the system's performance.

The SNR results show that the proposed system is feasible without a significant end-to-end SNR loss; in these particular tests, the losses were never greater than 2.4 dB.

For the same signals, an EVM analysis was conducted. The results are presented in Table 7.2, and it is shown that for the tested signal modulations and symbol rates, the EVM along the measure points, and particularly at the system's output, is relatively low and always inferior to 1.5 %.

**TABLE 7.1:** Comparative SNR values

| Signal modulation | Symbol rate | Measure Point SNR (dB) A | B | C | D | SNR loss (dB) |
|---|---|---|---|---|---|---|
| QPSK | 2 Mbps | 40.4 | 39.9 | 39.4 | 38.8 | 1.6 |
| | 5 Mbps | 40.1 | 39.3 | 38.3 | 37.7 | 2.4 |
| | 10 Mbps | 38.6 | 38.3 | 36.2 | 36.4 | 2.2 |
| 64-QAM | 2 Mbps | 40.0 | 39.4 | 38.6 | 37.6 | 2.4 |
| | 5 Mbps | 39.4 | 38.7 | 37.9 | 37.2 | 2.2 |
| | 10 Mbps | 38.0 | 37.0 | 36.0 | 35.7 | 2.3 |

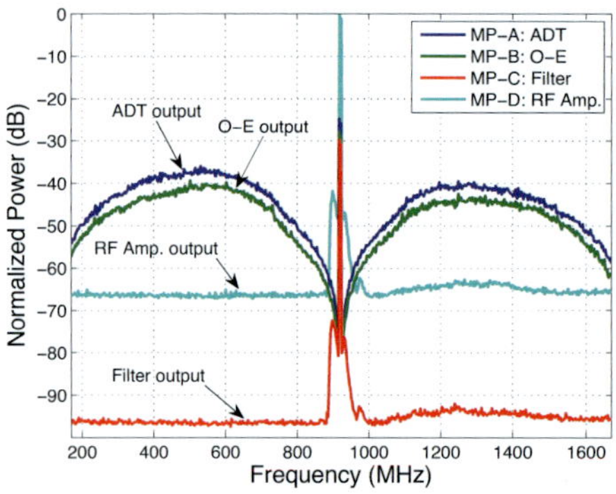


**FIGURE 7.4:** Signal spectra from the VSA for the measurement points defined in Fig. 7.2 centered at 920 MHz with a 1500 MHz frequency span, 30 kHz resolution bandwidth, and 1 kHz video bandwidth.

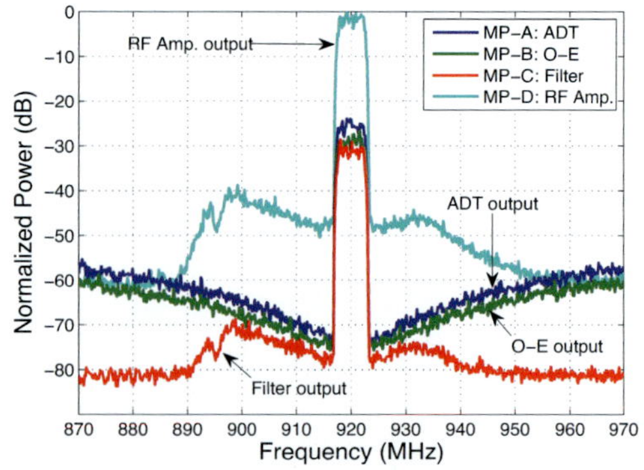


**FIGURE 7.5:** Signal spectra from the VSA for the measurement points defined in Fig. 7.2 with a 100 MHz frequency span, 100 kHz resolution bandwidth, and 5 kHz video bandwidth.

**TABLE 7.2:** Comparative EVM values

| Signal modulation | Symbol rate | Measure Point EVM (%) A | B | C | D |
|---|---|---|---|---|---|
| QPSK | 2 Mbps | 0.83 | 0.86 | 0.93 | 1.00 |
| | 5 Mbps | 0.89 | 0.93 | 1.06 | 1.14 |
| | 10 Mbps | 1.01 | 1.06 | 1.37 | 1.34 |
| 64-QAM | 2 Mbps | 0.85 | 0.92 | 1.01 | 1.12 |
| | 5 Mbps | 0.92 | 1.01 | 1.10 | 1.21 |
| | 10 Mbps | 1.08 | 1.23 | 1.40 | 1.44 |

### 7.2.4 Key Remarks

In this section, a new digital RoF transmission architecture was proposed and proven to be feasible. The use of an ADT allows the generation of a binary digital signal at RF frequencies that can be used to reconstruct the analog RF signal at the RRH side with reduced hardware complexity and low cost. The digital signal over the fiber can benefit from the more favorable propagation characteristics of digital modulations for RoF systems.

## 7.3 Optical Upconversion for Delta-sigma-over-fiber

In this section, two optical upconversion architectures are shown and analyzed. The first one is called single-carrier modulated optical upconversion, and the second one is called dual-carrier modulated optical upconversion. Both demonstrate that the combination of delta-sigma modulation with optical heterodyning is a promising candidate for the upcoming 60 GHz RoF front haul. As shown in Fig. 7.6, by using delta-sigma modulation, an all-digital transmitter can be employed, resulting in the consequent benefit of digital radio transmission over fiber. In addition, optical heterodyning and simple filtering can be

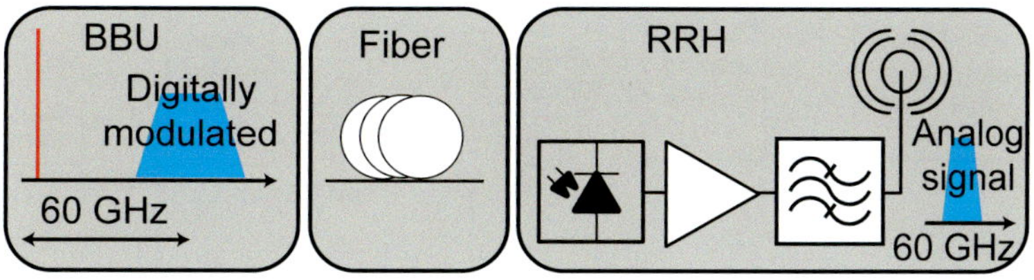


**FIGURE 7.6:** Schematic of the proposed fronthaul architecture. By using delta-sigma modulation, an unmodulated carrier and digital data are transmitted from the BBU to the RRH. The RF signal is created by heterodyning both signals in a photodiode, followed by amplification and filtering; thus, reducing complexity in the RRH.

employed to generate the mm-wave signal, thereby reducing the complexity of the RRH, also called RU. Ultimately, these Delta-sigma-over-fiber solutions combine the benefits of both digital and analog radio-over-fiber solutions.

### 7.3.1 Single Carrier Modulated Optical Upconversion Architecture

In the single-carrier modulated optical upconversion architecture, a real-time implementation of a ΔΣM was used to show that the QAM of 16, 64, and 256 QAM signals is only increased from about 2.9 % to about 3.7 % from the FPGA output to the mm-wave generated signal after 20 km of Standard Single-Mode Fiber (SSMF) transmission.

### 7.3.2 Single Carrier Modulated Optical Upconversion Architecture—FPGA-Based Delta-Sigma Modulator

The architecture for the all-digital transmitter used in the single carrier modulated optical upconversion architecture was introduced in [46]. Its fully parallel and pipelined design enabled the achievement of one of the highest-speed FPGA-based ΔΣMs reported in the literature at the time. As depicted in Fig. 7.7, the all-digital transmitter implementation was divided into the baseband interface, the polyphase interpolation filter, the ΔΣM, the digital up-conversion chain, and the multi-gigabit transceiver. In the baseband interface, I/Q signals sampled at 200 Msamples/s were transmitted from the computer to the FPGA memory. In our experiment, we emulated single-carrier data shaped by a root-raised cosine filter with a roll-off factor of 0.2%. However, the FPGA is capable of quantifying any signal with a similar bandwidth. Afterwards, the polyphase FIR filter interpolated and filtered the incoming data to get an oversampling factor of $N = 16$ for each I and Q sample. As shown, the de-interleaving module rearranges the data with the aim of ensuring that the feedback loop processes $K = 64$ contiguous samples of the signal. In each ΔΣM module, the quantization error was shaped by the filter: $1 - 2z^{-1} + z^{-2}$. Additionally, we employed a parallel error-feedback structure to achieve a short critical path, enabling high sampling rates. As presented in [46] and shown in Fig. 7.7, the propagation of state registers allowed for the reduction of the impulsive noise caused by the discontinuity between each block of the ΔΣM. This method reduced the number of discontinuities from every K samples to every $K \cdot N$ samples. After the ΔΣMs, the interleaving module arranged the samples with vertical temporal continuity. Then, the signal was forwarded to the digital up-conversion stage where the I and Q data are combined in the form of $[I, \overline{Q}, \overline{I}, Q]$, shifting the signal from baseband to one-fourth (3.2 GHz) of the MGT sampling rate (12.8 Gb/s). Finally, the MGT was responsible for the high-speed serialization of the signal. The insets in Fig. 7.7 show the evolution of the temporal and frequency representation of the signal along the ADT

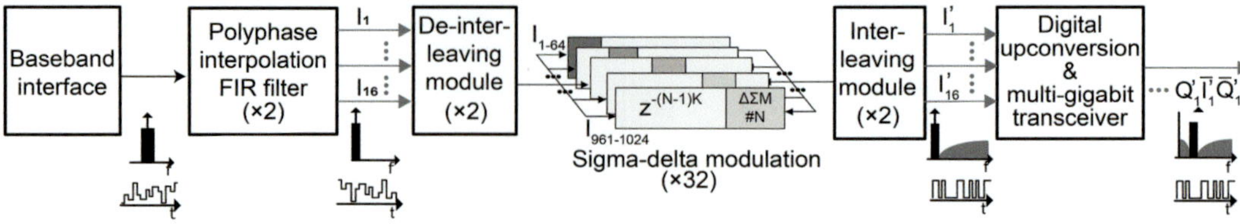


**FIGURE 7.7:** Block diagram of the FPGA (Virtex Ultrascale XCVU095) implementation.

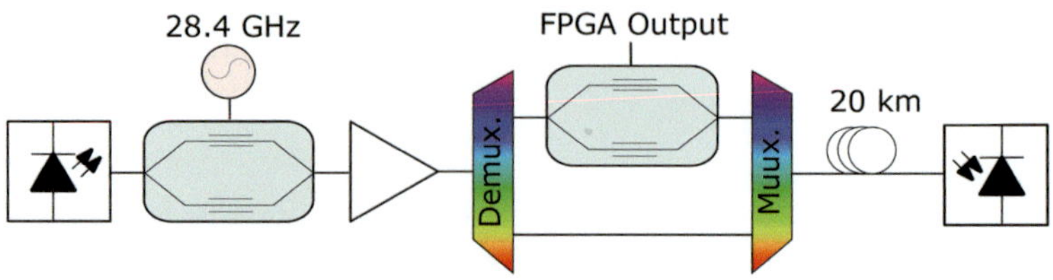


**FIGURE 7.8:** Schematic of the experimental setup.

stages. The output of the MGT consisted of binary data in which the analog information is centered at 3.2 GHz, and most of the quantization noise is out of band.

### 7.3.3 Single Carrier Modulated Optical Upconversion Architecture–Experimental Setup

The experimental setup is shown in Fig. 7.8. A laser at 1550.6 nm with 13 dBm output power was modulated by a 28.4 GHz tone in a Mach-Zehnder Modulator (MZM). By biasing the modulator at the null point, two phase-locked carriers spaced by 56.8 GHz were generated, while the carrier was suppressed by more than 20 dB compared to those carriers. To compensate for the modulator's loss, the generated carriers were amplified by an Erbium-Doped Fiber Amplifier (EDFA) to about 13 dBm and split into different arms using a standard 100 GHz add-drop multiplexer. One of the carriers was modulated in a second MZM with the signal generated in the FPGA. The modulated and the unmodulated carriers were combined before being launched into the 20 km SSMF transmission. By using the mentioned implementation, the system was not affected by chromatic dispersion. The spectrum of the signal at the fiber input is shown in Fig. 7.9(top). The power ratio between both carriers was about 9 dB, due mainly to the modulator's loss. However, it was decided not to equalize them, thereby avoiding the need for a second amplifier and maintaining a simple experimental setup.

After the 20 km SSMF transmission, the mm-wave signal was generated by electro-optical heterodyning in a 100 GHz bandwidth photodetector. The detected signal was amplified before being sampled with a real-time oscilloscope operating at 200 Gsample/s and with 70 GHz bandwidth. The spectrum of the measured mm-wave signal is shown in Fig. 7.9(bottom), which shows the typical shape of delta-sigma modulation with large quantization noise in the out-of-band region. The offline processing consisted of frequency downconversion, downsampling, filtering, and a least-mean-square equalizer with 5 taps. Afterwards, EVM was normalized to the mean power, taking into account the transmitted data. For comparison purposes, we also measured the signal transmitted through 20 km of SSMF and modulated the laser directly with the FPGA output as an on-off keying signal. In this case, the photodetected signal power was approximately -0.5 dBm.

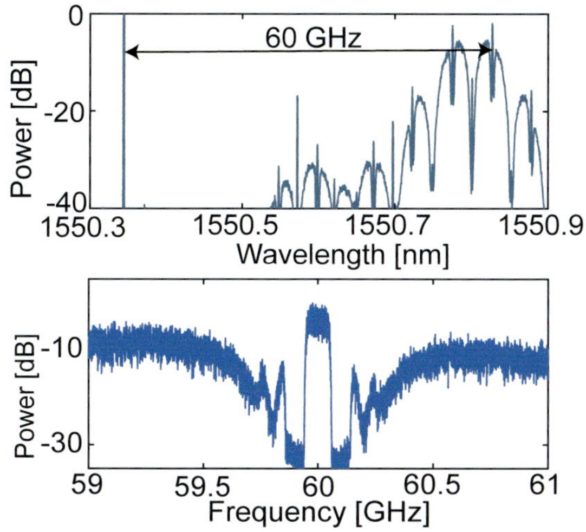


**FIGURE 7.9:** Spectrum measured at the fiber input (top). For visualization purposes, the unmodulated carrier power has been reduced. Spectrum of the detected signal (bottom).

**TABLE 7.3:** Measured QAM for different modulation formats at the FPGA output, after 20 km transmission at the intermediate frequency carrier (3.2 GHz) and up-converted to 60 GHz.

| | 16-QAM | 64-QAM | 256-QAM |
|---|---|---|---|
| FPGA | 2.89 % | 2.91 % | 2.96 % |
| 3.2 GHz | 3.44 % | 3.35 % | 3.57 % |
| 60 GHz | 3.73 % | 3.58 % | 3.70 % |

### 7.3.4 Single Carrier Modulated Optical Upconversion Architecture - Experimental Results

Table 7.3 shows the measured QAM for three different modulation formats. In the cases of the signal provided by the FPGA and transmitted through 20 km SSMF without up-conversion, similar QAMs were measured in offline processing and in a VSA. The QAM was degraded by about 0.55 % when an optical carrier was modulated and transmitted compared to the FPGA's output. The variations of the QAM's EVM among the different modulation formats were due to changes in the bias conditions. When the signal was up-converted to a frequency of 60 GHz, the QAM's EVM was further degraded by 0.25 %. The measured EVM of 3.7 % for the 256-QAM when upconverted to 60 GHz is only 0.2 % over the required for LTE signals with 256-QAM; however, the signal used had a larger bandwidth. In addition, it should be emphasized that the amplifier used to boost the detected signal was specified to have a bandwidth of 40 GHz; therefore, its performance is expected to have been one of the causes of the signal degradation. The constellation diagrams of the 256-QAM signals at the FPGA output and transmitted and up-converted signals are shown in Fig. 7.10.

The dependence of the QAM's EVM with the received power was also studied by varying the output power of the EDFA. As shown in Table 7.4, the signal quality deteriorated when the output power was reduced. Again, it is expected that the use of a RF amplifier suited to the 60 GHz band could mitigate this degradation. It also should be noted that balancing the power of the unmodulated and modulated carriers could lead to a larger sensitivity at the expense of a more complex transmitter.

**TABLE 7.4:** Measured QAM vs power levels at the photodetector for the 16-QAM signal.

| P. [dBm] | 4 | 3.2 | 2 | 0.8 | -0.2 | -1.4 |
|---|---|---|---|---|---|---|
| EVM [%] | 3.73 | 3.89 | 4.05 | 4.18 | 4.40 | 4.85 |

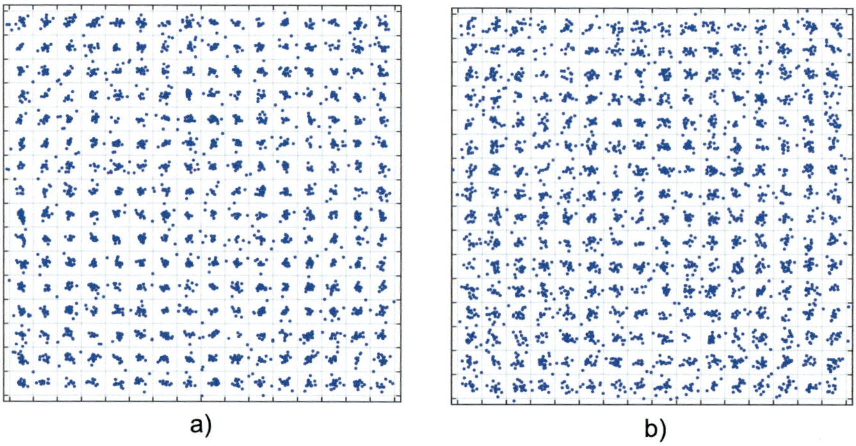


**FIGURE 7.10:** Constellation diagrams of 100 Mbaud 256-QAM at the FPGA output a) and generated after 20 km transmission and upconverted to 60 GHz b).

### 7.3.5 Single Carrier Modulated Optical Upconversion Architecture - Discussion

By using delta-sigma modulation, simple one-bit quantizers can be used at the expense of oversampling the signal. In this case, an oversampling factor of 16 was applied to each input I and Q sample. The signal was upconverted to an intermediate frequency of 3.2 GHz using the mentioned $[I, \overline{Q}, \overline{I}, Q]$ scheme. To make the overall system more attractive, architectures with lower oversampling (i.e., higher bandwidth for the same output sampling rate) should be utilized. Offline studies have, however, shown that lower oversampling can be utilized by adapting the signal cardinality to the quantization noise [213], which can be achieved in an OFDM system.

Despite having performed the experiments with single carrier signals, the architecture is independent of the modulation format, and similar results are expected with other multi-carrier or OFDM signals.

### 7.3.6 Single Carrier Modulated Optical Upconversion Architecture—Key Remarks

The combination of delta-sigma modulation and optical heterodyning was demonstrated to combine the benefits of digital transmission and analog processing at the RRH location. By employing delta-sigma modulation, 1-bit quantizers could be used while reducing quantization noise in the signal's band. The use of optical heterodyning for signal upconversion reduces complexity at the RRH, which could be of great importance for 5G and beyond applications.

### 7.3.7 Dual Carrier Modulated Optical Upconversion Architecture

For the dual carrier modulated optical upconversion architecture, the same principle applies, where ADTs with photonics-based optical mixers are combined to create a flexible solution based on off-the-shelf components. This architecture presents low complexity and large flexibility regarding carrier frequency, modulation format, and signal bandwidth. The ADT is once again implemented in a FPGA, and the architecture is based on the one presented in [46], where a 1-bit $\Delta\Sigma$M with a bandwidth of up to 120 MHz and an intermediate frequency of 3.2 GHz was demonstrated. Then, optical heterodyning is used, allowing the up-conversion of the FPGA's output to frequencies ranging from 27 up to 63 GHz with low EVM penalty in a fully flexible transmitter. The system is evaluated with single-carrier 16-ary quadrature amplitude modulation (16-QAM) signals with symbol rates of 20, 50, and 100 Mbaud. A maximum EVM of 4.2% is measured for the 100 Mbaud signal up-converted to 63.2 GHz. The results show the potential of combining ADTs with optical heterodyning for flexible frequency up-conversion.

### 7.3.8 Dual Carrier Modulated Optical Upconversion Architecture–Parallel Delta-Sigma Modulator

The ADT presented here is implemented using a parallel $\Delta\Sigma$M that was first introduced in [46]. In it, an error-feedback structure is employed due to its short critical path, which enables higher sampling rates. This structure is characterized by a STF and a NTF, where $\text{STF}(z) = 1$ and $\text{NTF}(z) = 1 - 2z^{-1} + z^{-2}$. From these two transfer functions, it can be seen that the signal passes through the modulator unchanged, while the noise undergoes high-pass shaping, ensuring that the baseband signal is minimally affected by noise, resulting in the so-called low-pass $\Delta\Sigma$M.

This architecture was chosen due to its ability to reach higher sampling frequencies while maintaining a relatively low FPGA clock rate (200 MHz in this case). The underlying idea is to slice the signal into blocks of length K (64 in this case), which then feed the N (16 per I and Q component) parallel modulators. This division allows each modulator to operate at a frequency N times slower than the signal's sampling frequency. However, a new problem arises due to discontinuities between different blocks. In the conventional architecture, shown in Fig. 7.11a, the $\Delta\Sigma$Ms suffer from discontinuities every K samples. In [212], a solution to this has been presented in the form of the propagation of state registers. To achieve this, delay registers are introduced before and after each modulator, as depicted in Fig. 7.11b, as well as an entirely new block referred to as "synchronous start-up control logic". This last block is responsible for monitoring the final state registers of every modulator and passing them to the adjacent modulators. Ultimately, this solution can reduce discontinuities from every K sample to every K $\times$ N sample, thereby improving the overall performance of the ADT.

The baseband interface module is responsible for transferring data to the ADT. In particular, data is received from the PC and stored in block RAMs for later processing. Then, the samples are read from memory and sent to the polyphase interpolation FIR filter, which performs an interpolation of the signal. Afterwards, the output of the filter needs to be rearranged to correctly enter the $\Delta\Sigma$Ms. The de-interleaving block is responsible for rearranging the signal to ensure temporal continuity within each phase. After the $\Delta\Sigma$M, the inverse operation is performed by the interleaving block, which ensures a temporal continuity between each phase.

Afterwards, the digital upconversion stage module is responsible for shifting the carrier frequency from baseband to an intermediate frequency. This is done by replicating and combining the I and Q samples in the form of $[I\ \overline{Q}\ \overline{I}\ Q]$. This is formally equivalent to

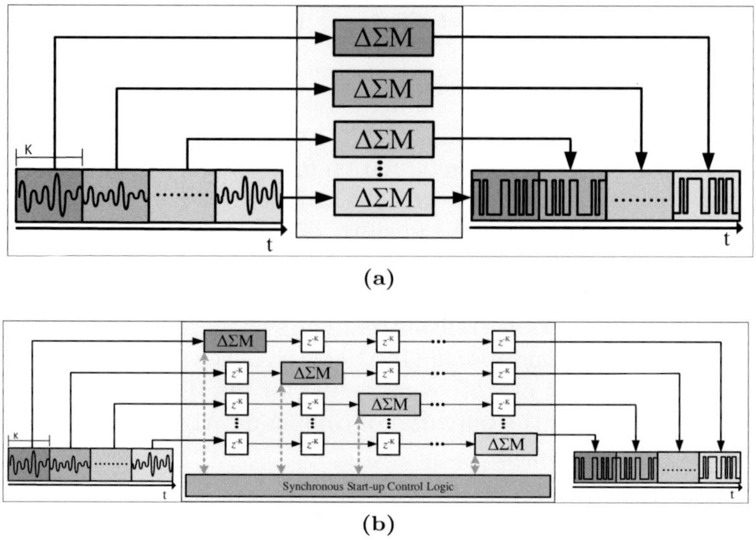


**FIGURE 7.11:** Comparison between the conventional architecture presented in [37] (a), and the architecture presented in [46] and used in our work (b).

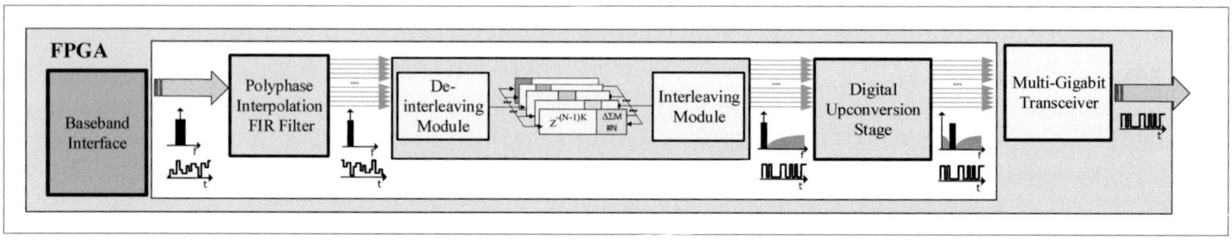


**FIGURE 7.12:** Block diagram of our ADT implementation with propagation of stage registers.

the digital upconversion to one-fourth of the MGT operating frequency. In our case, as the MGT is working at 12.8 Gb/s, the signal is upconverted to 3.2 GHz. Finally, the output word is serialized by the MGT. The final ADT architecture can be seen in Fig. 7.12.

### 7.3.9 Dual Carrier Modulated Optical Upconversion Architecture—Optical Up-Conversion

#### 7.3.9.1 Dual Carrier Modulated Optical Upconversion Architecture—Experimental Setup

The experimental setup is shown in Fig. 7.13a. A laser with 13 dBm output power and centered at 1550.1 nm was modulated in a Mach-Zehnder modulator by a tone using carrier-suppressed technique to obtain two carriers with a frequency separation twice the tone frequency. In our experiments, the tone frequency values were 15.1 GHz, 20 GHz, and 30 GHz. Carrier suppression of about 20 dB was achieved in all experiments. The power of the generated carriers was about -1 (-5) dBm each for the case of the 15.1 (30) GHz tones. These frequency values were chosen to demonstrate the flexibility of the proposed scheme. Afterwards, another Mach-Zehnder modulator was used to modulate both carriers as on-off keying signals using the electrical signals provided by the FPGA. In our studies, the FPGA output quantified single-carrier signal modulated as 20, 50, and 100 Mbaud 16-QAM signals,

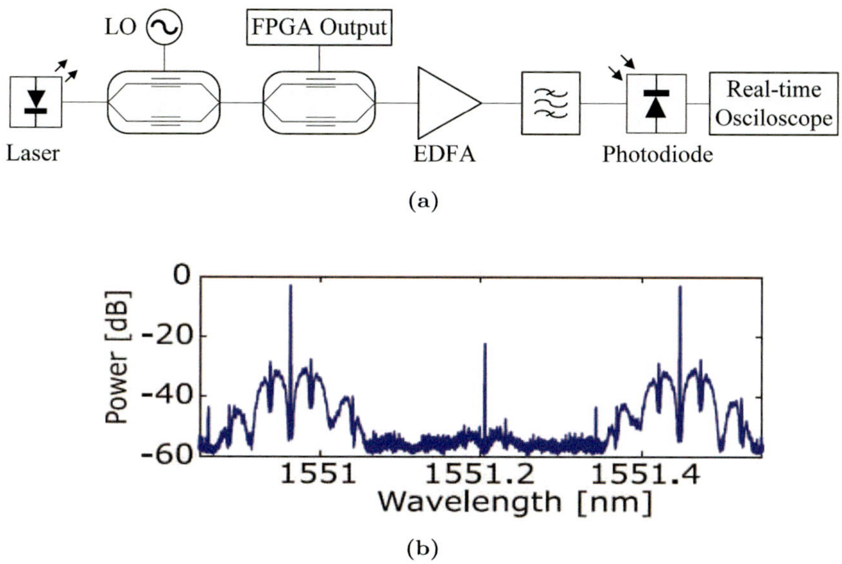


**FIGURE 7.13:** Block diagram of the experimental setup (a) and optical spectrum after the second modulator (b).

shaped as a Nyquist filter with a roll-off factor of 0.2. The spectrum of the signal after the second MZM for the case of 100 Mbaud 16-QAM signal and with 60 GHz carrier separation between both carriers is depicted in Fig. 7.13b. In the spectrum, the two main carriers are observable. A third carrier, due to the suppression ratio of the first modulation, can also be seen. This third carrier is also modulated, but the modulation cannot be seen since it is below the noise floor of the optical spectrum analyzer. After the modulator, the signal was amplified and filtered with a band-pass filter with 200 GHz bandwidth. Finally, the signal was detected in a photodiode with a bandwidth of approximately 100 GHz. The optical power incident on the photodiode was approximately 6 dBm. After the photodiode, the signal was sampled in a real-time oscilloscope operating at 200 GS/s and with an analog bandwidth of 70 GHz. In Fig. 7.14a), the spectrum of a 100 Mbaud 16-QAM signal up-converted to 60 GHz is shown. As can be seen, the spectrum has the typical shape of $\Delta\Sigma$M, with the quantization noise shaped in the out-of-band region. This spectrum shape can also be observed in Fig. 7.13b. The captured signal showed modulated data at 3.2 GHz frequency [not shown in the Fig. 7.14(a)], and the frequency given by twice the frequency of the synthesizers plus/minus the intermediate frequency since we are using a double-sideband modulation. This could be avoided by using optical filters to filter out only the desired upper or lower sideband for each carrier, at the expense of a more complex setup.

The captured samples were processed offline using MATLAB. The offline processing consisted of estimating the frequency prior to frequency down-conversion, matched filtering, and downsampling to 2 samples per symbol, followed by a least-mean-square equalizer with 5 taps. Afterwards, the EVM was normalized to the peak power. In an actual transmitter, the signal should go through a band-pass filter so that the analog signal can be reconstructed and radiated.

#### 7.3.9.2 Dual Carrier Modulated Optical Upconversion Architecture - Experimental Results

Table 7.5 shows the measured EVM for the signal generated at the FPGA output (3.2 GHz carrier frequency) and for the signal converted to different frequencies using optical

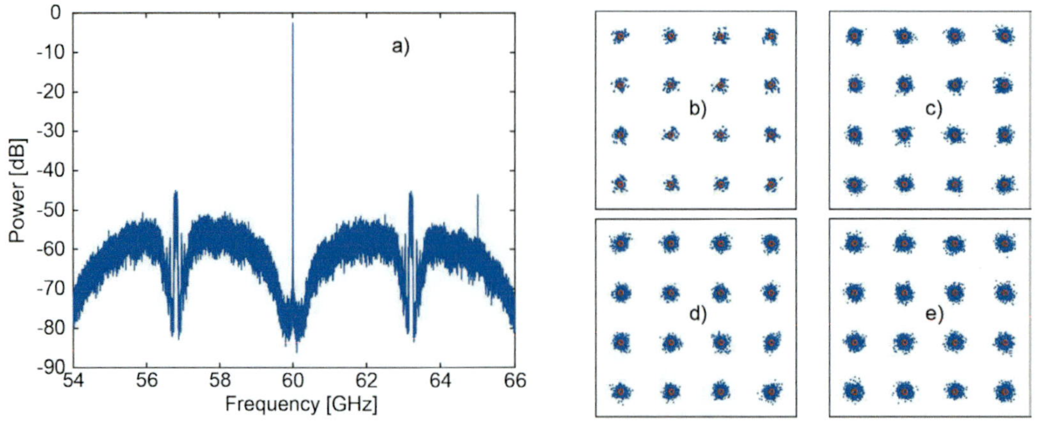


**FIGURE 7.14:** Spectrum of the 100 Mbaud signal centered at 56.8 GHz and 63.2 GHz. Constellation diagrams of the 100 Mbaud signal provided by the FPGA (b), up-converted to 27 GHz (c), to 43.2 GHz (d), and to 63.2 GHz (e).

heterodyning. These frequencies were obtained by selecting the lower sideband (27 GHz) or upper sideband (43.2 GHz and 63.2 GHz) of the signal sampled in the oscilloscope. As can be seen, the EVM is slightly degraded when using the photonics-based up-conversion scheme. One can see that the higher the symbol rate, the higher the increase in EVM. For example, the EVM increases about 1 % for the 20 Mbaud signal when up-converted to 63.2 GHz, whereas it increases about 1.5 % and 2 % for the 50 and 100 Mbaud signals, respectively. Figures 7.14(b)-7.14(e) show the constellation diagrams for the cases of 100 Mbaud 16-QAM signals with different carrier frequencies. In agreement with the EVM results, we can observe that the signal provided by the FPGA suffers from less noise than the other constellations. The degradation of the constellations is due to several factors: noise added by the modulator's drivers, imperfect bias settings of the modulator, noise added by the EDFA and the photodetector, and the bit resolution of the oscilloscope used. To fully understand which one of these noise contributions is the dominant one, a more thorough investigation should be performed.

Table 7.5 also demonstrates that, for the same symbol rate, there is an EVM increase for higher carrier frequencies. This is caused by the lower bit resolution of the oscilloscope at higher frequencies and the filtering effects resulting from the oscilloscope's bandwidth, indicating that this is one of the effects affecting photonics-based up-conversion. Nevertheless, in a real application scenario, this effect would not impact the proposed system, as discussed earlier, since a bandpass filter should be placed after the photodetector before transmitting

**TABLE 7.5:** EVM results for different symbol rates and carrier frequencies. The results at 3.2 GHz correspond to the 16-QAM signal provided by the FPGA. The results at other carrier frequencies correspond to the use of optical up-conversion by the heterodyning technique.

| Carrier Frequency / Symbol Rate | 3.2 GHz | 27 GHz | 43.2 GHz | 63.2 GHz |
|---|---|---|---|---|
| 20 Mbaud | 1.32% | 1.80% | 2.18% | 2.69% |
| 50 Mbaud | 1.72% | 2.54% | 2.77% | 3.21% |
| 100 Mbaud | 2.30% | 3.35% | 3.62% | 4.19% |

the delta-sigma signal at the mm-wave carrier frequency. Another effect we have observed that degrades the constellations slightly is the phase noise from our RF clock, which, as expected, increases with the frequency. To mitigate this effect, it was required to increase the step-size parameter of the least-mean-square algorithm to compensate for the phase noise of the clock.

### 7.3.10 Dual Carrier Modulated Optical Upconversion Architecture - Key Remarks

In this dual-carrier modulated optical upconversion architecture, photonics-based frequency upconversion is used to shift the signal to frequencies ranging from 27 GHz to 63 GHz. We have evaluated this architecture with 20, 50 and 100 Mbaud 16-QAM signals. The EVM results show that EVM increases by 1 % for the 20 Mbaud signal, 1.5 % for the 50 Mbaud signal, and 2 % for the 100 Mbaud signal when performing up-conversion to 63.2 GHz. Despite this degradation, for all the studied cases, the EVM remains below 4.2 %, providing a good margin compared to the specified EVM of 12.5 % for LTE signals. Therefore, this work demonstrates that photonic-based mm-wave generation is a promising technology for combining with an all-digital transmitter.

### 7.3.11 Key Remarks

In this section two architectures for optical upconversion of ADT's signals were analyzed. Both use the first MZM to generate the two tones separated by the desired final upconversion frequency. From here, the single-carrier modulated optical upconversion architecture utilizes an optical demultiplexer to separate the two tones, allowing only one of them to be modulated. In the second MZM, the carrier is modulated using the signal from the ADT. In contrast, the dual carrier modulated optical upconversion architecture uses the two tones directly in the second MZM. This means that both will be modulated. This difference is crucial and is reflected in an increase in the overall noise in the final signal after the photodiode for the dual-carrier modulated optical upconversion architecture. In the end, a trade-off between complexity and lower noise can be observed when comparing both architectures, with the dual-carrier modulated one being simpler; however, it also presents a higher level of noise, which is reflected in an increase in the final EVM.

---

## 7.4 All-Digital Transceivers

In this section, a new all-digital RF transceiver architecture, fully integrated into a single FPGA chip, will be presented and evaluated. The entire signal processing from the baseband up to the RF stage for the transmitter, and the opposite for the receiver, is completely performed in the digital domain. In this sense, the proposed system does not utilize conventional Analog-to-Digital Converter (ADC) and DAC converters, nor any type of analog mixing device, thereby enabling the creation of a much more compact and integrated radio.

### 7.4.1 State of the art

SDR transceiver architectures will be fundamental in future radio communication systems to endow the physical layer with high flexibility, spectral, and energy efficiency [130]. Therefore, this subsection addresses the state-of-the-art on SDR transceivers focusing on all-digital platforms, due to the enormous advantages that such architectures can provide.

#### 7.4.1.1 Transmitters

All-digital transmitter architectures typically employ pulsed-modulation approaches to enable the full digitalization of the radio system, which represents a significant step toward the complete software description of RF signals proposed in SDR.

Recent advances involving pulsed RF transmitters include the development of novel all-digital transmitter architectures where the radio datapath is digital from the baseband up to the RF stage. These transmitters perform a conversion from an $m$-bit digital representation to a 2-level signal, usually done by means of $\Delta\Sigma$M or PWM schemes. State-of-the-art pulsed transmitters can use amplitude and phase modulation of the RF carrier using PWM [140, 147, 148, 150], band-pass $\Delta\Sigma$M [101, 103] or low-pass $\Delta\Sigma$M with digital up-conversion [63, 64, 86, 90, 98]. To improve signal quality, many works have focused on the increase of the signal OSR. Most of pulsed-RF transmitters achieve higher OSR by optimizing the digital design for a higher frequency [24]. However, digital switching logic has frequency limits that bound the achievable OSR to a maximum value, depending on the technology used. An alternative approach for achieving higher OSR is to utilize parallelized processing followed by dedicated high-speed logic, such as a multiplexer or serializer [36, 54, 81]. This eases the design of higher-bandwidth transmitters operating at reduced logic frequencies and facilitates competitive integration with lower-cost technologies.

The functionality and flexibility of the transmitter are also of great importance since the purpose of a digitally controlled transmitter is to easily adapt the radio to its environment and user needs. Carrier frequency flexibility means that a single digital chip can be a wide-band radio with multiple channels and standards. In this section, we present a transmitter capable of covering signal bands from 400 MHz to 2.4 GHz.

#### 7.4.1.2 Receivers

Traditional RF receivers are commonly based on homodyne or heterodyne architectures, usually performing analog down-conversion up to the baseband or up to an IF stage, where the signal is sampled by an I/Q ADC or by an IF ADC. These architectures exhibit a high dependence on analog components, which poses problems such as limited bandwidth due to the frequency response limitations of these components. Alternatively, the IF ADC can be replaced by an RF ADC, allowing sampling the spectrum directly at the RF stage and therefore performing the entire signal processing in the digital domain, as it was envisioned as the ideal SDR receiver [130], Fig. 7.15. These architectures offer high flexibility and high bandwidth, allowing for the advantage of highly efficient and accurate digital signal processing techniques, thereby avoiding analog impairments and mismatches [119].

Modern SDR receiver performance and flexibility are susceptible to the ADC's bandwidth as well as its linearity. Current commercial RF ADCs have reached sampling frequency and analog input bandwidth of a few GHz. However, they are expensive and have a significant energy consumption, which may reduce the overall system's efficiency.

Recently new possibilities to build RF ADCs based on pulsed-modulation converters were proposed, such as; $\Delta\Sigma$M [12], Pulse-Frequency Modulation (PFM) [87, 89] or PWM [123, 161]. All these types of ADCs are single-bit ADCs, i.e., all deal with a digital pulsed representation of the analog signal. The $\Delta\Sigma$M ADCs are often used in audio applications due to their high resolution, which they provide [119]. Several research works have presented RF $\Delta\Sigma$M ADCs, such as [12], where a high-resolution ADC was built; however, it had a low analog input bandwidth. A PFM ADC is based on a Voltage-Controlled Oscillator (VCO) to create a PFM representation of the input analog signal [87]. These

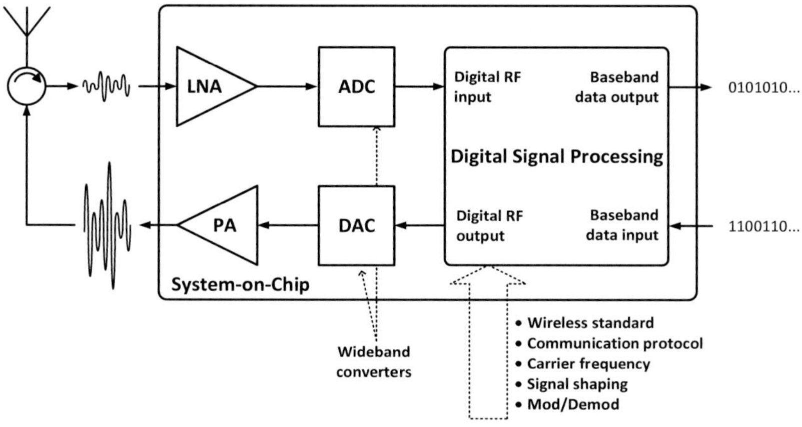


**FIGURE 7.15:** Ideal SDR transceiver architecture.

types of ADCs present high undesirable nonlinear behavior related to the VCO that may be improved using techniques such as the ones presented in [89]. A PWM ADC is made of a single comparator, whose inputs are the analog signal and a given reference wave (usually a triangular wave), generating a PWM representation of the analog signal [228]. They are commonly used in low-frequency applications [228]; nevertheless, they are also being applied in the RF world, as presented in [123, 161]. In [123], high-speed comparators to generate the PWM representation are used, and in [161] a single FPGA chip is used to build the ADC and the remaining receiver, presenting a flexible and promising architecture. However, none of these works presents an exact method to obtain the best reference waveform for optimizing the receiver figures of merit. In this work, the authors present an algorithm that optimizes the frequency of a reference triangular wave to improve receiver performance.

### 7.4.2 Proposed architecture

This subsection presents and explains the proposed architecture for the all-digital transceiver. For the sake of clarity, the transmitter and receiver architectures will be presented separately, but they will be implemented and tested together.

#### 7.4.2.1 Proposed Transmitter Architecture

The proposed architecture for the all-digital transmitter will be presented and explained. The digital transmitter architecture used is shown in Fig. 7.16. The baseband complex signal is divided into two components, the in-phase (I) and quadrature (Q) signals. It should be noted that for the purpose of this work, no particular protocol or standard is considered, i.e., the digital baseband processing can be performed as in conventional transmitter architectures for any particular standard use.

The proposed architecture uses $\Delta\Sigma$M with a two-level output, as shown in Fig. 7.17. The delta-sigma modulator is a signal-shaping stage of the modulator that converts the I and Q signals with N-bit resolution into single-bit representation signals at the trade-off of a higher resolution in time. The signal is initially up-sampled to improve the OSR using a digital interpolation filter. The filter inside the delta-sigma loop shapes the quantization noise inserted by the 1-bit quantizer at the output by feeding back the error

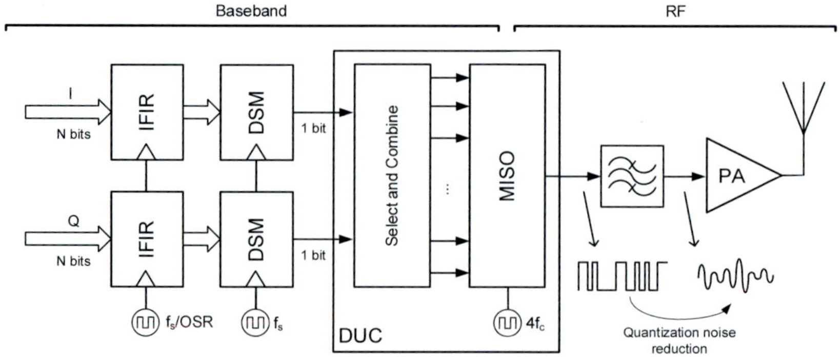


**FIGURE 7.16:** ΔΣM-based ADT architecture. Baseband delta-sigma clock frequency is $f_s$ while the digital up-conversion works at 4 times the carrier frequency $f_c$.

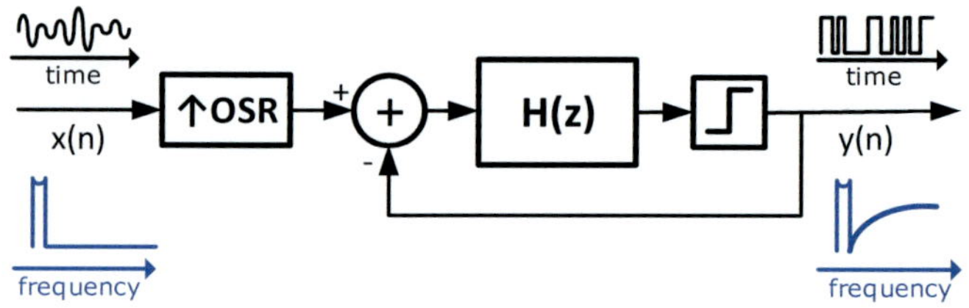


**FIGURE 7.17:** Delta-sigma modulation loop with loop filter $H(z)$.

signal. The output of the modulator is a binary signal with added out-of-band quantization noise.

The ΔΣM loop filter will define the transmitted signal resolution and consequent signal quality in terms of SNR. Higher-order ΔΣM allows higher noise rejection for the signal's band at the same OSR. However, the higher in-band noise rejection increases the noise in the adjacent bands. In [179], an extensive study was made on the influence of the ΔΣM's order on the SQNR. An approximation of the SQNR value given by the Eq. (7.1) and equivalent signal resolution in number of bits is shown in Fig. 7.18 for a first-, second-, and third-order ΔΣMs. In the presented equation, $OSR$ is the over-sampling ratio and $L$ is the ΔΣM order. Higher ΔΣM orders require lower values of OSR to achieve the same performance when compared to lower ΔΣM orders, and therefore lower sampling frequencies could be used. Hence, it would be desirable to use a higher-order modulator to improve the transmitter SNR.

$$SQNR_{dB} = 3.01 \times log_2(OSR) \times (2L+1) - 9.36L - 2.76 \tag{7.1}$$

In RF applications, this becomes a problem since additional out-of-band noise reduces the transmitter efficiency. The relationship between in-band power and total transmitted power is the modulator $C_{eff}$. It can be defined as the signal channel's power $P_s$ over the total transmitted power $P_t$ [71], as it is shown in Eq. (7.2).

$$CE = \frac{P_s}{P_t} \tag{7.2}$$

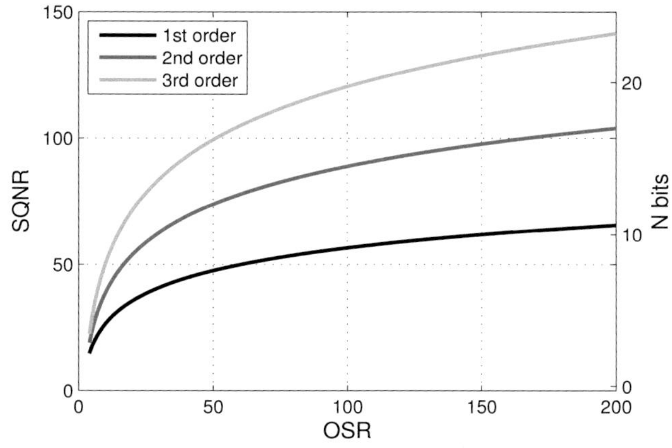


**FIGURE 7.18:** Signal to quantization noise ratio and effective number of bits for two-level delta-sigma modulation with different orders of modulation.

The $C_{eff}$ of $\Delta\Sigma$M varies subject to the nature of the input signal, particularly its PAPR, and is typically lower than 30%. However, higher-orders of modulation will decrease the $C_{eff}$ of the transmitter even further. Fig. 7.19 shows a representative graph of $C_{eff}$ measurements for an all-digital transmitter with both first- and second-order modulators for different PAPR signals. It is evident that for common wireless communication standards with PAPR varying from 3 dB to 12 dB, the $C_{eff}$ is considerably low, especially for higher-order modulators.

Since higher-order $\Delta\Sigma$M limits the transmitter's $C_{eff}$, increasing the modulator's order is not a satisfactory solution to improve the signal SNR. A possible alternative is to increase the OSR of the modulator by increasing its operation speed. However, a key limitation in digital $\Delta\Sigma$M is the typically low bandwidth of the signal modulator, due to frequency restrictions on the digital logic clock resulting from loop-filter propagation delays. To overcome this, a time-interleaved $\Delta\Sigma$M was used to improve the maximum modulator sampling frequency. This parallel $\Delta\Sigma$M architecture used was initially presented in [36] to improve the modulator OSR by allowing the processing of multiple parallel delta-sigma samples in a single clock period.

The $\Delta\Sigma$M used is a first-order modulator because it has a higher $C_{eff}$ for the transmitted signal. Additionally, a simpler loop filter can be implemented with fewer logic elements. Fewer logic gates allow for a shorter logic delay path, thereby enabling a higher sampling rate and signal bandwidth.

DUC is achieved by time-interleaving the delta-sigma-modulated two-level samples (I and Q) in time with their inverted versions. The resulting wave, $w$ in Eq. (7.3), is equivalent to the multiplication of the I and Q baseband components with two square waves with a 90-degree phase shift between them. To achieve this, a parallel-to-series block with a switching frequency four times higher than the desired carrier frequency $f_c$ can be used.

$$w = [I; Q; \overline{I}; \overline{Q}] \tag{7.3}$$

The resulting output signal has the baseband complex signal up-converted to $f_c$ with added quantization noise in the sidebands, which is introduced by the delta-sigma modulation. This noise can be filtered, and the RF signal can be recovered.

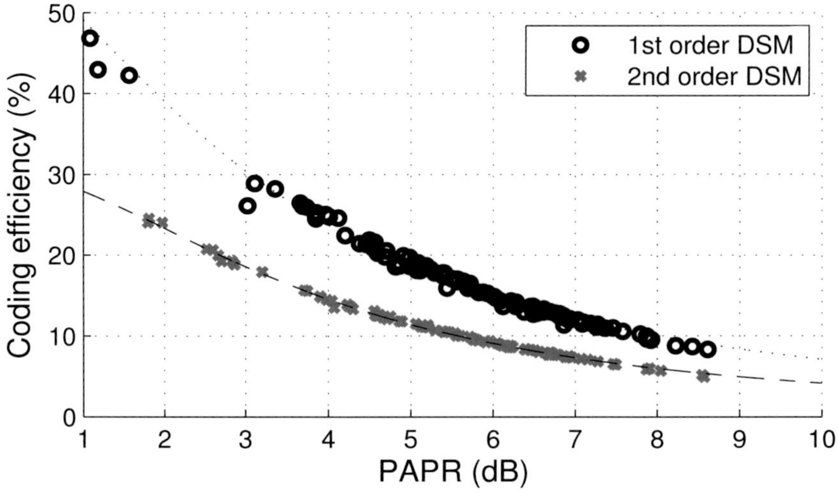


**FIGURE 7.19:** Coding efficiency variation with input signal for first- and second-order ΔΣM. The measured PAPR and $C_{eff}$ values are from random phase and amplitude multi-sine signals.

#### 7.4.2.2 Proposed Receiver Architecture

The receiver architecture will be presented and explained, considering a PWM-based topology, which was chosen mainly due to the high analog input bandwidth and flexibility that such a topology can provide.

The generic block diagram architecture for the proposed receiver is presented in Fig. 7.20, where the signal $x(t)$ represents the RF signal, centered at $f_c$ with bandwidth $BW$, received at the antenna after filtering and low noise variable gain amplification (LNA/VGA). The $x(t)$ signal is one of the inputs of a comparator, while the other is a reference signal ($r(t)$), of frequency $f_r$, which is usually a triangular wave. At the output of this comparator, there is a continuous-time PWM signal $p(t)$, that contains the information of the $x(t)$ signal. The $p(t)$ signal is a natural sampling PWM representation of $x(t)$ [196], which implies that no quantization noise is added into the signal's band. Following the comparator, there is a single-bit register responsible for sampling and discretizing the continuous-time PWM signal $p(t)$ at a sampling rate of $f_s$, with ($f_s > 2f_c$ and $f_s >> BW$). Thereafter, the signal is available for further processing, including digital downconversion and filtering, to recover the information in the baseband signal.

The analog-to-digital conversion stage, which is the key element of this receiver, like any other converter, can be separated into two processes: sampling and quantization, which are related to discretization in both time and amplitude, respectively. The proposed method for implementing the conversion (PWM) is similar to the stochastic-ergodic converter (SEC) presented in [20], which was a popular idea in the 1960s to build low-cost and low-frequency ADCs. However, at the time, due to the lack of enabling technology, the idea was abandoned and sigma-delta converters were preferred [20]. This converter primarily operates by adding a dithering wave to an analog signal and then feeding it to a single-bit quantizer. The theoretical demonstration of this converter can be developed based on statistical quantization theory, which shows that if the dithering signal presents uniform distribution, the input signal will be equally quantized, i.e., the mean of the input signal will be contained in the output quantized signal, allowing its recovery. A common selection for the dithering wave is a triangular or sawtooth wave, since both present uniform amplitude distributions [20,217].

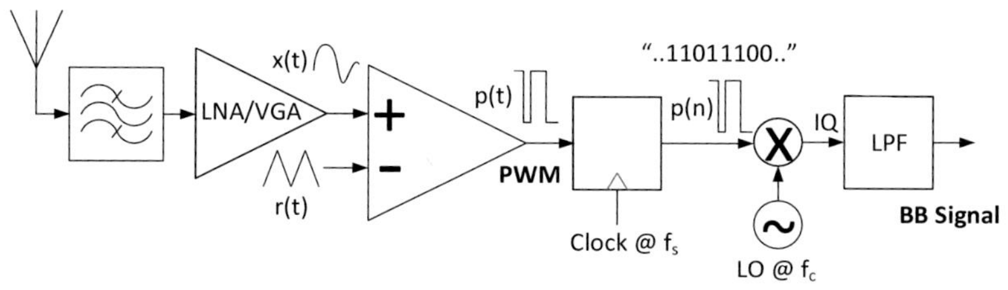


**FIGURE 7.20:** PWM-based all-digital receiver architecture.

This dithering wave can be directly transposed for the PWM RF ADC, playing the same role as the reference signal $r(t)$, which should also present uniform amplitude distribution.

In addition to the reference signal amplitude distribution, two other important characteristics are the reference frequency and the sampling frequency of the sampler after the comparator, which define the maximum bandwidth of the signal that can be acquired. The latter two variables will also impose the converter's resolution, since the ratio between the effective sampling frequency ($f_s$) and the reference signal frequency ($f_r$) indicates the number of PWM levels and, therefore, the number of bits of the converter Eq. (7.4) and its SNR Eq. (7.5), as follows:

$$N_{bits} = log_2(N_{levels}) = log_2 \frac{f_s}{f_r} \tag{7.4}$$

$$SNR = 6.02N_{bits} + 1.76 \tag{7.5}$$

The previous relations indicate that the higher the ratio between the sampling frequency and the reference signal, the higher the converter's effective resolution will be. Therefore, we may consider that the limit for $f_s$ will be imposed by the maximum frequency allowed by a given technological process or by a set of requirements, such as resolution or power consumption, of a certain application. Nonetheless, and without loss of generality, to simplify the digital architecture and achieve the maximum resolution, $f_s$ should be fixed to the maximum achievable value, while maintaining the previous assumptions: $f_s > 2f_c$ and $f_s >> BW$. Thereafter, considering an analog signal occupying a given bandwidth ($BW$), it is important to select the minimum reference signal frequency to provide the highest resolution. While selecting a minimum reference frequency, $f_r$, of $2BW$ for a sawtooth waveform or $BW$ for a triangular waveform allows achieving the same resolution, it creates a different harmonic content in the PWM spectrum. The sawtooth wave puts the first distortion band closer to the interest band, making the triangular wave a preferable way to acquire the signal and to provide a relaxation of the filtering requirements in the Digital Down-Convert (DDC) chain.

At the register's input, ideally, there is a perfect square wave with varying duty cycle of infinitesimal resolution, which presents an infinite power spectrum, as shown in the spectrum before sampling in Fig. 7.21. Then, the single-bit register running at $f_s$ samples the continuous-time $p(t)$ waveform, generating a discrete-time $p(n)$ PWM waveform. As already mentioned, the relation $f_s > 2f_c$ imposes the signal of interest $x(t)$ to be in the first NZ. However, the sampling process causes a frequency overlap at the first NZ due to the aliasing of PWM harmonics coming from higher NZs, as shown in Fig. 7.21 after sampling. This aliasing is unavoidable since an anti-aliasing filter cannot be included between the comparator and the register, in order to maintain a 2-level signal. The use of a binary signal allows for simplifying the sampling hardware to a single-bit register. Therefore, an

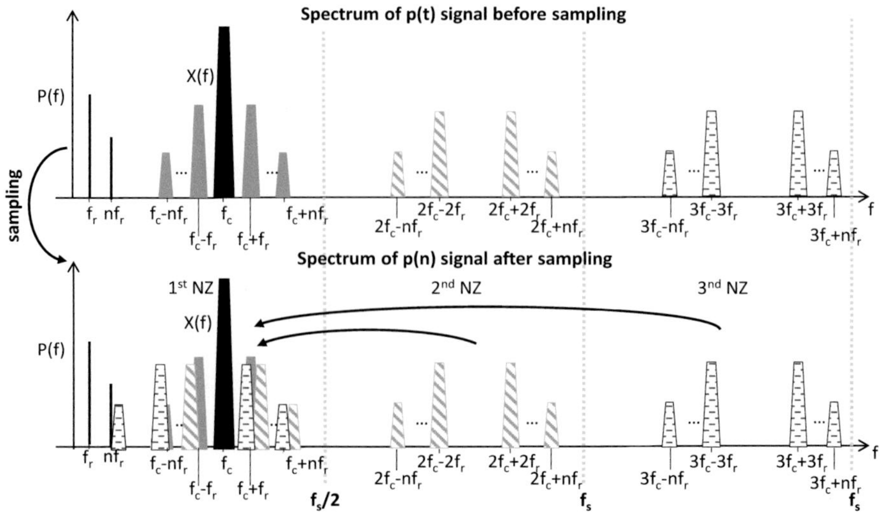


**FIGURE 7.21:** Conceptual representation of the PWM signal spectrum, in the first three NZ, before and after sampling.

analysis of the $p(t)$ PWM spectrum signal should be considered, in order to understand how this overlapping can be avoided in the signal's band.

The PWM modulation process (7.6), which is the comparison between the RF signal, $x(t)$, and the reference signal, $r(t)$, can be described by the sign function Eq. (7.7).

$$p(t) = \text{sign}(x(t) - r(t)) \tag{7.6}$$

$$\text{sign}(x(t)) = \begin{cases} -1, & \text{if } x(t) < 0 \\ +1, & \text{if } x(t) > 0 \end{cases} \tag{7.7}$$

In [217], a general mathematical model to approximate (7.6) is presented in (7.8), and is valid for every $x(t)$ signals:

$$p(t) = x(t) + \frac{2}{\pi} \sum_{\substack{n=-\infty \\ n\neq 0}}^{+\infty} \frac{1}{n} e^{jn\omega_r t} sin\left(\frac{n\pi}{2}(1 + x(t))\right) \tag{7.8}$$

where $\omega_r$ is the angular reference signal frequency and $x(t)$ is the RF signal that can be described as $x(t) = x_{BB}(t)sin(\omega_c t)$, where $x_{BB}(t)$ is the baseband signal's envelope modulated by a carrier centered at $\omega_c$. Then, replacing $x(t)$ by the previous description and applying some well-known trigonometric identities Eq. (7.8) can be simplified into Eq. (7.9):

$$p(t) = x(t) + \sum_{n=1}^{+\infty} \frac{4}{n\pi} cos(n\omega_r t) sin\left(\frac{n\pi}{2} x_{BB}(t) sin(\omega_c t) + \frac{n\pi}{2}\right). \tag{7.9}$$

Consequently, applying the sum-to-product trigonometric identity, Eq. (7.9) becomes Eq. (7.10):

$$p(t) = x(t) + \sum_{n=1}^{+\infty} \frac{4}{n\pi} cos(n\omega_r t) sin\left[A(t)_{n\,even} + A(t)_{n\,odd}\right] \tag{7.10}$$

where, $A(t)_{n\,even}$ and $A(t)_{n\,odd}$ are respectively given by Eqs. (7.11) and (7.12).

$$A(t)_{n\,even} = cos(\frac{n\pi}{2})sin(\frac{n\pi}{2}x_{BB}(t)sin(\omega_c t)) \tag{7.11}$$

$$A(t)_{n\,odd} = sin(\frac{n\pi}{2})cos(\frac{n\pi}{2}x_{BB}(t)sin(\omega_c t)) \tag{7.12}$$

To know the spectrum of the PWM signal $p(t)$, a Fourier transform must be applied on Eq. (7.10). However, the Fourier transform of all the elements is not trivial since it implies calculating the transform of a function in the format $sin(\beta sin(\theta))$ and $cos(\beta sin(\theta))$. It can be further simplified using the Jacobi-Anger expansion, which uses the first-order kind Bessel function Eqs. (7.13) and (7.14) [143]:

$$sin(\beta sin(\theta)) = 2\sum_{k=1}^{+\infty} J_{2k-1}(\beta)sin((2k-1)\theta) \tag{7.13}$$

$$cos(\beta sin(\theta)) = J_0(\beta) + 2\sum_{k=1}^{+\infty} J_{2k}(\beta)cos(2k\theta) \tag{7.14}$$

where $J_k, k \geq 0$ represents the first-order kind Bessel function. Therefore, by applying Eqs. (7.13) and (7.14) on Eqs. (7.11) and (7.12), a new approximation for $A(t)_{n\,even}$ and $A(t)_{n\,odd}$ can be achieved in the form of Eqs. (7.15) and (7.16):

$$A(t)_{n\,even} = cos\left(\frac{n\pi}{2}\right) 2\sum_{k=1}^{\infty} J_{2k-1}\left(\frac{n\pi}{2}x_{BB}(t)\right) sin((2k-1)\omega_c t) \tag{7.15}$$

$$A(t)_{n\,odd} = sin\left(\frac{n\pi}{2}\right)\left[J_0\left(\frac{n\pi}{2}x_{BB}(t)\right) + 2\sum_{k=1}^{\infty} J_{2k}\left(\frac{n\pi}{2}x_{BB}(t)\right) cos(2k\omega_c t)\right] \tag{7.16}$$

Considering now Eq. (7.10) with the new expressions for $A(t)_{n\,even}$ and $A(t)_{n\,odd}$, it is possible to reach an approximation of the central frequency bins where the PWM distortion will fall. In fact, it is not relevant to solve the entire Fourier transform of Eq. (7.10), since the idea is just to choose a reference frequency $r(t)$ to avoid the distortion overlap in the first NZ and consequently optimize the receiver SNR. Thus, Eqs. (7.17) and (7.18) present the center frequencies of the PWM distortion bands:

$$P^*_{n\,even}(f) = \sum_{n=1}^{\infty} \frac{2}{n\pi}\delta(f - nf_r) * \left[\sum_{k=1}^{\infty} |J_{2k-1}(B(t))|_{max}\delta(f - (2k-1)f_c)\right] \tag{7.17}$$

$$P^*_{n\,odd}(f) = \sum_{n=1}^{\infty} \frac{2}{n\pi}\delta(f - nf_r) * \left[|J_0(B(t))|_{max} + \sum_{k=1}^{\infty} |J_{2k}(B(t))|_{max}\delta(f - 2kf_c)\right] \tag{7.18}$$

where, $P^*_n(f)$ are the central bins of the positive part of the $p(t)$ spectrum, $\delta(f)$ is the Dirac delta function, $|J_k(B(t))|_{max}$ with, $B(t) = \frac{n\pi}{2}x_{BB}(t)$, is the maximum absolute value of the amplitude response for a given distortion bin.

At this point, it is important to note that Eqs. (7.17) and (7.18) give the information of the central frequencies of each PWM harmonic, as shown in the spectrum before sampling in Fig. 7.21. Thus, considering a certain level of required or imposed SNR, it should be guaranteed that the PWM harmonics do not fall over the interest band in the first NZ. Therefore, the folding frequency bins placement in the first NZ should be calculated using Eq. (7.19):

$$f_{fold} = \left|f_c - \left\lfloor\frac{f_c}{f_s}\right\rceil f_s\right| \tag{7.19}$$

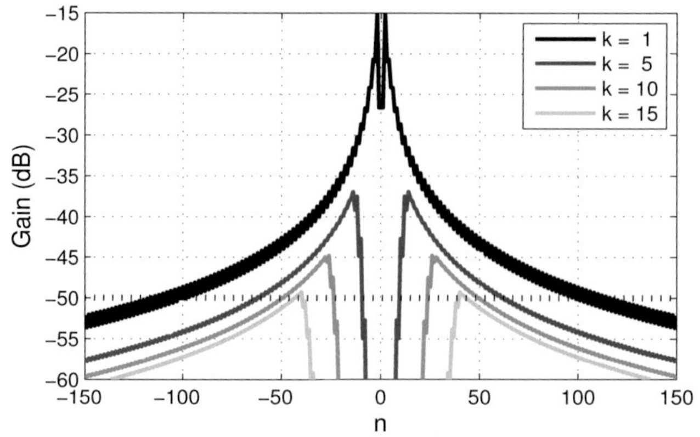


**FIGURE 7.22:** Magnitude of the amplitude response of $P^*(f)$.

where $f_{fold}$ is the folding frequency in the first NZ, $f_c$ is the carrier frequency, $f_s$ is the sampling frequency, $\lfloor ... \rceil$ is the rounding operation toward nearest integer and $|...|$ is the absolute value. Additionally, the amplitude response of $|J_k(B(t))|_{max}$ must be analyzed, up to an $n$ and $k$ value that guarantees the calculation of all distortion bands above the SNR limit.

The maximum $k$ and $n$ values from Eqs. (7.17) and (7.18) can be extrapolated from an imposed limit of the amplitude response of $P_n^*(f)$. In Fig. 7.22, a limit of -50 dB is considered for the amplitude gain, which immediately imposes a maximum value of 15 for the maximum carrier frequency harmonic $k$. Then, after obtaining the $n$, $k$ that will be considered, for the calculation of the folding spectrum version using Eq. (7.19), this information must be used in an algorithm to obtain the best reference frequency (Fig. 7.23).

The proposed algorithm starts with a set of inputs which are the target SNR, the carrier frequency, $f_c$ and the bandwidth, $BW$, of the signal $x(t)$, the sampling frequency, $f_s$, and a step for the reference frequency ($\Delta f_r$). It should be noted that $BW$ here considered is the channel bandwidth, not the signal bandwidth, in order to allow for a small frequency margin for filtering, typically a small percentage of the signal bandwidth, such as 30%.

Regarding the algorithm, it begins by computing the $k$ and both the minimum and maximum $n$ values to ensure the target SNR, as previously described. Then, it calculates the minimum reference frequency, which is given by $2BW$. Following, there is the calculation of the folded version of the spectrum and the verification of the presence of any overlap that may jeopardize the SNR. If there are no overlaps, the reference frequency calculation is complete; otherwise, a new reference frequency is tested, calculated as the previous frequency plus a given step ($\Delta f_r$). This step should be small enough to allow a small loss of resolution and should have an irrational relation to $f_s$ in order to avoid successive occurrences of overlaps from various NZs. Therefore, this step was fixed to be a sub-multiple of $\pi$. An alternative should also be to vary the $f_s$, however, that would increase the complexity of the DDC chain due to the need for irrational decimation filters. This procedure will enable the obtaining of a $f_r$ that generates a sampled signal obeying the target SNR limit.

In order to exemplify the proposed algorithm, Fig. 7.24 presents the simulation results of the proposed receiver considering an ideal comparator. The simulation was performed considering a carrier frequency of 1650 MHz and a 16-QAM signal with 5 MHz of symbol rate. The minimum reference is 14 MHz, and the optimized reference after computing the proposed algorithm is 30.9 MHz. As it is possible to verify in Fig. 7.24, the signal's spectrum using the minimum reference frequency presents a degraded SNR when compared with the

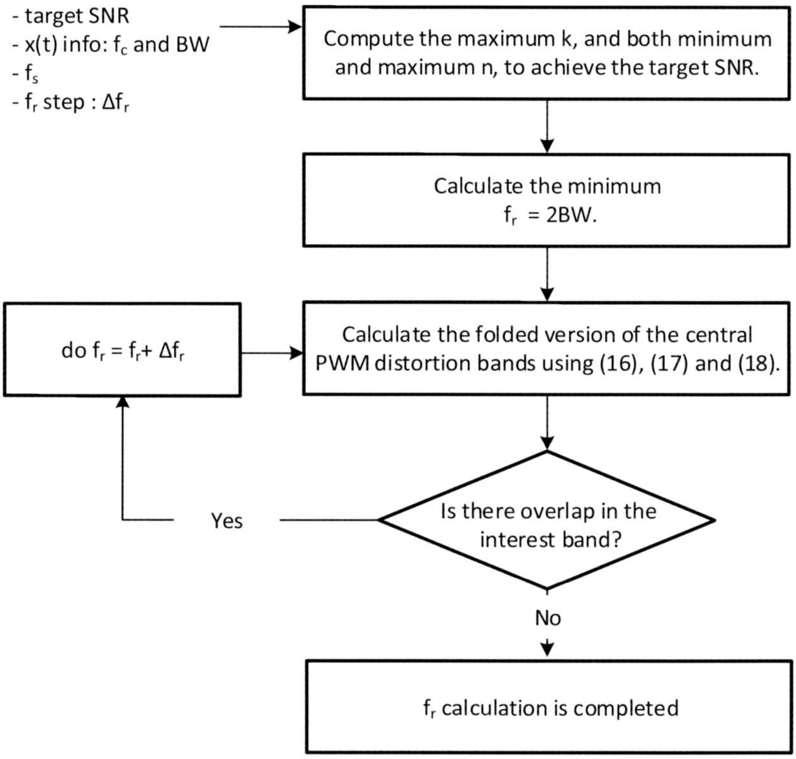


**FIGURE 7.23:** Flowchart diagram of the proposed algorithm to find the best reference frequency to improve the SNR.

spectrum with the best reference. In fact, after computing the EVM of both situations, it was verified an SNR improvement of 9 dB, which validates the proposed technique.

Finally, during laboratory measurements, new spectral bins, differing from those predicted by Eqs. (7.17) and (7.18), were detected. In fact, these new distortion bins are associated with a DC offset error $\alpha$ at the input of the comparator, which can be modeled by Eq. (7.8) considering that $x(t)$ is now given by $x(t) + \alpha$. Therefore, applying the mathematical steps presented previously, a new expression for $P_n^*(f)$ considering the mismatch $\alpha$, can be reached Eqs. (7.20) and (7.21).

$$P^*_{n\,even}(f) = \sum_{n=1}^{\infty} \frac{n\pi}{2}\delta(f - nf_r)* \left[ |sin(\frac{n\pi}{2}\alpha)||J_0(B(t))|_{max} + |cos(\frac{n\pi}{2}\alpha)| \sum_{k=1}^{\infty} |J_{2k-1}(B(t))|_{max}\delta(f-(2k-1)f_c) + |sin(\frac{n\pi}{2}\alpha)| \sum_{k=1}^{\infty} |J_{2k}(B(t))|_{max}\delta(f-(2k)f_c) \right] \quad (7.20)$$

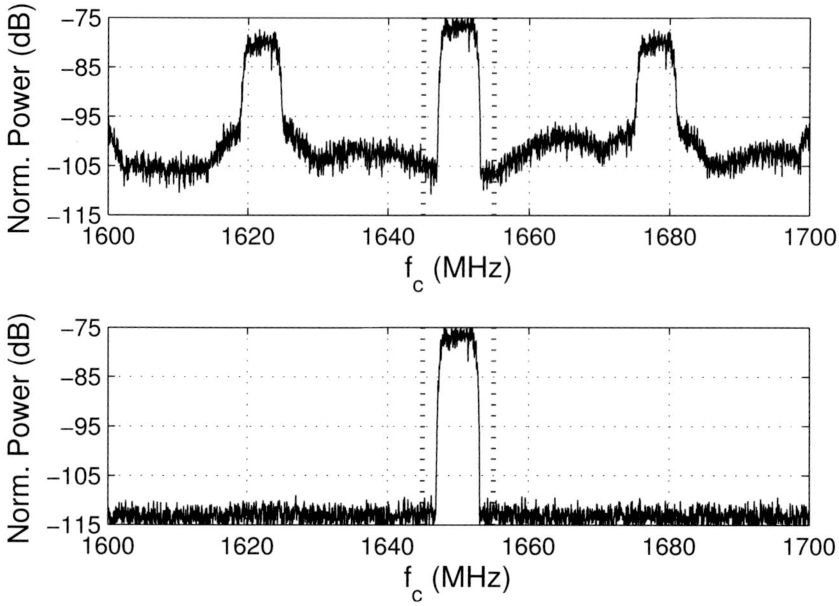


**FIGURE 7.24:** Receiver simulation results with and without best reference calculation, top and bottom plot respectively.

$$P^*_{n\,odd}(f) = \sum_{n=1}^{\infty} \frac{n\pi}{2} \delta(f - nf_r) * \left[ |cos(\frac{n\pi}{2}\alpha)||J_0(B(t))|_{max} + |cos(\frac{n\pi}{2}\alpha)| \sum_{k=1}^{\infty} |J_{2k}(B(t))|_{max} \delta(f - 2kf_c) - |sin(\frac{n\pi}{2}\alpha)| \sum_{k=1}^{\infty} |J_{2k-1}(B(t))|_{max} \delta(f - (2k-1)f_c) \right] \quad (7.21)$$

#### 7.4.2.3 FPGA Implementation

A proof-of-concept prototype was developed using the proposed transmitter and receiver architectures presented in the previous sub-sections. The hardware design was implemented in a KC705 board from Xilinx with a Kintex-7 XC7K325T FPGA. FPGA design enables a completely integrated approach to both the all-digital transmitter and receiver, thanks to the high logic capacity of modern FPGAs. Additionally, their inherent re-configurability enables rapid adjustment of the radio hardware to different purpose scenarios, which aligns closely with the concept of SDR. The FPGA-based transceiver architecture, shown in Fig. 7.25, can be separated into its two main functional blocks, the transmitter (ADT) and the receiver (ADR). The implemented system has an Ethernet connection to a local PC running a Matlab routine responsible for the baseband signal generation and demodulation, transceiver carrier control, and reference signal frequency calculation. The transmitter interpolation filter defines the datapath sampling rate. Higher sampling frequencies will improve the signal OSR in the $\Delta\Sigma$M and therefore the transmitted signal quality. However, the sampling rate must be adjusted depending on the carrier frequency. For carriers between 400 MHz and 1 GHz the sampling frequency is equal to $f_c$, and the RF waveform is the following parallel word $w_{low}$ on Eq. (7.22). The use of repeated samples in Eq. (7.22) reduces the carrier

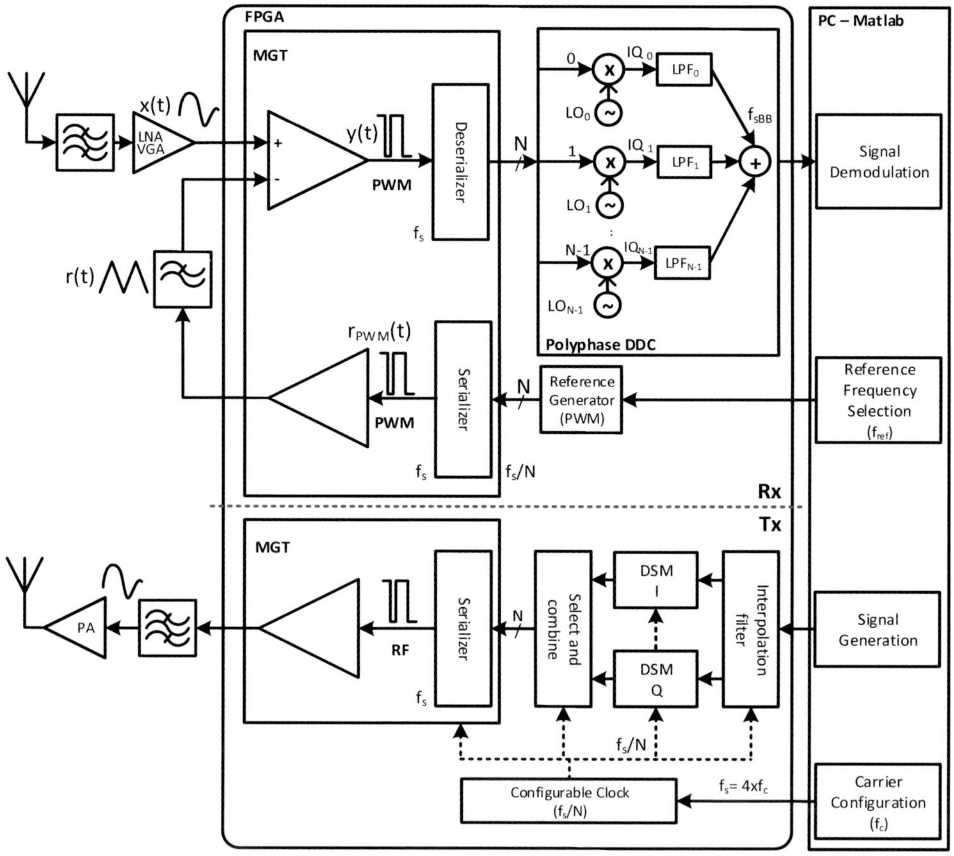


**FIGURE 7.25:** Block diagram representation of the proposed FPGA-based all-digital transceiver architecture.

frequency to half while maintaining the transceiver clock within the FPGA Phase Locked Loop (PLL) frequency lock range. This allows the use of this ADTx architecture for carrier frequencies between 400 MHz and 2.4 GHz.

$$w_{low} = [I_0; I_0; Q_0; Q_0; \overline{I_1}; \overline{I_1}; \overline{Q_1}; \overline{Q_1}; I_2; I_2; Q_2; Q_2; \overline{I_3}; \overline{I_3}; \overline{Q_3}; \overline{Q_3}] \tag{7.22}$$

The input I/Q signal is up-sampled to an equivalent sampling frequency of $f_s = f_c$ using a polyphase interpolation filter with four parallel paths. This enforces a digital hardware clock frequency of $f_c/4$ (between 100 MHz and 250 MHz depending on $f_c$) for the interpolation filter, $\Delta\Sigma$M, and select and combine logic. For carriers above 1 GHz and up to 2.4 GHz the sampling frequency is half of $f_c$, and the RF waveform is given by the word $w_{high}$ as in Eq. (7.23).

$$w_{high} = [I_0; Q_0; \overline{I_0}; \overline{Q_0}; I_1; Q_1; \overline{I_1}; \overline{Q_1}; I_2; Q_2; \overline{I_2}; \overline{Q_2}; I_3; Q_3; \overline{I_3}; \overline{Q_3}] \tag{7.23}$$

In this case, the equivalent sampling rate varies between 500 MHz and 1200 MHz (four parallel paths), and the logic clock will be between 150 MHz and 300 MHz depending on

$f_c$. The ΔΣM stage is constructed with two parallel modulators, one for each of the I and Q datapaths, featuring a polyphase implementation of a first-order modulator with four parallel paths. The ΔΣM hardware operates at the same clock speed as the interpolation filter, allowing it to process up to 1200 Msps. The output will be a 4-bit word for each modulator, corresponding to 4 consecutive time samples. The up-conversion to the carrier frequency is done by combining the ΔΣM outputs accordingly to a specific order as stated in equations (20) or (21), subject to the carrier frequency. The word $w$ is then serialized accordingly to the specified order. The serialization of the word is using an integrated high-speed transceiver capable of data rates up to 10 Gbps. The transceiver's clock frequency will be either $4f_c$ or $8f_c$, depending on whether the word at the input is $w_{low}$ or $w_{high}$. At the output of the serializer, the two-level signal will be centered in $f_c$, surrounded by quantization noise from the ΔΣM in the sidebands. To remove quantization noise, an output filter can be used at the target frequency band.

Similarly to the RF output signal, the proposed transceiver also uses a binary output to produce the reference waveform for the receiver. In this case, a baseband PWM is used to generate a triangular wave at the specified reference frequency, varying between 4 MHz and 30 MHz. The signal is filtered with a low-pass filter to recover the required waveform by removing the PWM quantization noise. The reference signal is compared with the input RF signal using the FPGA high-speed differential input buffers as a comparator. This comparator is followed by a register working at 10 GHz of sampling frequency, generating a digital bitstream that will be deserialized by a factor of N as indicated in Fig. 7.25. Therefore, the sampled PWM waveform arrives at the remaining FPGA logic at $f_s/N$, with N=64. Then, the RF data must be down-converted to baseband, filtered, and decimated. The down-conversion and filtering are performed using a polyphase DDS and a polyphase filter, in this case with N=64 parallel branches. After the DDC the signal is transferred to the local PC via an Ethernet connection, where the baseband data is demodulated and analyzed for its EVM.

### 7.4.3 Measurement Results

The transceiver performance was measured in terms of signal quality for both the transmitted and received signals. The figure of merit used to evaluate the signal quality is the EVM using two different signals 16-QAM and 64-QAM, with 2 MHz and 5 MHz of symbol rate, both generated using a root raised cosine filter with a roll-off of 0.22, which implies a channel bandwidth of 2.44 MHz and 6.1 MHz. The performance results for the isolated transmitter (Figs. 7.27 and 7.28b), the isolated receiver (Fig. 7.30), and the transceiver point-to-point configuration will now be presented.

#### 7.4.3.1 Transmitter

The transmitter performance was measured in terms of SNR, performing a sweep over the carrier frequencies between 400 MHz and 2.425 GHz with a step of 25 MHz. The measurement setup was assembled as depicted in Fig. 7.26. In Figs. 7.27a and 7.27b, the EVM results are shown for 2 MHz and 5 MHz symbol rate. The results show that for the tested modulations and symbol rates, the EVM along the measured carrier frequencies is always inferior to 2.5%.

The ACPR was measured for both the 2 MHz and 5 MHz channels along the tested carrier frequencies. The measured ACPR values are shown in Fig. 7.28. The maximum and minimum values correspond to the best and worst ACPR scenarios, considering the tested modulations of 16-QAM and 64-QAM. Both the lower and upper channels were measured

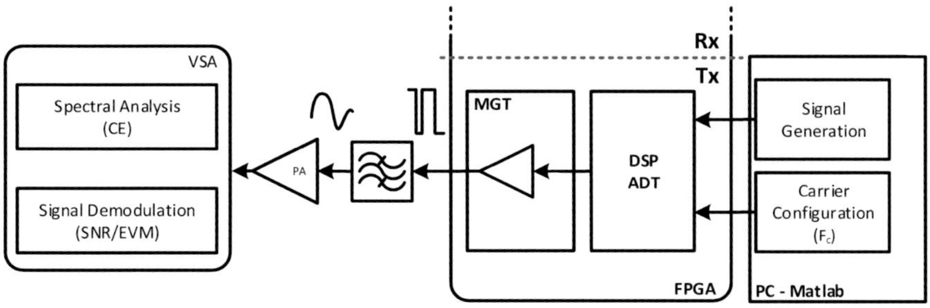


**FIGURE 7.26:** Block diagram of the transmitter measurement setup.

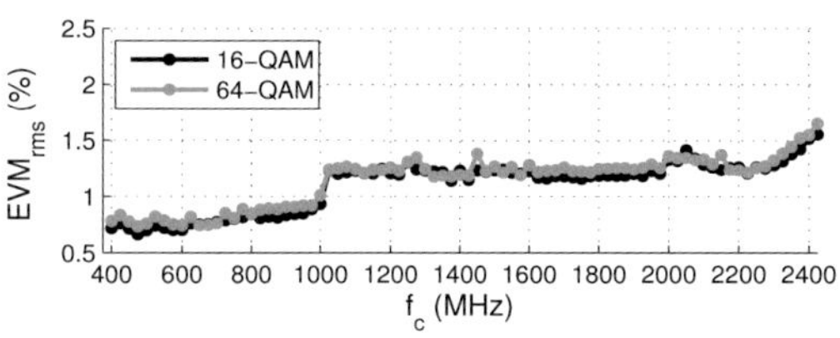


(a)

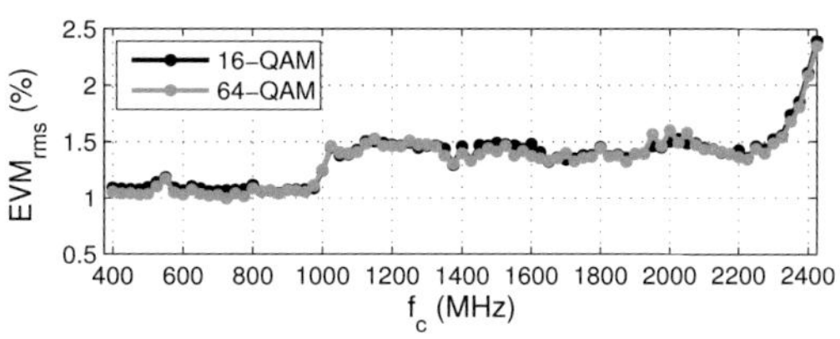


(b)

**FIGURE 7.27:** Transmitter EVM using 16-QAM and 64-QAM signals. (a) 2 MHz bandwidth signal. (b) 5 MHz bandwidth signal.

for every frequency, and the worst case was considered as the ACPR value for that carrier frequency.

The $C_{eff}$ for the transmitted signal of 16-QAM and 64-QAM was also measured and was found to vary between 15 % and 22 % depending on the PAPR of the signal. The EVM variation around 1 GHz occurs due to the change from $w_{low}$ to $w_{high}$, which presents the worst jitter behavior. The EVM change around 2.4 GHz, however, appears because the ΔΣM reaches its maximum clock operation and starts to present errors. Nonetheless, these results demonstrate the viability of this transmitter for use over the measured frequencies.

#### 7.4.3.2 Receiver

The receiver performance was measured by performing a sweep over the carrier frequency from 400 MHz up to 2400 MHz with a step of 25 MHz, considering the setup presented in the block diagram of Fig. 7.29. A two-channel AWG (AWG7000A from Tektronix) is directly connected to the receiver and is responsible for generating both the reference and RF signal. For each carrier frequency, the reference signal frequency was previously optimized, considering the algorithm described in the previous subsection.

The EVM sweep results for 2 MHz and 5 MHz symbol rate signals are depicted in the Figs. 7.30a and 7.30b. In both of these sweeps, it is possible to see the points that were measured using the minimum reference and optimized reference. The use of the optimized reference allows for the EVM to maintain a relatively constant value for the swept carrier frequencies, despite the PWM folded harmonics. Regarding the 2 MHz signal, it was possible to reach an EVM below 2.7% for the entire bandwidth.

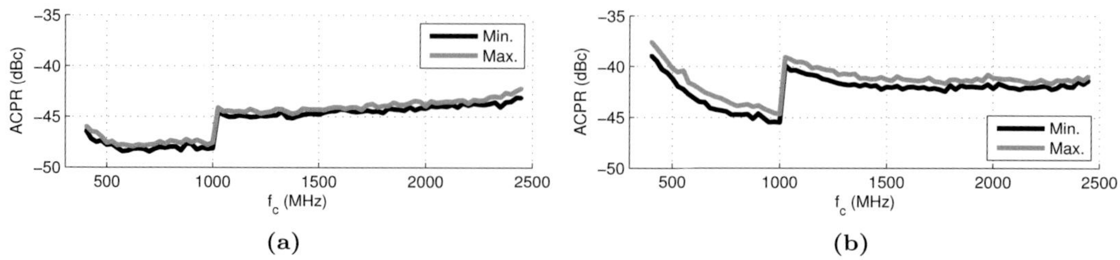


**FIGURE 7.28:** Transmitter ACPR measurement. (a) 2 MHz bandwidth channels. (b) 5 MHz bandwidth channels.

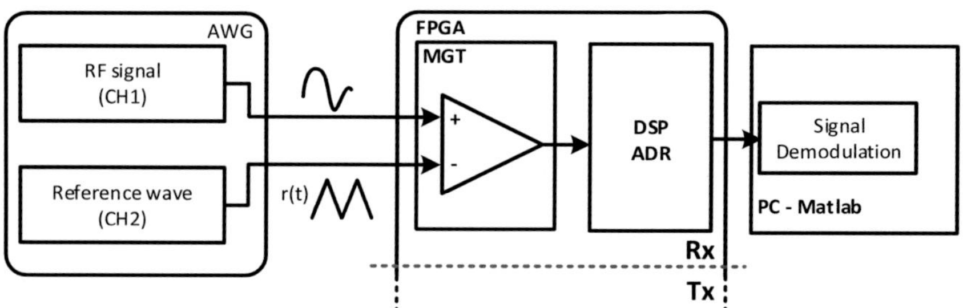


**FIGURE 7.29:** Block diagram of the receiver measurement setup.

#### 7.4.3.3 Full System Transceiver

The implemented all-digital transceiver was tested in a Tx/Rx point-to-point configuration for some frequency bands of interest and feasible with the proposed architecture. For this measurement, the transmitted signal is filtered by a band-pass filter and is sent directly to the receiver, which acquires the signal. Then, after the DDC, the signal quality is analyzed in terms of EVM. Simultaneously, the signal is also analyzed and demodulated in a VSA to obtain a reference EVM measurement for the transmission. The measurement setup is depicted in Fig. 7.31. Fig. 7.32 presents a photo of the measurement setup. Additionally, Fig. 7.33 shows the spectrum of the transmitted and the received signals at 1850 MHz. In the latter, it is possible to observe the PWM distortion around the carrier frequency.

The test signals are 64-QAM with 2 MHz and 5 MHz sampling rate, varying the carrier frequency over the selected bands with a step of 5 MHz. Fig. 7.34 shows a sweep over $f_c$ in the 900 MHz ISM band starting on 903 MHz and up to 923 MHz. Fig. 7.35 shows a sweep over $f_c$ in the 1800 MHz LTE band starting on 1827 MHz and up to 1881 MHz. The presented measurements demonstrate the feasibility of the proposed transceiver architecture for certain common frequency bands.

Finally, an overall comparison between the proposed architecture and other similar state-of-the-art transceivers is warranted. However, since this is the first time such an architecture is presented, this comparison can only be made with the transmitter and receiver independently, using their respective state-of-the-art architectures.

The proposed transmitter employs a polyphase architecture, as presented in [36], which aligns with state-of-the-art figures of merit in terms of bandwidth, SNR, and $C_{eff}$. However,

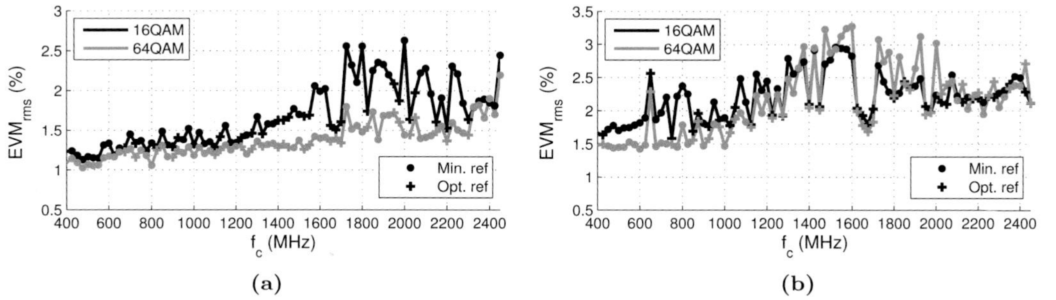


**FIGURE 7.30:** Receiver EVM measurement. (a) 2 MHz symbol rate signal. (b) 5 MHz symbol rate signal.

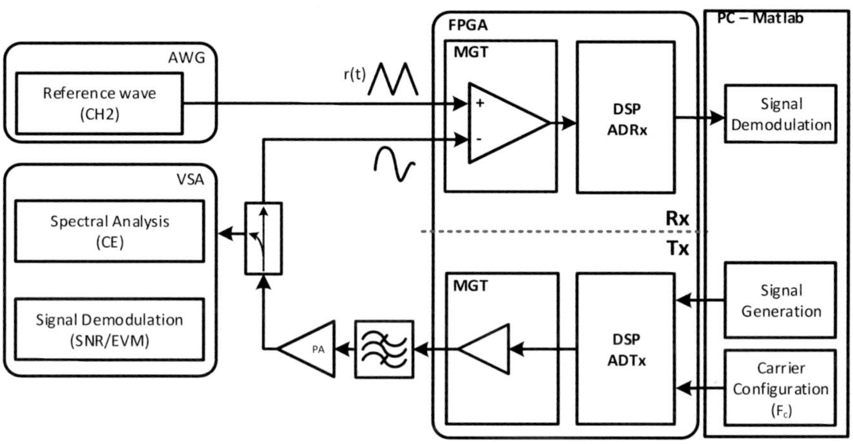


**FIGURE 7.31:** Block diagram of the transceiver loop-back measurement setup.

the transmitter architecture was designed to allow for fast carrier frequency flexibility; in fact, this is the first time such a large set of carrier frequencies has been tested, demonstrating the feasibility and functionality of this transmitter for these frequencies. On the receiver side, the same carrier frequencies were also tested, presenting reasonable results. When comparing this system with [123], focusing on the 5 MHz signals, there were improvements in terms of SNR/EVM. Despite the fact that in [123] the PWM sampling is two times superior to the 10 GHz used in this work, the proposed technique allowed to obtain an EVM lower than 3.5%, whereas in [123] the presented EVM is about 6%. Moreover, when comparing the current work with [161], an improvement was also obtained. Both the previous improvements are due to the new technique for selecting the reference frequency.

To consider a receiver evaluation in terms of sensitivity and dynamic range, the LNA/VGA should be mandatorily included, which is outside the scope of this paper. However, it is possible to get an estimation of the instantaneous dynamic range by transposing the EVM values into SNR by applying the formulation presented in [76]. Therefore, considering that the best and the worst obtained EVM values are respectively 1% and 3.5%, by

**FIGURE 7.32:** Measurement setup photo.

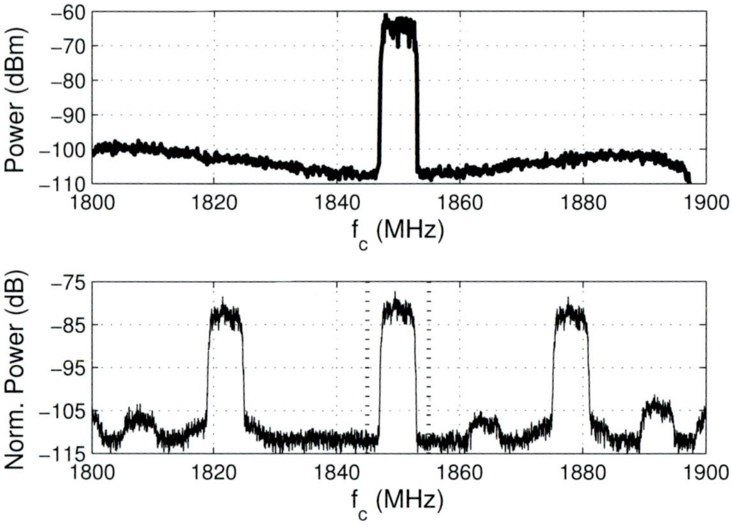


**FIGURE 7.33:** Spectral capture of the transmitted signal with a VSA (top plot) and the RF received signal at the all-digital receiver (bottom plot).

applying the previous formulation, it is possible to get an SNR of 40 dB and 30 dB. Hence, this leads to an indicative value of the receiver's instantaneous dynamic range.

### 7.4.4 Key Remarks

In this section, a fully integrated single FPGA-based all-digital RF transceiver was presented, which marks a breakthrough in digital RF system design. The entire signal processing from the baseband up to the RF stage for the transmitter, and the opposite for the receiver, is completely performed in the digital domain. The transmitter chain of the radio is implemented according to $\Delta\Sigma$M modulation, and the receiver is based on PWM modulation, which was optimized with a non-linear analysis of its behavior.

The proposed system presents high flexibility in a bandwidth of almost 2.5 GHz, while maintaining reasonable EVM results for 2 MHz and 5 MHz symbol rate signals. The obtained results make this system highly suitable for future radio transceivers, specifically

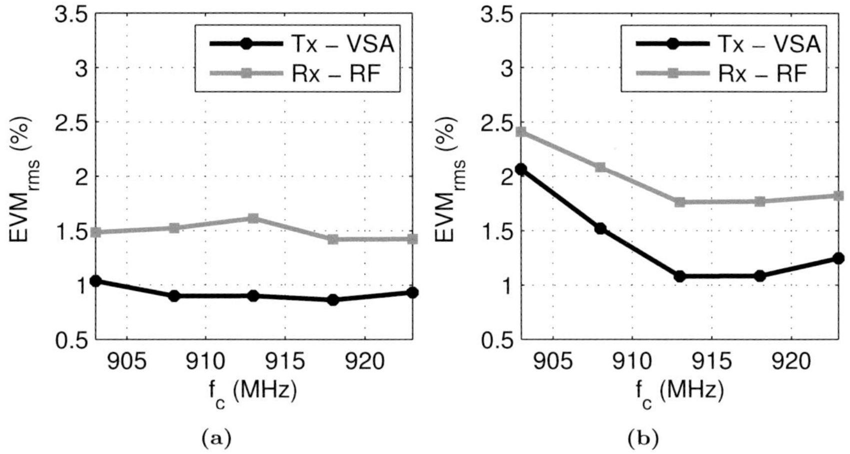


**FIGURE 7.34:** Measurement over the 900 MHz ISM band. (a) Using a 2 MHz 64-QAM signal. (b) Using a 5 MHz 64-QAM.

focusing on the integration of previous standards that operated in sub-2.5 GHz bands. An FPGA was used as a proof of concept for implementation due to its faster verification and deployment capabilities; nevertheless, an ASIC could be built following the proposed architecture. However, due to the FPGA's reconfigurability capabilities, it adds an increased degree of flexibility required by SDR applications.

## 7.5 Designing Highly-integrated Antenna Array Systems

Both scientific and industrial communities have been investing significant efforts in the development of novel antenna technologies, with antenna arrays at their core, and integration and massification being the underlying watchwords. However, it is widely known that, despite the vast opportunities that massive antenna arrays enable, several issues, such as complexity, cost, and energy efficiency, become particularly problematic. Thus, in this section, this research line is addressed and a promising architecture for the design of scalable and fully integrated RF front-ends for antenna arrays is proposed.

This section is supported by the conference paper [51] and by the journal [50]. The chapter presents a BB-stage ADT architecture that combines eight independent ADTs to perform the direct driving of an antenna array. By removing the need for external DACs and external upconversion stages, the reported solution presents a high level of scalability.

### 7.5.1 Contextualization

The typical design methodology of an antenna array involves the integration of $M$ dedicated RF front-ends to interface with $M$ radiating elements. However, this leads to heavy, bulky, and complex systems, which, ultimately, preclude the integration of a massive number of RF front-ends. The use of a single RF front-end with multiplexing techniques [Time-Division Multiplexing (TDM) and Code-Division Multiplexing (CDM)] has been proposed by some

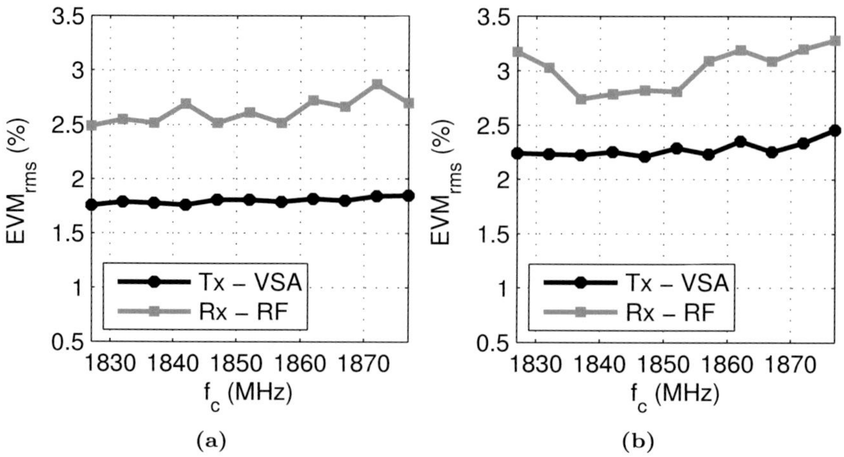


**FIGURE 7.35:** Measurement over the 1800 MHz LTE band. (a) Using a 2 MHz 64-QAM signal. (b) Using a 5 MHz 64-QAM.

authors [131,209]. However, these techniques exhibit limited available bandwidth, high noise levels, and increased system complexity [131]. Other alternatives propose the use of hybrid beamforming architectures for designing large-scale antenna arrays [79, 194]. The aim is to reduce the number of RF chains without decreasing the beamforming gains. However, even though some design requirements are relaxed, there is a higher level of system complexity in designing and operating the digital and analog beamformers properly.

#### 7.5.1.1 Proposed Architecture

The proposed idea falls under the exploration of reconfigurable all-digital transmitters integrated into a single FPGA to implement antenna array systems with a massive number of elements, known as the concept of All-Digital Antenna Array Transmitter (ADAAT). The high number of FPGA's I/O pins, jointly with FPGA-based reconfigurable all-digital transmitter techniques, can be explored, together with miniaturized antennas and bandpass filters, to implement scalable and compact massive antenna array transmitters. The target is to minimize the number of external components. Taking into account the relevant literature, one can verify that the FPGA's resource utilization factor for implementing an ADT is usually very low [189]. Therefore, as far as scalability is concerned, there are enough logic resources to include more stages in the digital datapath. In particular, the main objective is to minimize the inherent complexity of the beamforming network, while maximizing the digital physical subsystems built into a single FPGA. In addition to the appealing scenario of having the majority of the chains integrated into a single chip, the potential of having software-defined antenna array transmitters is very promising and may be a key to enabling a seamless migration path for the future large-scale antenna arrays.

The conceptual model of this ADAAT is depicted in Fig. 7.36 and intends to demonstrate the physical interaction between all the subsystems in the final system.

#### 7.5.1.2 Block Diagram

The general block diagram is depicted in Fig. 7.37, and one can see that it is based on the parallelization of multiple ADTs.

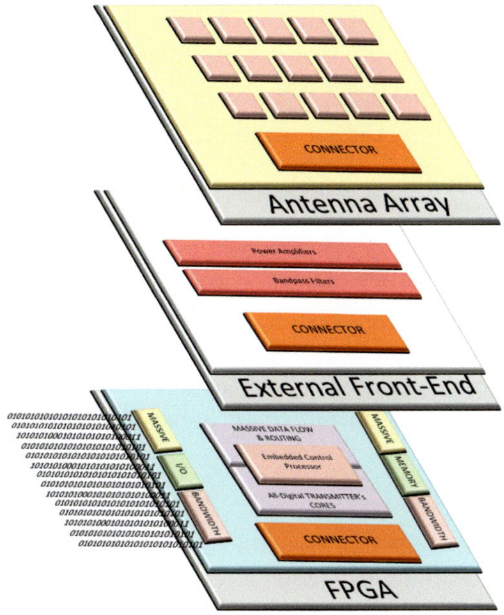


**FIGURE 7.36:** Conceptual model for the ADAAT.

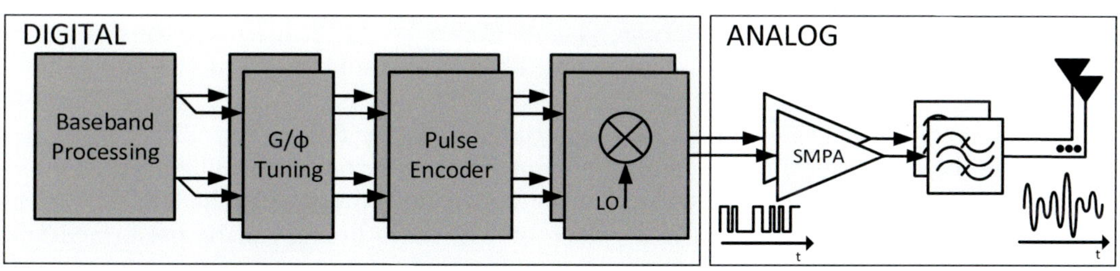


**FIGURE 7.37:** Conceptual block diagram of an ADAAT.

The system comprises $M$ independent ADTs, each one performing the direct driving of a single radiating element. In addition to the typical ADT subsystems, a new one was included: the *G/$\phi$ Tuning* module. This module operates at low sampling rates and is responsible for performing amplitude scaling as well as the BB phase shifting of each ADT's output. Its core (depicted in Fig. 7.38) is based on two digital gain blocks and a COordinate Rotation DIgital Computer (CORDIC) in rotation mode. The two modules have different functionalities. On the one hand, the digital gain blocks enable inter-lane amplitude calibration, as well as the real-time amplitude scaling of each radiant element. On the other hand, the CORDIC enables the inter-lane phase calibration, as well as the phase shifting of each radiant element, also in real-time. This real-time reconfigurability is essential to enable the inclusion of beamsteering algorithms. It is essential to note that for the transmission of wide bandwidths, the narrowband assumption is no longer valid, and the CORDIC must be replaced with a digital filter.

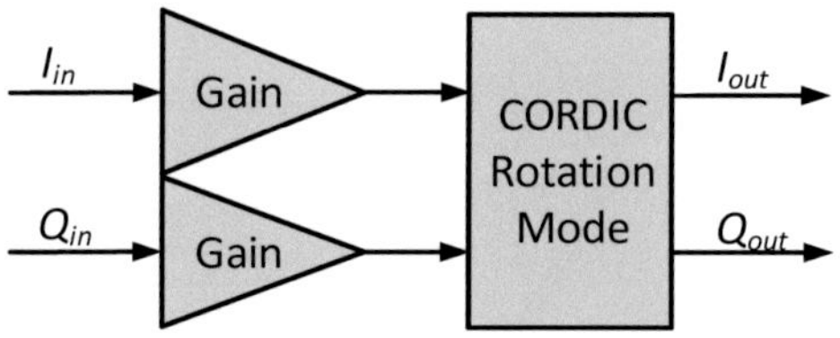


**FIGURE 7.38:** Block diagram of the digital $G/\phi$ *Tuning* module.

#### 7.5.1.3 Calibration Procedure

The existence of a calibration procedure for this type of system is mandatory due to the inherent timing mismatches between the different MGT lanes, which are worsened by the use of different cables, connectors, and other components with varying electrical lengths. Additionally, all the transceivers' output power must be equivalent to avoid unbalances.

**Phase Calibration**

For the sake of simplicity, one can restrain the chosen calibration procedure to just two elements before moving to the $M$ elements, and one can also consider the continuous-time domain. If a low-frequency complex exponential tone ($x(t) = e^{j(\omega_{IF}t)}$) feeds two ADTs, after being modulated by the LP-$\Delta\Sigma$Ms and upconverted to a given RF frequency, the two outputs will be $y_1 = Ae^{j(\omega_{RF}t+\phi)} + n_1(t)$ and $y_2 = Be^{j(\omega_{RF}t+\theta)} + n_2(t)$, where $\phi$ and $\theta$ are the phase deviation specific to each transmitter, $A$ and $B$ represent some gain/attenuation inherent to the DUC, and $n_1(t)$ and $n_2(t)$ are the quantization noise generated by the respective modulators. For this analysis, $A$ and $B$ can be ignored, as well as the out-of-band noise, given the noise-shaping property. Thus, the combination of the two outputs can be simply expressed as $y \simeq e^{j(\omega_{RF}t+\phi)} + e^{j(\omega_{RF}t+\theta)} = 2\cos\left((\phi-\theta)/2\right) e^{j(\phi+\theta)/2} e^{j\omega_{RF}t}$. Taking into account that the maximum magnitude occurs for $\theta = \phi$, an indirect procedure to estimate the phase deviation between two lanes could rely on performing a phase sweep in one of the BB incoming signals, while monitoring the output power in the bins of interest. Moreover, as the output waveform is *a priori* known, curve fitting can be performed with just a few iterations, thereby decreasing the calibration time.

An external $M$ to 1 power combiner can be used to accommodate the $M$ MGTs. Even though all the outputs are directly connected, just two can be turned on at the same time, and one of them must be used as a reference to ensure a relative calibration. Without loss of generality, by considering the first port as the reference, the result is: $e = \begin{bmatrix} 0 & p_{12} & \dots & p_{1M} \end{bmatrix}$, where each variable represents the estimated phases between each two lanes. After all the power combiner's phase unbalances are characterized in the Vector Network Analyzer $\left(c = \begin{bmatrix} p_{1s} & p_{2s} & \dots & p_{Ms} \end{bmatrix}\right)$, each CORDIC reference input must be updated according to: $p = (e - c)\,\%\,2\pi$. A flowchart summarizing the implemented phase calibration method is depicted in Fig. 7.39.

**Magnitude Calibration**

The most straightforward way to correct the amplitude mismatches, without external amplification circuitry at RF, is to adjust the BB gain of each input signal. To ensure an automated calibration procedure, all the $M$ MGT's outputs can be connected to a power combiner, and just one element can be turned on at each moment. After comparing the output power from each lane, a low-complex control loop can be implemented to attenuate all the BB input signals until the same output power is present in all the outputs. Depending on

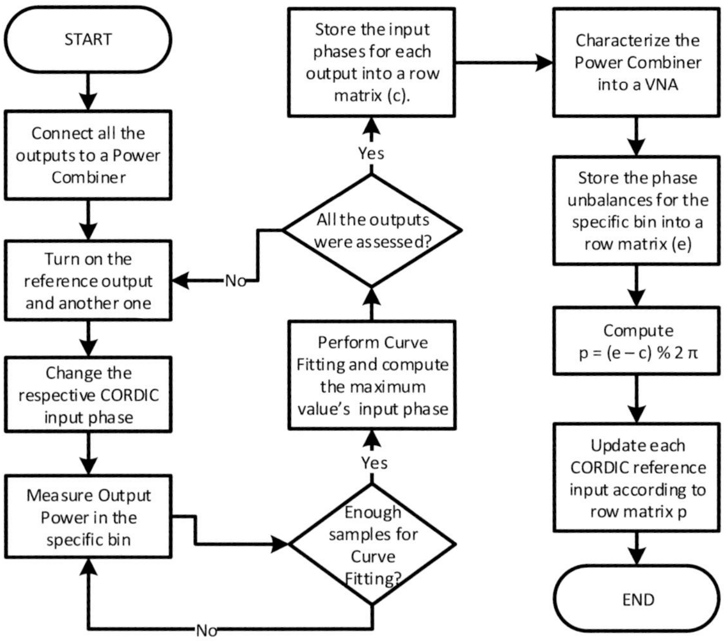


**FIGURE 7.39:** Flowchart of the implemented phase calibration method.

the case, the power combiner's insertion losses must also be assessed to verify whether they are negligible or if their correction needs to be performed in the digital domain. A flowchart summarizing the implemented magnitude calibration method is depicted in Fig. 7.40.

### 7.5.2 Prototype Validation

To validate the proposed idea, the proposed block diagram from Fig. 7.41 was fully implemented into a Virtex Ultrascale XCVU095 FPGA using the Xilinx VCU1287 Characterization Kit. The experimental setup is depicted in Fig. 7.42.

For the purpose of this proof-of-concept, eight independent ADTs are fully integrated into the same FPGA. Even though more elements could be included in the same FPGA, board interface limitations precluded the integration of more than eight elements.

After integrating the eight independent ADTs in the same FPGA, the BB test signals are stored in the FPGA's internal memory. Afterwards, the BB signals feed all the *G/$\phi$ Tuning* subsystems. After that, a Polyphase Interpolation FIR Filter increases the sampling rate by a factor of five, with the samples divided into five different branches. Therefore, the signals feed a polyphase equivalent 5-path LP-$\Delta\Sigma$M sampled at a clock frequency of 125 MHz, resulting in a global sampling rate of 625 Msps. The polyphase equivalent $\Delta\Sigma$M was designed according to the theory discussed in [205]. The choice of the five phases was made taking into account the trade-off between the number of phases and the overall critical path [36]. As a $\Delta\Sigma$M contains a feedback loop, polyphase techniques may significantly impact the critical path. Thus, the number of paths must be chosen as a trade-off between global clock frequency and LP-$\Delta\Sigma$M's complexity. The chosen LP-$\Delta\Sigma$M is based on an error-feedback structure previously depicted in Fig. 3.52.

As usual, the corresponding samples are then combined and replicated to generate a parallel output word with the four components of the RF signal ($x_I[n]$, $x_Q[n]$, and inverted versions) that is serialized by the FPGA's MGTs.

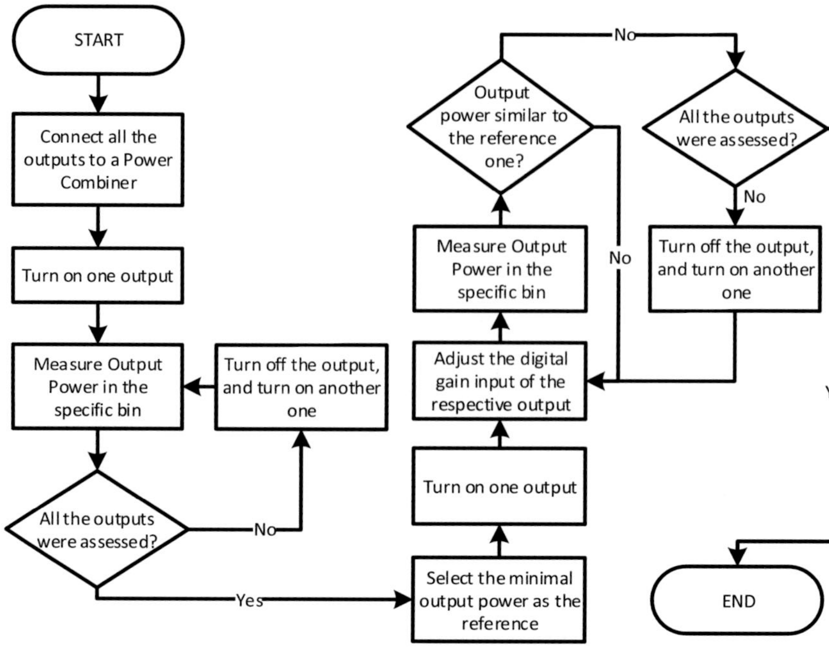


**FIGURE 7.40:** Flowchart of the implemented magnitude calibration method.

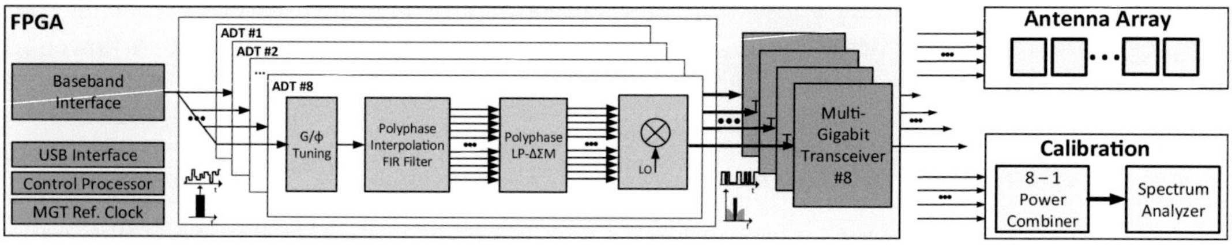


**FIGURE 7.41:** Block diagram of the implemented FPGA-based ADAAT.

The calibration procedure introduced in the previous section was implemented. It is worth noting that the proposed calibration procedure does not require any additional modules or subsystems beyond those already present in the design. In particular, the *G/ϕ Tuning* module is used for both the calibration procedure and for the beamsteering operation. The functionality of this module is controlled by the *Control Processor*. The Power Combiner was a ZN8PD1-53S+ from Mini-Circuits. It is essential to note that, depending on the system's constraints, incorporating an external power combiner may not be feasible. Thus, the same calibration procedure was successfully validated in far-field conditions within a controlled environment, using a spatial power combination to replace the external power combiner. Regardless of the scenario, phase and magnitude calibrations were performed in a straightforward manner, without the need for manual interaction.

All the measurements were performed in an anechoic chamber with a distance of four meters between the ADAAT (depicted in Fig. 7.41) and the horn measurement antenna (to ensure the far-field region). The ADAAT comprises an in-house developed 14.47 dBi linear 8-element microstrip patch antenna array matched for 2.5 GHz that was developed to demonstrate the capability to control the amplitude/phases of each transmitter in a

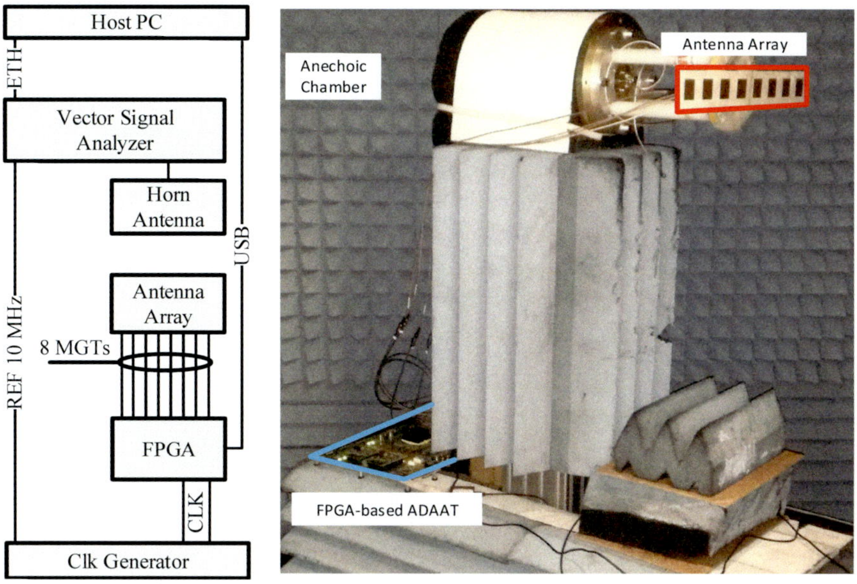


**FIGURE 7.42:** Experimental setup in the anechoic chamber: the ADAAT connected to one in-house developed linear 8-element microstrip patch antenna array.

real-life scenario. Connected to the measurement antenna is a VSA R&S FSW 8. In particular, two different experimental results were obtained:

- Radiation pattern diagrams based on peak gain measurements (Figs. 7.43 and 7.44);
- Measured EVM and MER values for different symbol rates in different steering angles (Table 7.6).

The calibration procedure is validated in Fig. 7.43, where the measured radiation pattern matched the simulated one.

To assess the beamsteering capability, the typical progressive phase method with constant amplitude was applied, and each computed value updated the respective CORDIC phase value. Normalized radiation patterns are depicted in Fig. 7.44 for azimuth angles in the range of -40° to 40° with emphasis on the 1° resolution. Due to the use of a 16-bit register for the CORDIC module, the authors believe that sub-degree resolution can be achieved. However, limitations in the measurement procedure precluded its experimental validation.

All experimental results present EVM values below 5 % for a distance of 4 meters (without external amplification circuitry), which resulted in a well-defined constellation (Table 7.6). As usual, for higher symbol rates, the available bandwidth of the patches (around 25 MHz, as assessed in the patch's $S_{11}$ parameters) together with the LP-$\Delta\Sigma$M's processing bandwidth starts to deteriorate the signal integrity.

Table 7.7 reports the ADAAT resources usage. The higher percentage of resources ($\simeq$ 73% of DSP slices) is being used in the non-optimized polyphase interpolation FIR filters. However, this percentage can be lowered by resorting to multiplier-less FIR and CIC filters. In most cases, a simple average FIR filter can be used. This is one of the most resource-efficient topologies because no arithmetic operations are required. Thus, from a system designer's point of view, novel filtering methodologies should be employed to ensure the integration of more RF front-ends. Taking into account the used FPGA chip, the resources

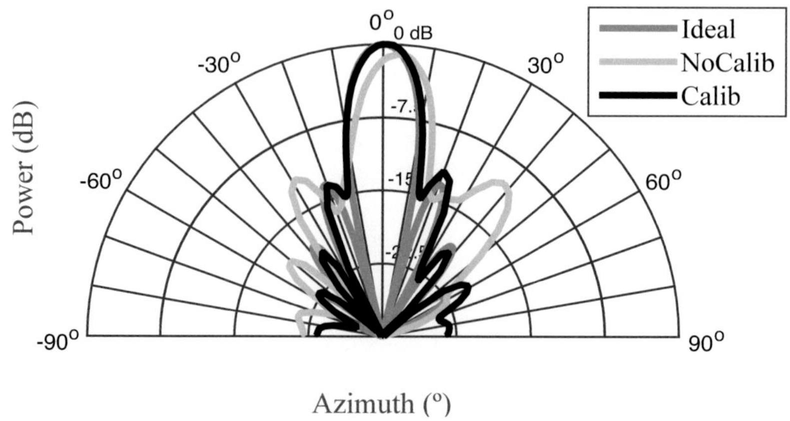


**FIGURE 7.43:** Impact of the calibration procedure on the radiation pattern.

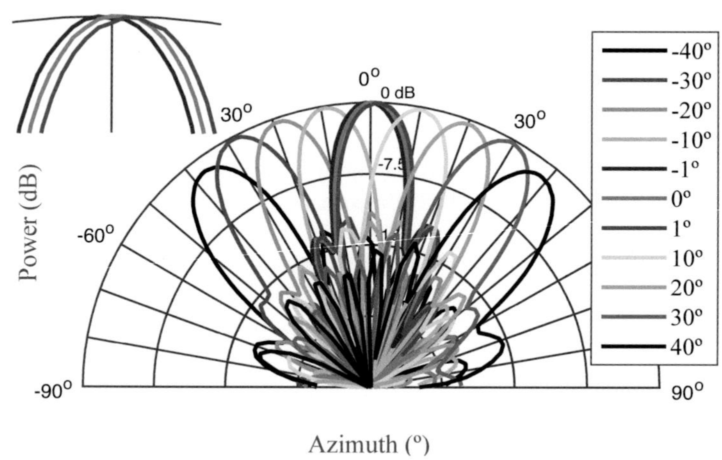


**FIGURE 7.44:** Normalized radiation patterns for different steering angles. The 1-degree resolution is shown in the left corner.

usage from the proposed implementation, and an interpolation based on a ZOH (0% of DSP occupation), one can estimate that fifty-four parallel ADTs can be easily integrated into the same chip.

### 7.5.3 Key Remarks

In this section, a novel architecture for an FPGA-based All-Digital Antenna Array RF Transmitter was proposed and validated. In particular, eight complete digital RF transmitters were optimized and integrated into the same FPGA.

The proposed concept presents a lower complexity than the state-of-the-art counterparts. By enabling the design of antenna array transmitters without external DACs and external upconversion stages, the overall system's layout can be highly simplified, and consequently, more radiating elements can be integrated.

**TABLE 7.6:** Measured performance with a 16-QAM in far-field (4 m). Symbol rate is in Msps, EVM in % and MER in dB.

| | Steering Angles (°) | | | | | | | |
|---|---|---|---|---|---|---|---|---|
| | -40 | | -20 | | 20 | | 40 | |
| SRate | EVM | MER | EVM | MER | EVM | MER | EVM | MER |
| 3.125 | 1.93 | 34.4 | 1.63 | 35.7 | 1.67 | 35.6 | 1.92 | 34.4 |
| 7.8125 | 2.56 | 31.9 | 2.40 | 32.4 | 2.54 | 32.5 | 2.57 | 31.9 |
| 12.5 | 3.10 | 30.1 | 3.02 | 30.4 | 3.00 | 30.5 | 3.13 | 30.1 |
| 15.625 | 3.44 | 29.3 | 3.42 | 29.6 | 3.42 | 29.6 | 3.43 | 29.3 |
| 25 | 4.77 | 26.4 | 4.84 | 26.2 | 4.78 | 26.4 | 4.87 | 26.2 |

**TABLE 7.7:** ADAAT occupied resources.

| Logic Resources | Used | Total | Percentage (%) |
|---|---|---|---|
| Flip Flops | 30401 | 1075200 | 2.83 |
| LUTs | 23701 | 537600 | 4.41 |
| Memory | 256 | 1728 | 14.81 |
| DSP | 560 | 768 | 72.92 |
| GTHs | 8 | 28 | 28.57 |
| GTYs | 0 | 26 | 0 |

This novel concept has been successfully implemented and validated on an FPGA-based transmitter. Radiation patterns and the transmission of modulated signals for various steering angles demonstrate the validity of the proposed architecture.

## 7.6 ΔΣ-based All-Digital Beamforming Transmitters

Massive MIMO systems are anticipated to play a crucial role in the evolution of next-generation communication systems. These types of systems are essential for improving overall network efficiency and capacity. However, achieving the full potential of massive MIMO comes with significant challenges, including complex calibration, optimal beamforming, and managing the high resource demands associated with such systems. Addressing these issues is vital for the advancement of next-generation communication technologies. By overcoming these hurdles, we can unlock the numerous advantages that massive MIMO provides, such as enhanced data rates, improved signal quality, and greater user capacity, ultimately contributing to the development of more reliable and efficient wireless networks. Moreover, continued research and innovation in algorithms, hardware design, and signal processing techniques will be instrumental in realizing the transformative impact of massive MIMO in future communication systems.

This section is supported by the journal paper [8]. The section presents an innovative architecture that combines modular ΔΣ-based ADTs with digital TTD techniques for the calibration of antenna array channels, offering scalable and digital beamforming control. This integration facilitates low-latency management of individual channel delays, enabling dynamic adjustments in beam direction and enhancing transmission integrity throughout the antenna array. By leveraging these advanced digital techniques, the proposed system illustrates how modular ADTs, when paired with digital TTD, can deliver a flexible and high-performance solution tailored to the requirements of next-generation wireless communications.

### 7.6.1 Contextualization

The rapid evolution of wireless communication systems and especially the anticipated demands of the next generation of mobile communications have intensified the need for scalable, efficient, and flexible transmitter architectures that support massive MIMO applications [4]. These systems require high-throughput data handling, enhanced spectral and energy efficiency, and sophisticated beamforming capabilities, all pushing conventional analog and hybrid RF systems to their limits. ADTs offer a promising solution for antenna array systems. By converting multi-bit data streams directly into single-bit digital signals, ΔΣ-based ADTs simplifies hardware, potentially reduces power consumption, and fully leverages SoC-based digital processing capabilities. This approach creates an entirely digital RF path from baseband to the antenna [35, 116, 146, 154].

In [51], an ADAAT is used to generate beamforming signals by combining multiple ADT-cores that process baseband signals and perform the phase shift and gain required for each of the channels before the ΔΣ-encoding, thus being able to generate and control the beamforming signal. Also, [41], demonstrated that by combining various ADT-cores, it was possible to achieve 1.25 GHz bandwidth signals using an antenna array. The author in [164] uses multi-core ΔΣ modulation to enable wider fractional bandwidths, achieving up to 50% in some designs. This approach relies on polyphase filtering and a multi-core ΔΣ setup to handle high data rates and broad bandwidths effectively, as demonstrated by prototypes capable of 2.5 GHz center carrier with 1.25 GHz bandwidth.

While these works explore different approaches to ADAATs, they all depend on multiple ADTs cores, which demand significant resources. This dependence restricts their scalability for large-scale antenna array implementations and limits both the performance and flexibility of these systems. In [7], a modular tile-based ΔΣ-based ADT architecture was developed and proposed, introducing a new level of flexibility and scalability. It proposed an architecture capable of dynamically adjusting the available user channels in the RU, and also combining individual channels to achieve better performance, reduced quantization noise, and higher carrier frequencies and bandwidths.

Calibration and phase management of antenna array systems remain challenging, particularly in maintaining signal integrity and effectively leveraging massive MIMO for precise beamforming. In [9], a highly flexible and scalable implementation with very low resource consumption, a digital TTD approach was proposed for the calibration of various MGT. Furthermore, the TTD module offers a delay control of one MGT sample period, enabling a finer phase control of each channel for higher sampling speeds on the MGT.

This section introduces an innovative architecture that combines modular ΔΣ-based ADTs with digital TTD techniques, enabling scalable digital beamforming control for modern and next-generation mobile communication antenna array systems. This integration enables low-latency control over individual channel delays, allowing for dynamic beam direction adjustments and improved transmission integrity across the antenna array. By combining these advanced digital techniques, the proposed system demonstrates how modular ADTs, paired with digital TTD, can provide a flexible, high-performance solution for the demands of next-generation wireless communications.

### 7.6.2 Architecture Description

The proposed architecture leverages a modular ΔΣ-based ADT design [7], integrated with digital TTD techniques [9] for generation and control of beamforming in antenna array

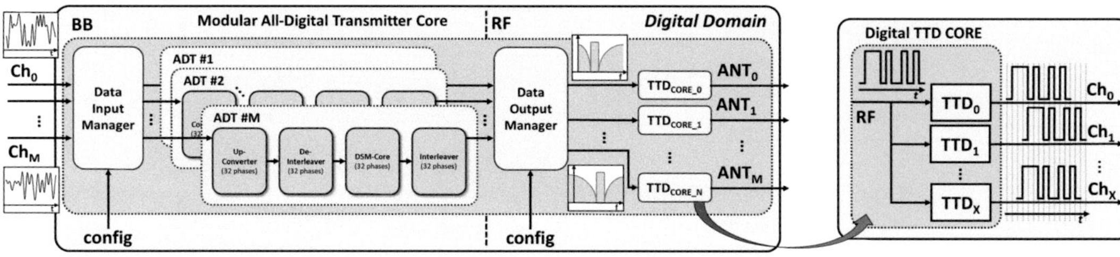


**FIGURE 7.45:** Proposed architecture of a scalable $\Delta\Sigma$-based Modular ADT for antenna arrays.

systems. This innovative system offers significant flexibility and scalability, allowing the dynamic adjustment of available resources in the RU to meet network demands. By integrating multiple ADT cores, the architecture supports both independent multi-channel operation and combined high-performance signal generation, achieving higher carrier frequencies and bandwidths. Each channel has a digital TTD core block, enabling beamforming generation and control for each antenna group. This approach not only simplifies hardware complexity but also potentially reduces power consumption, making it a robust solution for next-generation wireless communication systems. Fig. 7.45 shows a diagram of the proposed architecture.

The modular ADT core is composed of $M$ individual ADTs, each capable of independently encoding different $M$ baseband signals into RF signals at different user-defined carrier frequencies. Each ADT consists of four main cores, Up-Converter, De-Interleaver, $\Delta\Sigma$M-Core and Interleaver, each with 32 internal parallel processing stages to achieve higher performance. The up-converter performs the up-conversion from a baseband IQ signal to the desired output carrier frequency using Direct Digital Synthesis, resulting in an up-converted signal in real format. This block can be adapted in real-time, allowing a greater degree of freedom in the system's real-time operation. The De-Interleaver is responsible for sorting the output samples of the previous core into the correct input phases of the $\Delta\Sigma$M-Core, to align them sequentially in each phase. The $\Delta\Sigma$M-Core processes a sequence of 32 consecutive time real samples in each of its 32 input phases and performs the $\Delta\Sigma$-encoding. The Interleaver then performs the reverse operation of the De-Interleaver, but on a bit level, sorting each of the 32-time consecutive $\Delta\Sigma$-encoded output samples (bits) from each of the previous core's output phases into sequential order by phase, thus being time-sorted and organized for direct sampling by the MGT core.

In addition, depending on the configuration provided to the individual ADT cores and data input/output manager blocks, it is possible to combine several individual ADTs to generate $Y$ number of independent signals, where $Y$ is less than $M$, for higher carrier frequencies with better SNR performance. This feature allows the system to seamlessly adapt to a wide range of operational scenarios, supporting higher carrier frequencies with improved SNR when required, and ensures efficient resource utilization and superior signal quality.

Each output signal, $ANT_M$, represents a different ANTenna with up to $X$ number of elements. These signals are generated in the $TTD_{CORE_M}$ blocks. The internal architecture of these blocks can be seen in Fig. 7.45, where it can be seen that the $\Delta\Sigma$ waveform generated as mentioned above, is connected to up to $X$ digital TTD blocks, each of which is connected to a single MGT channel, $Ch_X$, thus allowing the connection to an antenna array of $X$ elements. These digital TTD blocks perform the digital phase control by combining $N$ flip-flops for coarse delay control and a barrel shifter for finer delay control, as shown in Fig. 7.46. Each TTD block receives $N$ bits of $\Delta\Sigma$-encoded signal, corresponding to the

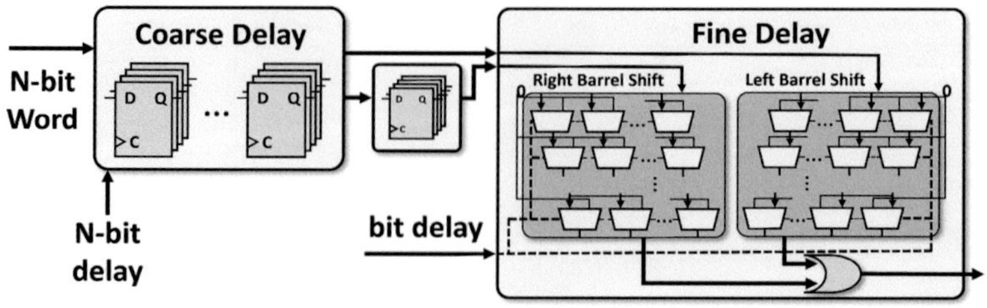


**FIGURE 7.46:** Internal digital TTD block design.

number of internal phases configured on the previous Modular ADT. Depending on the desired delay to be set on each $Ch_X$, the TTD block uses the internal sequence of $N$ flip-flops if the delay is greater than or equal to $N$, and the barrel shifter for finer control from 0 up to $N-1$ bits of delay. This provides the ability to generate and control beamforming signals in any connected antenna array.

This architecture can encode multiple user channels and generate and control beamforming signals for single or multiple antenna arrays within a single SoC. The resolution of the beamforming control is based on the sampling frequency of the MGT, $f_s^{MGT}$ and to better understand this, we must first consider Eq. 7.24 from [14]. This equation determines the required delay for each channel to achieve the desired $\theta$ angle in degrees for the beamforming signal.

$$delay = \frac{d \times sin(\theta)}{c} \tag{7.24}$$

Note that $d$ is the distance between the elements of the antenna array and $c$ is the speed of light. By resolving the equation in terms of the angle $\theta$, we obtain Eq. 7.25.

$$\theta = \arcsin\left(\frac{delay \times c}{d}\right) \tag{7.25}$$

Here we see that by controlling the *delay* we can control the angle of the beamsteering, since $c$ is a universal constant and $d$ is constant for the antenna array in question. The smallest delay that we can impose with the proposed architecture is equal to a single bit delay through the TTD blocks, which is equal to $1/f_s^{MGT}$. Thus, the higher the sampling frequency of the MGT, the smaller the time delay, and thus the finer the resolution of the beam steering control. To better understand how phase control is implemented in each antenna array element, let's consider the following examples.

Example 1: Consider a 1-D antenna array with a phase delay ratio equal to $d$. Each of the digital TTD blocks will receive a signal indicating the delay to be implemented, with a ratio between them equal to $d$. In other words, let us consider an $X$ number of channels, as previously discussed, connected to $TTD_0$, $TTD_1$ ... $TTD_X$. The first channel $TTD_0$ would get a delay of 0, $TTD_1$ a delay of $d$, up to $TTD_X$ with a delay of $d \times (X-1)$. This would generate a radiated signal with a beamforming angle determined by Eq. 7.25.

Example 2: Consider a 2-D antenna array. Everything remains the same, both equations are valid, and each digital TTD is still connected to an antenna array element. However, the array now has two dimensions, X and Y, and the delays are applied in both dimensions. The only additional step is to add the X and Y delays to be implemented in each digital TTD using the same methodology presented in Example 1.

### 7.6.3 Prototype Validation

To implement and validate the proposed architecture, the Xilinx VCU129 Evaluation Kit was used, which provides several MGT interfaces used for serial connections with speeds up to 28 Gbps in a non-PAM4 configuration.

The antenna array used in this work consists of 8 elements in a horizontal arrangement (Fig. 7.47) and is designed for a carrier frequency of 2.45 GHz. Using Eq. 7.24, we obtain a delay of $d = 3.49\ ps$ for a $\theta = 1°$. However, since our control is done in a fixed way, in multiples of one sample period of the MGT, using Eq. 7.25 we get an angle of $\theta = 14.48°$ for a delay of $d = 50\ ps$. Although the presented results are quite rough, the resolution in the delay was done considering a $f_s^{MGT} = 20\ GHz$. Considering a higher sampling rate, 56 GHz, as available in the Xilinx kit VPK180, in non-PAM4 configuration, we could get a $\theta = 5.12°$, demonstrating the capabilities of the proposed architecture with the current technology, and also leading to better results in the future with each technology leap.

The project used for the implementation and proof-of-concept was developed using a modular ADT with four individual channels capable of generating four individual data channels or one combined high-performance channel. To validate these results, only a single user was considered, so the results presented below are obtained using the combination of the four individual channels into one high-performance channel.

The digital TTD architecture can delay the modular ADT output signal in several channels, in this case, 8, since the antenna array consists of 8 elements. Each channel can be delayed by 256 periods of $1/f_s^{MGT}$. This provides a high degree of flexibility for the initial calibration of the system to align all channels within the $1/f_s^{MGT}$ period, as discussed down below. In addition to these main cores, extra logic was needed to control the system and the MGT cores used for the ADT.

Table 7.8 shows the utilization report for the implementation of the final prototype project, including the individual ADTs used within the Modular ADT. These ADTs are designed with $\Delta\Sigma$-cores to process 32 samples and 32 output phases. These cores can generate signals with bandwidths up to 100 MHz and carrier frequencies up to 2.5 GHz. Working together in a modular configuration, the ADTs can achieve bandwidths of up to 800 MHz and carrier frequencies of up to 10 GHz through additional logic in the Modular ADT that aligns the samples temporally between each core [7]. Combined with the required additional logic and the digital TTD control, we can see that compared to the previous approaches, with a single ADT for each channel, we only need a single high-performance ADT for the eight-element antenna array, making this approach more flexible and scalable. Also, the resource consumption is low as shown in Table 7.8, with a low percentage of occupation of the FPGA. If we take a closer look at the resources used, we can see that the TTD Control hierarchy requires a very small amount of resources, and it is worth remembering that it is composed of 8 digital TTD-Cores capable of controlling the digital phase delay of 8 individual antenna elements. By performing some calculations, we can determine that with a slight increase of about 10 % in LUTs and 1 % in FFs, we could control the digital phase of up to 80 antenna elements on a single chip, using the 32 GTY and 48 GTM available in the VCU129 FPGA kit.

Considering the power consumption of the proposed architecture, to obtain a more realistic power report using the power measurement tools provided by AMD in Vivado, a simulation of the system was performed with real signal stimulus and the switching of the different parts of the architecture was captured. In this way, the power values obtained for the three main hierarchies of the proposed architecture are: Modular ADT with 3.181 mW; GTM core with 3.914 mW; TTD Control with 70 mW. However, the GTM core comprises four duals that allow two transmitter MGTs each. If we look at the consumption of the MGT, we have 489,25 mW. The same is true for the TTD core, which implements the

**TABLE 7.8:** Implementation utilization report.

| | LUTs | FFs | BRAMs | DSPs | GTM Dual |
|---|---|---|---|---|---|
| **Individual ADT** | 18.689 | 8.702 | 32 | 353 | 0 |
| **Modular ADT** | 139.050 | 43.657 | 128 | 1.412 | 0 |
| **TTD Control** | 11.157 | 2.048 | 0 | 0 | 0 |
| **GTM Core** | 4.752 | 16.303 | 32 | 70 | 4 |
| **System Control Logic** | 5.469 | 17.791 | 2.5 | 0 | 0 |
| **Total** | **160.428** | **79.799** | **162.5** | **1.482** | **4** |
| **Available** | 1.728.000 | 3.456.000 | 2.688 | 12.288 | 24 |
| **Percentage Used (%)** | **9.28** | **2.31** | **6.05** | **12.06** | **16.67** |

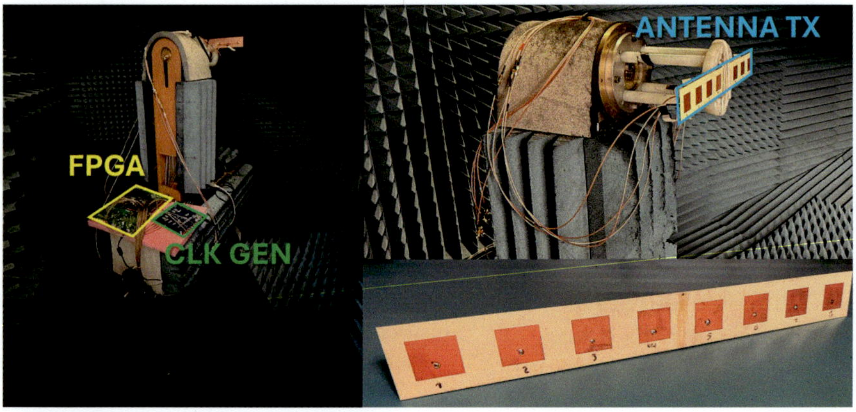


**FIGURE 7.47:** Laboratorial Setup.

digital TTD in eight channels. If we look at the channel, we have 8,75 mW. These results show the low power consumption of the proposed architecture and its potential scalability for antenna systems with many elements, using only a single system on chip.

Figs. 7.47 and 7.48 show the laboratory setup, composed of an external reference clock generation board for the MGTs, the FPGA running the proof-of-concept implementation and the antenna array used. The measurements were performed in an anechoic chamber.

**Prototype Calibration**

The first step required before performing the validation measurements is to calibrate the system so that all the channels corresponding to each of the elements of the antenna array are aligned within one $1/f_s^{MGT}$ period, where $f_s^{MGT} = 20$ GHz. A square wave is used as a reference signal for the digital TTD blocks present in the proof-of-concept implementation. Each time the system is powered up, the channels have different states concerning the phase of the rising edge of the reference square wave. Figure 7.49 shows the state of the different channels observed after powering up the system.

To align all the channels, a script was run using an oscilloscope to read the rising edges of all the channels, then detect the most delayed channel (**ch1** in Fig. 7.49) and calculate the necessary delays for the remaining channels to align them within the $1/f_s^{MGT}$ period.

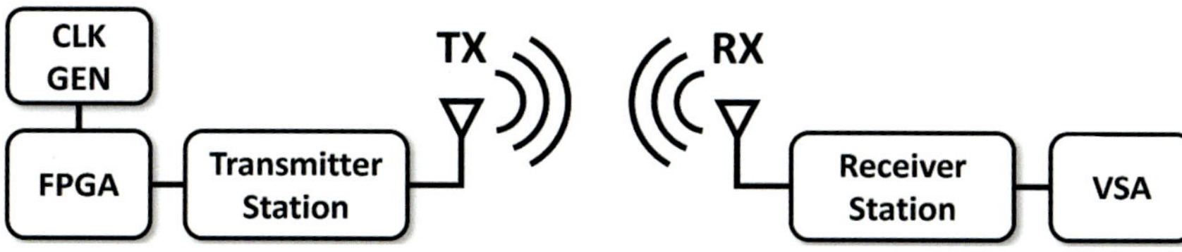


**FIGURE 7.48:** Block Diagram of Laboratorial Setup.

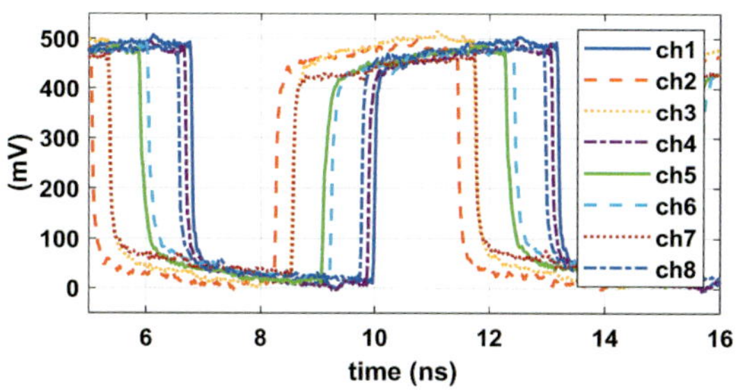


**FIGURE 7.49:** Individual channels pre-sync alignment.

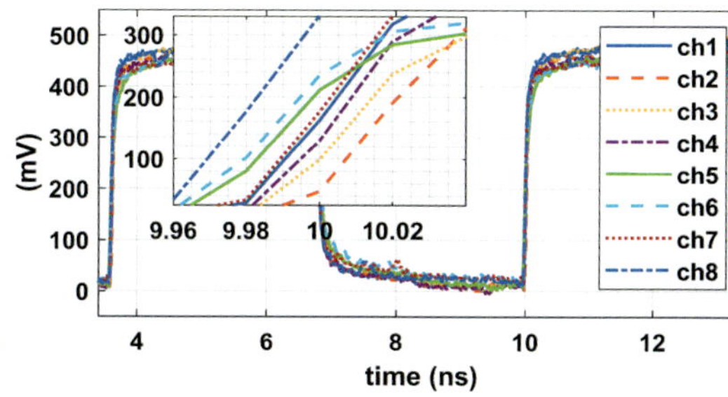


**FIGURE 7.50:** Individual channels post-sync alignment.

The complete result of the initial calibration process of the system can be seen in Fig. 7.50, where the alignment was achieved within about 40 ps, thus reaching our "aligned" state for all channels.

### Prototype Validation

With the system calibrated, all channels were connected to each individual element of the antenna array and the modular ADT was activated with a 16-QAM signal with 100 MHz bandwidth at a carrier frequency of 2.44 GHz. Fig. 7.51 shows the normalized radiation pattern for the system configuration with all channels in the "aligned" state, i.e., no additional delays beyond the initial calibration.

The normalized radiation pattern in Fig. 7.51 shows a peak at an angle of $-2°$. This was expected since the alignment between all the channels is not perfect, as seen in Fig. 7.50. This misalignment affects the stability of the beam steering, but can be improved by increasing the sampling frequency of the MGT, resulting in a smaller period of alignment, thus improving the steering and precision of the beam. Fig. 7.52 shows a sweep with all available beam steering options for the antenna array used with an MGT sampling frequency of 20 GHz.

Each positional beam steering angle is obtained by delaying each element by multiples of $1/f_s^{MGT}$. By delaying each channel by one $1/f_s^{MGT}$ period incrementally from channel 1 to channel 8, we get the 15° angle, by delaying by $2 \times 1/f_s^{MGT}$ periods, we get the 30° angle, and so on, and if we start delaying from channel 8 to channel 1, we get the other negative angles. We can see some discrepancies between the expected steering angles and those obtained in Fig. 7.52. This is again to be expected, as mentioned before, due to the misalignment under $1/f_s^{MGT}$ period in the "aligned" state, but overall the angles obtained are close to the expected ones within a margin of 3°, and in some cases precisely right, like at angles 30° and $-30°$.

Table 7.9 shows the measured SNR and EVM for the steering angles previously shown in Fig. 7.52 for a 100 MHz bandwidth signal at 2.44 GHz. Although some of the steering angles deviated from the expected ones by as much as 3° in some cases, the measured EVM is less than 0.72 % for all measured steering angles, and the SNR is greater than 41 dB.

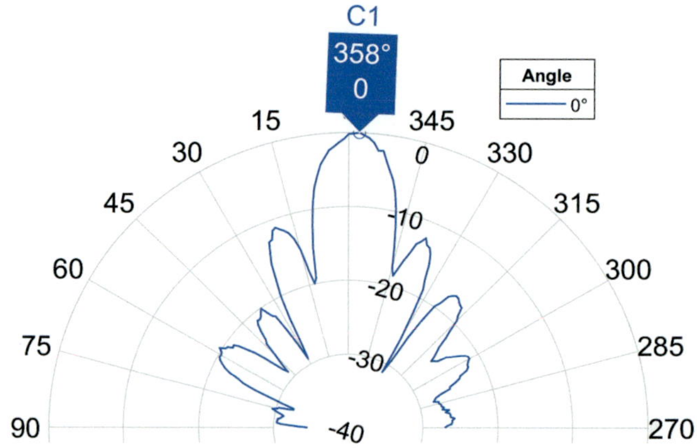


**FIGURE 7.51:** Normalize radiation pattern with no individual channels delay for a 100 MHz bandwidth signal at 2.44 GHz.

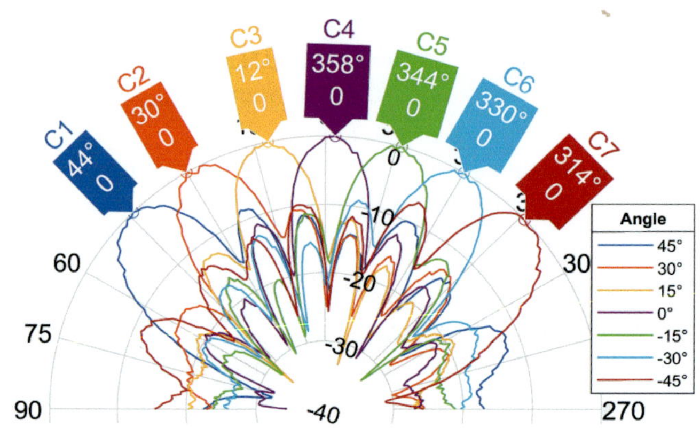


**FIGURE 7.52:** Normalized radiation patterns for different delays between the various individual channels for a 100 MHz bandwidth signal at 2.44 GHz.

The results demonstrate the proposed approach's viability for antenna array calibration and beam steering as the technology in MGT evolves and improves.

### 7.6.4 Key Remarks

This section explored the implementation and validation of a novel architecture for generating and controlling beamforming signals for wireless communication antenna arrays. This architecture integrates modular ADT with digital TTD techniques, providing a scalable and efficient approach to enhance beamforming capabilities for massive MIMO applications.

The calibration process, which involves the alignment of multiple channels within an antenna array, was successfully achieved with an accuracy of approximately 40 ps, ensuring synchronization within the sampling period of the MGT. This streamlined calibration demonstrates the system's ability to simplify complex processes while maintaining high accuracy. In addition, the potential to further refine this alignment by increasing MGT sampling rates highlights the scalability of the solution to support evolving performance requirements.

**TABLE 7.9:** Measured Performance for a 100 MHz bandwidth 16-QAM signal at 2.44 GHz.

| **Steering Angles** | $-45°$ | $-30°$ | $-15°$ | $0°$ | $15°$ | $30°$ | $45°$ |
|---|---|---|---|---|---|---|---|
| **SNR (dB)** | 41.31 | 41.70 | 42.16 | 43.29 | 42.44 | 41.76 | 40.98 |
| **EVM (%)** | 0.68 | 0.68 | 0.58 | 0.55 | 0.57 | 0.68 | 0.72 |

For beam control and steering, the architecture demonstrated effective beamforming over a range of angles ($-45°$, $-30°$, $-15°$, $0°$, $15°$, $30°$, and $45°$) with an error of $3°$ due to the calibration and digital TTD limitations as mentioned earlier. Despite minor misalignments, the system maintained robust performance, achieving consistently high SNRs above 41 dB and EVMs below 0.72%.

The results demonstrate the viability of the proposed approach for antenna array calibration and beam steering and show its potential to meet the high-performance requirements of future wireless communication technologies in specific use cases. Overall, the experimental validation highlights the effectiveness of the architecture in providing a scalable and less complex solution for next-generation wireless communications, with promising implications for improved performance as the technology evolves.

## 7.7 Conclusion

This chapter provided an overview of the various applications of ADT. From RoF solutions, passing through optical-upconversion, all-digital transmitters, and also ADAAT, it was shown that ADTs can not only be a viable solution to replace traditional transmitters, but they can also be implemented in several other scenarios, empowering them with flexibility and/or simplicity that could not be achieved through conventional transmitters.

# Chapter 8

# Conclusions

Throughout this book, we explored the evolution, architecture, and applications of All-Digital Transmitters (ADTs) and related receiver concepts as building blocks for next-generation wireless and optical communication systems. By digitizing the transmission chain from baseband to the antenna, all-digital RF front-ends address the growing limitations of conventional mixed-signal architectures in advanced semiconductor technologies, where analog circuits scale poorly relative to digital logic.

All-digital RF transmitters move the digital-to-analog conversion boundary closer to the antenna and implement modulation, upconversion, and envelope shaping in the digital domain. This enables a software-defined signal chain with reduced analog dependence and allows direct control of RF behavior through digital signal processing.

The work in this book consolidated architectural principles, modulation strategies, implementation constraints, and system-level impacts across several transmitter topologies, ranging from single-band stand-alone transmitters to multi-band and multi-element arrays, including intermediate optical transport and phased-array signal distribution.

Experimental results obtained from commercial FPGAs demonstrated the feasibility of one-bit ADTs using multi-gigabit transceivers as oversampled DACs or waveform sources, and showed that per-element amplitude and phase control can support real-time beam steering and calibration.

These implementations confirm that ADT concepts can be realized with available digital hardware and known resource, power, and clocking limitations and form a credible pathway toward ASIC realizations with higher clock frequencies, lower jitter, and improved energy efficiency.

Looking ahead, semiconductor technology scaling is expected to reinforce this trend. Advanced process nodes increasingly rely on FinFET and gate-all-around devices, which primarily improve digital logic density and speed. Analog circuits, in contrast, face reduced supply voltages and higher parasitic effects that degrade linearity, gain, and frequency response. This divergence underscores the practical advantages of fully digital transmitter and receiver architectures and motivates the work presented in this book.

## 8.1 Summary of the Contributions

The following subsections summarize the main contributions presented and discussed in this book.

### 8.1.1 Conceptual Unification and Theoretical-Practical Framework

A unified theoretical-practical foundation for ADTs was established, integrating oversampled converters, pulse-width and delta-sigma modulation, and switched-mode amplification within a single framework. The book developed an architectural view that covers:

DOI: 10.1201/9781003225713-8

- oversampled waveform generation using $\Delta\Sigma$ modulation, PWM, and hybrid quantization methods;
- digital upconversion and envelope shaping without analog mixing stages;
- direct switched-mode RF drive, optionally followed by simple analog filtering;
- applicability to single-element and multi-element transmitter configurations.

This synthesis clarified how quantization, digital upconversion, and envelope shaping can replace analog mixers and multi-bit DACs. It also consolidated prior work that treated $\Delta\Sigma$ modulators, PWM, and switched-mode RF techniques in separate contexts, highlighting their common characteristics and trade-offs. The resulting framework relates oversampling ratio, quantizer order, transition density, and spectral shaping to implementation cost, power, and performance.

### 8.1.2 Architectural Innovations

The book introduced and evaluated several practical transmitter configurations, including:

- hybrid $\Delta\Sigma$/PWM encoders, intended to reduce out-of-band noise with constrained oversampling and relaxed analog filtering requirements;
- LUT-based waveform synthesis, aimed at reducing logic usage and consequently enabling higher modulation oversampling ratio;
- time-interleaved and multicore $\Delta\Sigma$ modulators, used to extend output bandwidth at moderate clock rates, achieving multi-GHz bandwidth with adequate EVM;
- cascaded and state-propagated designs, intended to increase effective noise shaping without exponential resource growth and to relax the oversampling ratio versus bandwidth trade-off;
- pre-compensated multi-band modulation chains, used for controlled spectral placement of multiple carriers and for concurrent operation of independent bands.

Experimental validation confirmed that these architectures can be implemented on mid- and high-end FPGA platforms with measurable error vector magnitude, output spectra, and modulation bandwidths relevant to RF communication.

### 8.1.3 System-Level Methodologies and Figures of Merit

A consistent methodological framework was defined with quantitative *Figures of Merits (FoMs)*, namely Coding Efficiency, Spectral Utilization, and Power Efficiency, for evaluating ADT performance. The analysis showed how digital design can be jointly optimized across temporal and spectral domains, aligning hardware constraints with communication-theoretic limits. Applying these metrics demonstrated the existence of operating points that depend on target bandwidth, modulation format, frequency plan, available clock rate, and resource budget rather than a single universally optimal transmitter configuration.

Measurements demonstrated that FPGA-based prototypes can achieve EVM below 1%, SNR above 40 dB, and high spectral efficiency under controlled bandwidth conditions and moderate oversampling, with performance commensurate with conventional architectures for comparable conditions. These results span RF and optical front-ends, showing the feasibility of $\Delta\Sigma$-over-fiber systems. The same architectural concepts were also applied to phased arrays using per-element digital phase and amplitude control. These results confirm that ADT methods are not restricted to theoretical studies but can operate reliably in digital hardware with deterministic timing.

Collectively, these contributions treat the transmitter as a programmable digital system with explicit control over spectral and temporal characteristics, shifting the design emphasis from an analog-centric to a digital-centric RF implementation.

## 8.2 Challenges and Open Issues

Although the demonstrated systems validate feasibility, several technical issues must be addressed before large-scale deployment or ASIC-level commercial integration. Reported all-digital transceiver architectures have so far balanced trade-offs among carrier frequency agility, signal bandwidth, number of simultaneous bands, and number of array elements. Limited digital resources, clock rates, and I/O throughput force designs to optimize one parameter at the expense of others, often leading to zero-sum trade-offs. Overcoming these limitations is essential for advancing the technology readiness level of all-digital RF systems.

### 8.2.1 Bandwidth, Clock Distribution, and Power

High oversampling ratios are required for acceptable spectral shaping. Clock rates in commercial FPGAs and serializers restrict achievable carrier frequency and instantaneous bandwidth. Dynamic switching power increases with clock rate and transition density, which limits long-term scalability without specialized logic cells, serializers, and power-aware encoding. Maintaining deterministic timing across multiple MGTs also becomes more difficult as carrier frequencies increase.

### 8.2.2 Analog Interface and Filtering

Despite the objective of minimal analog circuitry, some analog filtering remains necessary to remove quantization noise and switching energy. The required filter complexity depends on the spectral mask, modulation order, and oversampling ratio, and interacts with the achievable out-of-band attenuation of the digital encoder. Efficient co-design between digital encoding and analog filtering is still needed to minimize total power and area while meeting regulatory masks.

### 8.2.3 Linearity, Quantization, and Spectral Purity

Maintaining spectral integrity in switched-mode operation remains challenging. Quantization noise folding, duty-cycle distortion, and interleaving mismatches introduce deterministic spurs. Adaptive digital predistortion is possible but not yet fully optimized for one-bit $\Delta\Sigma$ or PWM streams, especially in wideband multi-band scenarios. Practical schemes must also account for implementation impairments in the serializers and packages.

### 8.2.4 Synchronization and Determinism

Coherent phased-array operation requires stable relative timing and deterministic phase relationships among elements. This requirement is more stringent than absolute clock accuracy. As the number of elements or FPGAs increases, synchronization and calibration must remain tightly controlled even across physically separated nodes. This depends on hardware clock distribution networks, time-frequency synchronization protocols, and digital calibration algorithms that compensate for dynamic channel, temperature, and supply variations.

### 8.2.5 Scalability and Interconnect Bottlenecks

Massive MIMO and distributed antenna systems demand scalable interconnects with deterministic timing and high throughput. Conventional board-level interconnect approaches do not scale well with element count and bandwidth. Next-generation serializers, chiplet-based high-speed fabrics, and short-reach optical links are needed to avoid bandwidth bottlenecks between digital logic, converters, and antenna elements.

### 8.2.6 ADC DAC Reciprocity

Extending digitalization to receivers requires matched $\Delta\Sigma$ or other oversampled ADCs, ensuring reciprocity in all-digital transceivers. Achieving comparable noise shaping at high bandwidths is still difficult due to power, clock jitter, and analog front-end limitations. FPGA-based demonstrations using MGTs as oversampled samplers are affected by equalization, buffering, and protocol logic that are absent in custom designs, which can distort the received waveform.

### 8.2.7 Reliability and Security

Thermal cycling, voltage stress, and high toggle rates introduce reliability concerns for long-term operation. Reconfigurable platforms also introduce firmware-related vulnerabilities. Secure bitstream handling, tamper detection, and robust remote update mechanisms must be integrated into all-digital front-ends, especially in distributed deployments.

---

## 8.3 Future Research Directions

Based on the challenges and open issues discussed above, several future research directions can be identified.

### 8.3.1 Architectures for Joint Bandwidth, Agility, and Scalability

Future work should focus on architectures that jointly address bandwidth, carrier frequency agility, number of simultaneous bands, and array size without linear growth in power and area. This includes joint optimization of modulation order, oversampling ratio, resource mapping, and interconnect topology to achieve wideband and multi-band operation within realistic device constraints.

### 8.3.2 Leveraging Emerging FPGA and ASIC Features

Recent FPGA and ASIC developments provide new opportunities. Modern devices integrate AI engines and vector processors that can replace part of the conventional DSP fabric for parallel and adaptive signal processing. High-speed serial transceivers with multi-level signaling, such as PAM-4, increase effective I/O throughput between logic and front-end circuits. These features enable more complex polyphase filtering, true-time-delay beamforming, and multi-band operation with reduced resource usage and can support digital calibration and adaptation in real time.

### 8.3.3 ASIC Implementations

FPGA-based realizations remain essential for rapid prototyping and architectural validation, but their reliance on generic multi-protocol MGTs introduce non-idealities such as equalization, buffering, and protocol-specific logic that are not required in a dedicated ADT. Future work should therefore consider ASIC implementations of all-digital transmitters. Custom digital front-ends optimized for direct RF sampling and reconstruction could eliminate unnecessary overheads and provide a more accurate assessment of attainable figures of merit in all-digital transmitter systems.

### 8.3.4 Joint Digital Analog Filter Co-Design

Systematic methods are needed to balance the complexity of digital noise shaping and the capability of analog filtering. This includes analytical and optimization tools that treat the digital encoder, serializer, package, and analog filter as a single chain, with objectives such as minimizing total power, satisfying constrained group delay, and meeting spectral masks. Such methods would guide the choice of $\Delta\Sigma$ order, oversampling ratio, and filter topology.

### 8.3.5 AI-Assisted Encoder and Decoder Design

The exploration of AI-based end-to-end encoder models for all-digital transmitters and receivers is a promising research direction. Data-driven models could learn efficient mappings between digital baseband data and quantized RF waveforms subject to implementation constraints. If properly constrained, such models may replicate or extend existing ADT functionality using fewer logic resources and improved adaptability across operating conditions, while still remaining interpretable and verifiable.

### 8.3.6 Centralized and Distributed Digital Architectures

The trade-off between centralized and distributed digital architectures remains open. Centralized implementations simplify synchronization and calibration but scale poorly to large arrays and geographically separated nodes. Distributed approaches are essential for massive MIMO and radio access networks but require robust time and phase alignment across devices. Future work should explore clock distribution schemes, synchronization protocols, and calibration algorithms that maintain coherence while allowing distributed processing.

### 8.3.7 Digital Photonic Co-Design

Coupling all-digital RF front-ends with photonic integrated circuits allows direct digital drive of optical modulators and supports $\Delta\Sigma$-over-fiber links. Further work is needed on joint electronic-photonic design, including modulation formats, noise-shaping strategies, and packaging solutions for short- and medium-reach fronthaul and backhaul applications.

### 8.3.8 Cognitive and Sustainable Networking

Integrating ADT architectures with cognitive radio and network-level resource management frameworks is another relevant direction. All-digital front-ends with fast reconfiguration can support dynamic spectrum allocation and power-aware operation. Research is required on control algorithms and network protocols that exploit this capability while keeping energy consumption and hardware complexity within practical limits.

### 8.3.9 Education and Open Innovation

Finally, the development of open-source ADT frameworks, reference designs, and educational material can help disseminate the concepts and accelerate adoption. Publicly available architectures, measurement scripts, and design flows would also support reproducibility and enable fair comparisons across different approaches.

---

## 8.4 Final Remarks

All-digital RF transmitters represent a significant change in radio design practice. By exploiting digital logic, clocking, and oversampled modulation, they reduce the dependence on precision analog components and align RF front-end design with the scaling properties of advanced CMOS technologies. With higher computational density available in modern process nodes and reduced reliance on scaling-sensitive analog circuitry, all-digital transceivers can more effectively benefit from technology scaling to achieve higher performance and integration levels.

From a system viewpoint, this shift has several implications:

- **Scalability:** Functionality can be replicated and extended through code and reconfigurable hardware instead of fixed analog circuits.
- **Flexibility:** Reconfiguration through software or hardware updates replaces repeated analog redesign.
- **Determinism:** Performance becomes more strongly tied to algorithms and clock constraints than to analog component tolerances.

At the same time, increased digital content brings new requirements for secure configuration, reliable long-term operation, and energy-aware design. As all-digital transceiver architectures continue to evolve, progress will depend on integrated advances in circuit design, signal processing, packaging, synchronization, and system architecture.

The work presented in this book provides a foundation for these developments by demonstrating practical ADT implementations, identifying the key architectural trade-offs, and outlining realistic directions for future research and engineering.

# *Bibliography*


[1] 3GPP. Technical Specification Group Radio Access Network; NR; Base Station (BS) radio transmission and reception. Technical Specification (TS) 38.104, 3rd Generation Partnership Project (3GPP), 06 2018. Version 15.2.0.

[2] 3rd Generation Partnership Project (3GPP). NR; base station (BS) radio transmission and reception. Technical Specification (TS) 38.104, 3rd Generation Partnership Project (3GPP), jan 2018.

[3] Victor Adrian, Yin Sun, and Joseph S. Chang. Design of a 5 GS/s Fully-digital Digital-to-analog Converter. In *Proc. of IEEE Int. Symp. on Circuits and Syst.*, pages 1548–1551, 2014.

[4] Hüseyin Arslan, Ahmet Yazar, Seda Doğan-Tusha. 6g vision: An ultra-flexible perspective. *ITU Journal on Future and Evolving Technologies*, 1(9):1–20, 2020.

[5] Ian F. Akyildiz, Ahan Kak, and Shuai Nie. 6g and beyond: The future of wireless communications systems. *IEEE Access*, 8:133995–134030, 2020.

[6] Peter Alfke et al. *The Programmable Gate Array Data Book.* Xilinx, 1991.

[7] Luís Filipe Almeida, Samuel S. Pereira, Arnaldo S. R. Oliveira, and N. B. Carvalho. Multi-channel modular all-digital transmitter. In *2025 IEEE Radio and Wireless Symposium (RWS)*, pages 119–122, 2025.

[8] Luís Filipe Almeida, Samuel S. Pereira, Arnaldo S. R. Oliveira, and N. B. Carvalho. Scalable modular all-digital transmitter antenna array for next-generation mobile networks. *IEEE Wireless Communications Letters*, pages 1–1, 2025.

[9] Luís Filipe Almeida, Samuel Santos Pereira, Ivo Rodrigues, Ricardo Correia, Miguel Drummond, Arnaldo S. R. Oliveira, and Nuno Borges Carvalho. True time delay-based alignment for all-digital beamforming transmitters. In *2023 IEEE Radio and Wireless Symposium (RWS)*, pages 122–124, 2023.

[10] Kyu Hwan An, Ockgoo Lee, Hyungwook Kim, Dong Ho Lee, Jeonghu Han, Ki Seok Yang, Younsuk Kim, Jae Joon Chang, Wangmyong Woo, Chang-Ho Lee, Haksun Kim, and Joy Laskar. Power-combining transformer techniques for fully-integrated cmos power amplifiers. *IEEE Journal of Solid-State Circuits*, 43(5):1064–1075, 2008.

[11] Jeffrey G. Andrews, Stefano Buzzi, Wan Choi, Stephen V. Hanly, Angel Lozano, Anthony C. K. Soong, and Jianzhong Charlie Zhang. What will 5g be? *IEEE Journal on Selected Areas in Communications*, 32(6):1065–1082, 2014.

[12] Ahmed Ashry and Hassan Aboushady. A 4th Order 3.6 GS/s RF/spl Sigma//spl Delta/ADC With a FoM of 1 pJ/bit. *IEEE Transactions on Circuits and Systems I: Regular Papers*, 60(10):2606–2617, Oct. 2013.

[13] 5G Infrastructure Association et al. The 5G infrastructure public private partnership: The next generation of communication networks and services, 2015.

[14] Constantine A. Balanis. *Antenna Theory: Analysis and Design.* Wiley, fourth edition edition, 2 2016.

[15] S. Balasubramanian, S. Boumaiza, H. Sarbishaei, T. Quach, P. Orlando, J. Volakis, G. Creech, J. Wilson, and W. Khalil. Ultimate Transmission. *IEEE Microwave Magazine*, 13(1):64–82, Jan. 2012.

[16] S. Balasubramanian and W. Khalil. Direct Digital-to-RF Digital-to-Analogue Converter Using Image Replica and Nonlinearity Cancelling Architecture. *Electronics Letters*, 46(14):1030–1032, 2010.

[17] D. B. Barkin, A. C. Y. Lin, D. K. Su, and B. A. Wooley. A cmos oversampling bandpass cascaded d/a converter with digital fir and current-mode semi-digital filtering. In *2003 Symposium on VLSI Circuits. Digest of Technical Papers (IEEE Cat. No.03CH37408)*, pages 79–82, 2003.

[18] S. Bassam, W. Chen, M. Helaoui, and F. M. Ghannouchi. Transmitter Architecture for CA: Carrier Aggregation in LTE-advanced Systems. *IEEE Microwave Magazine*, 14(5):78–86, 2013.

[19] V. Bassoo, K. Tom, A.K. Mustafa, E. Cijvat, H. Sjoland, and M. Faulkner. Potential Architecture for Future Generation 'Green' Wireless Base Station. In *International Symposium on Wireless Pervasive Computing Digest*, pages 1–5, Feb 2009.

[20] Ivars Bilinskis. *Digital Alias-free Signal Processing.* J.W. & Sons, 2007.

[21] Nuno Borges Carvalho and Dominique Schreurs. *Microwave and Wireless Measurement Techniques.* Cambridge University Press, 2013.

[22] J. Chen, L. Rong, F. Jonsson, and L. R. Zheng. All-digital Transmitter based on ADPLL and Phase Synchronized Delta Sigma Modulator. In *IEEE RFIC Symp. Dig.*, pages 1–4, Baltimore, MD, USA, Jun. 2011.

[23] Jian Chen, Liang Rong, Fredrik Jonsson, Geng Yang, and Li-Rong Zheng. The design of All-digital Polar Transmitter based on ADPLL and Phase Synchronized $\Delta\Sigma$ Modulator. *IEEE Journal of Solid-State Circuits*, 47(5):1154–1164, 2012.

[24] Jian Chen, Liang Rong, Fredrik Jonsson, and Li-Rong Zheng. All-digital transmitter based on adpll and phase synchronized delta sigma modulator. In *2011 IEEE Radio Frequency Integrated Circuits Symposium*, pages 1–4, 2011.

[25] Shuli Chi, Christian Vogel, and Peter Singerl. The Frequency Spectrum of Polar Modulated PWM Signals and the Image Problem. In *Proceedings of the IEEE International Conference on Electronics, Circuits, and Systems*, pages 679–682, Athens, Greece, Dec 2010.

[26] Kyung Ho Cho and Henry Samueli. A frequency-agile single-chip QAM modulator with beamforming diversity. *IEEE Journal of Solid-State Circuits*, 36(3):398–407, 2001.

[27] Wu Chou, Ping Wah Wong, and Robert M. Gray. Multistage Sigma-delta Modulation. *IEEE Transactions on Information Theory*, 35(4):784–796, 1989.

[28] Pei-Jung Chung. Green Radio - The Case for More Efficient Cellular Base Stations. Powerpoint presentation, 2004.

[29] S. Chung, R. Ma, S. Shinjo, K. Yamanaka, and K. H. Teo. A Concurrent Triple-Band Digital Transmitter Using Feedforward Noise Cancellation for Delta-Sigma Modulation. In *Proceedings of the European Microwave Conference*, pages 1–4, Nuremberg, Germany, Oct. 2017.

[30] SungWon Chung, Rui Ma, Shintaro Shinjo, Hideyuki Nakamizo, Kieran Parsons, and Koon Hoo Teo. Concurrent Multiband Digital Outphasing Transmitter Architecture Using Multidimensional Power Coding. *IEEE Transactions on Microwave Theory and Techniques*, 63(2):598–613, Feb. 2015.

[31] SungWon Chung, Rui Ma, Shintaro Shinjo, and Koon H. Teo. Inter-band Carrier Aggregation Digital Transmitter Architecture with Concurrent Multi-band Delta-sigma modulation using Out-of-band Noise Cancellation. In *IEEE MTT-S International Microwave Symposium Digest*, pages 1–4, Phoenix, AR, USA, May 2015. IEEE.

[32] R. F. Cordeiro, André Prata, Arnaldo S. R. Oliveira, José M. N. Vieira, and N. B. De Carvalho. Agile All-Digital RF Transceiver Implemented in FPGA. *IEEE Transactions on Microwave Theory and Techniques*, 2017.

[33] Rui F. Cordeiro, Arnaldo S. R. Oliveira, and José M. N. Vieira. All-digital Transmitter with RoF Remote Radio Head. In *Proceedings of the IEEE/MTT-S International Microwave Symposium*, pages 1–4, Jun. 2014.

[34] Rui F. Cordeiro, Arnaldo S. R. Oliveira, and José M. N. Vieira. Relaxing All-digital Transmitter Filtering Requirements Through Improved PWM Waveforms. In *Proceedings of the IEEE/MTT-S International Microwave Symposium*, pages 1–4, Phoenix, AZ, USA, May. 2015.

[35] Rui F. Cordeiro, Arnaldo S. R. Oliveira, and José M. N. Vieira. All-Digital Transmitter With a Mixed-Domain Combination Filter. *IEEE Transactions on Circuits and Systems II: Express Briefs*, 63(1):4–8, 2016.

[36] Rui F. Cordeiro, Arnaldo S. R. Oliveira, José M. N. Vieira, and Nelson V. Silva. Gigasample Time-Interleaved Delta-Sigma Modulator for FPGA-Based All-Digital Transmitters. In *Proceedings of the Euromicro Conference on Digital System Design*, pages 222–227, Aug. 2014.

[37] Rui F. Cordeiro, Arnaldo S. R. Oliveira, José M. N. Vieira, and Tomás O. Silva. Wideband All-digital Transmitter Based on Multicore DSM. In *Proceedings of the IEEE MTT-S International Microwave Symposium*, pages 1–3, San Francisco, CA, USA, May. 2016.

[38] Rui Fiel Cordeiro. *Novel Architectures for Flexible and Wideband All-digital Transmitters*. PhD thesis, Universidade de Aveiro (Portugal), 2018.

[39] Lawrence F. Cygan, Curtis M. Williams, and Andrew M. Khan. Method and Apparatus for Direct Digital to Radio Frequency Conversion. US Patent 7,825,724, November 2 2010. US Patent 7,825,724.

[40] Özlem Tugfe Demir and Emil Björnson. Channel estimation under hardware impairments: Bayesian methods versus deep learning. In *2019 16th International Symposium on Wireless Communication Systems (ISWCS)*, pages 193–197. IEEE, 2019.

[41] D. C. Dinis, R. Ma, K. H. Teo, P. Orlik, A. S. R. Oliveira, and J. Vieira. An fpga-based multi-level all-digital transmitter with 1.25 ghz of bandwidth. In *2018 IEEE/MTT-S International Microwave Symposium - IMS*, pages 659–662, 6 2018.

[42] Daniel C. Dinis, R. Cordeiro, Arnaldo S. R. Oliveira, and José M. N. Vieira. Towards an All-Digital Antenna Array Transmitter. In *IEEE Field Programmable Logic Digest*, pages 1–2, Lausanne, CH, Aug. 2016.

[43] Daniel C. Dinis, Rui Cordeiro, Filipe M. Barradas, Arnaldo S. R. Oliveira, and José M. N. Vieira. Agile Single- and Dual-band All-Digital Transmitter based on a Precompensated Tunable Delta-Sigma Modulator. *IEEE Transactions on Microwave Theory and Techniques*, 64(12):4720–4730, Dec. 2016.

[44] Daniel C. Dinis, Rui Cordeiro, Arnaldo S. R. Oliveira, and José M. N. Vieira. Tunable Delta-Sigma Modulator for Agile All-Digital Transmitters. In *IEEE MTT-S International Microwave Symposium Digest*, pages 1–4, San Francisco, CA, USA, May. 2016.

[45] Daniel C. Dinis, Rui Cordeiro, Arnaldo S. R. Oliveira, José M. N. Vieira, and T. O. Silva. A fully parallel architecture for designing frequency-agile and real-time reconfigurable fpga-based rf digital transmitters. *IEEE Transactions on Microwave Theory and Techniques*, 66(3):1489–1499, Mar. 2018.

[46] Daniel C. Dinis, Rui Cordeiro, Arnaldo S. R. Oliveira, José M. N. Vieira, and Tomás O. Silva. Improving the Performance of All-Digital Transmitter based on Parallel Delta-Sigma Modulators through Propagation of State Registers. In *Proceedings of IEEE International Midwest Symposium on Circuits and Systems*, pages 1133–1137, Boston, MA, USA, Aug. 2017.

[47] Daniel C. Dinis, Rui Ma, Shintaro Shinjo, Koji Yamanaka, Koon H. Teo, Philip Orlik, Arnaldo S. R. Oliveira, and José M. N. Vieira. A Real-time Architecture for Agile and Concurrent FPGA-based Triple-Band All-Digital RF Transmission. *IEEE Transactions on Microwave Theory and Techniques*, 66(11):4955–4966, Aug. 2018.

[48] Daniel C. Dinis, Rui Ma, Koon H. Teo, Philip Orlik, Arnaldo S. R. Oliveira, and José M. N. Vieira. An FPGA-based Multi-level All-Digital Transmitter with 1.25 GHz of Bandwidth. In *IEEE MTT-S International Microwave Symposium Digest*, pages 1–4, Philadelphia, PA, USA, Jun. 2018.

[49] Daniel C. Dinis, Rui Ma, Koon H. Teo, Philip Orlik, Arnaldo S. R. Oliveira, and José M. N. Vieira. Towards Analog Filter-Free All-Digital Transmitters Through Hybrid Estimation and Cancellation of Delta-Sigma Modulation's Quantization Noise. In *IEEE MTT-S International Microwave and Radar Conference Symposium Digest*, pages 1–4, Poznan, Poland, May. 2018.

[50] Daniel C. Dinis, Arnaldo S. R. Oliveira, and José Vieira. Recent advances on All-Digital Antenna Array Transmitters. *Submitted to IEEE Microwave Magazine*, pages 1–6, 2018.

[51] Daniel C. Dinis, Arnaldo S. R. Oliveira, and José M. N. Vieira. All-Digital Transmitter Based Antenna Array with Reduced Hardware Complexity. In *Proceedings of IEEE MTT-S International Microwave Symposium*, pages 153–156, Honolulu, HI, USA, Jun. 2017.

[52] Daniel C. Dinis, Arnaldo S. R. Oliveira, and José M. N. Vieira. All-Digital Transmitter Based on Cascaded Delta-Sigma Modulator for Relaxing the Analog Filtering Requirements. In *Proceedings of IEEE MTT-S International Microwave Symposium*, pages 145–148, Honolulu, HI, USA, Jun. 2017.

[53] Jr. Donald K. Weaver. A Third Method of Generation and Detection of Single-Sideband Signals. *Proceedings of the IRE*, 44(12):1703–1705, 1956.

[54] M. M. Ebrahimi, S. A. Bassam, M. Helaoui, and F. M. Ghannouchi. Feedback-based digital predistorter for multi-bit delta-sigma transmitter. In *Proceedings of IEEE 54th Int Circuits and Systems (MWSCAS) Midwest Symp*, pages 1–4, 2011.

[55] Mohammad Mojtaba Ebrahimi and Mohamed Helaoui. Reducing Quantization Noise to Boost Efficiency and Signal Bandwidth in Delta–sigma-based Transmitters. *IEEE Transactions on Microwave Theory and Techniques*, 61(12):4245–4251, 2013.

[56] Mohammad Mojtaba Ebrahimi, Mohamed Helaoui, and Fadhel M. Ghannouchi. Time-interleaved Delta-sigma Modulator for Wideband Digital GHz Transmitters Design and SDR Applications. *Progress in Electromagnetics Research B*, 34:263–281, Sept. 2011.

[57] Mohammad Mojtaba Ebrahimi, Mohamed Helaoui, and Fadhel M. Ghannouchi. Delta-sigma-based Transmitters: Advantages and Disadvantages. *IEEE Microwave Magazine*, 14(1):68–78, 2013.

[58] ETSI ETR. ETR 290 (v 1.2. 1):"DVB Measurement Guidelines for DVB systems.", May 1997.

[59] R. Feger, C. Pfeffer, and A. Stelzer. A Frequency-division MIMO FMCW Radar System using Delta-sigma-based Transmitters. In *IEEE MTT-S International Microwave Symposium Digest*, pages 1–4, June 2014.

[60] D. Ferling, T. Bohn, D. Zeller, P. Frenger, I. Gódor, Y. Jading, and W. Tomaselli. Energy Efficiency Approaches for Radio Nodes. In *Proc. Future Network and Mobile Summit*, pages 1–9, 2010.

[61] Oriol Font-Bach, Nikolaos Bartzoudis, Xavier Mestre, David Lopez-Bueno, Philippe Mege, Laurent Martinod, Vidar Ringset, and Tor André Myrvoll. When SDR meets a 5G Candidate Waveform: Agile Use of Fragmented Spectrum and Interference Protection in PMR Networks. *IEEE Wireless Communications*, 22(6):56–66, 2015.

[62] Brecht François, P. A. J. Nuyts, Wim Dehaene, and Patrick Reynaert. Extending Dynamic Range of RF PWM Transmitters. *Electronics Letters*, 49(6):430–432, 2013.

[63] A. Frappe, A. Flament, B. Stefanelli, A. Kaiser, and A. Cathelin. An All-Digital RF Signal Generator Using High-Speed Delta Sigma Modulators. *IEEE Journal of Solid-State Circuits*, 44(10):2722–2732, 2009.

[64] Antoine Frappe, Axel Flament, Bruno Stefanelli, Andreia Cathelin, and Andreas Kaiser. All-digital RF Signal Generation for Software Defined Radio. In *4th European Conference on Circuits and Systems for Communications (ECCSC)*, pages 236–239. IEEE, 2008.

[65] Antoine Frappe, Axel Flament, Bruno Stefanelli, Andreia Cathelin, and Andreas Kaiser. All-digital RF signal generation for software defined radio. In *2008 4th European Conference on Circuits and Systems for Communications*, pages 236–239, 2008.

[66] Antoine Frappé, Axel Flament, Bruno Stefanelli, Andreas Kaiser, and Andreia Cathelin. An All-Digital RF Signal Generator Using High-Speed $\Delta\Sigma$ Modulators. *IEEE J. Solid-State Circuits*, 44(10):2722–2732, Oct 2009.

[67] C. Fulton, M. Yeary, D. Thompson, J. Lake, and A. Mitchell. Digital Phased Arrays: Challenges and Opportunities. *Proceedings of the IEEE*, 104(3):487–503, March 2016.

[68] Steven Gao, Neil F. Chamberlain, and Y. Jay Guo. Guest Editorial: Antennas for Satellite Communications. *IEEE Transactions on Antennas and Propagation*, 63(4):1186–1190, 2015.

[69] F. T. Gebreyohannes, A. Frappé, and A. Kaiser. A Configurable Transmitter Architecture for IEEE 802.11ac and 802.11ad Standards. *IEEE Transactions on Circuits and Systems II: Express Briefs*, 63(1):9–13, Jan 2016.

[70] F. M. Ghannouchi. Power Amplifier and Transmitter Architectures for Software Defined Radio Systems. *IEEE Circuits and Systems Magazine*, 10(4):56–63, 2010.

[71] F. M. Ghannouchi. Power Amplifier and Transmitter Architectures for Software Defined Radio Systems. *IEEE Circuits and Systems Magazine*, 10(4):56–63, 2010.

[72] F. M. Ghannouchi, S. Hatami, P. Aflaki, M. Helaoui, and R. Negra. Accurate Power Efficiency Estimation of GHz Wireless Delta-Sigma Transmitters for Different Classes of Switching Mode Power Amplifiers. *IEEE Transactions on Microwave Theory and Techniques*, 58(11):2812–2819, 2010.

[73] Fadhel M. Ghannouchi. Power Amplifier and Transmitter Architectures for Software Defined Radio systems. *IEEE Circuits and Systems Magazine*, 10(4):56–63, 2010.

[74] Fadhel M. Ghannouchi, Safar Hatami, Pouya Aflaki, Mohamed Helaoui, and Renato Negra. Accurate Power Efficiency Estimation of GHz Wireless Delta-sigma Transmitters for Different Classes of Switching Mode Power Amplifiers. *IEEE Transactions on Microwave Theory and Techniques*, 58(11):2812–2819, 2010.

[75] K. M. Gharaibeh, K. G. Gard, and M. B. Steer. Accurate Estimation of Digital Communication System Metrics - SNR, EVM and rho; in a Nonlinear Amplifier Environment. In *Proc. of IEEE ARFTG Microwave Measurements Conference*, pages 41–44, Dec 2004.

[76] K. M. Gharaibeh, K. G. Gard, and M. B. Steer. Accurate estimation of digital communication system metrics - snr, evm and /spl rho/ in a nonlinear amplifier environment. In *64th ARFTG Microwave Measurements Conference, Fall 2004.*, pages 41–44, 2004.

[77] C. Giancarlo and C. Sodini. A Slope Adaptive Delta Modulator for VLSI Signal Processing systems. *IEEE Transactions on Circuits and Systems*, 33(1):51–58, 1986.

[78] Ulf Gustavsson, Thomas Eriksson, Hossein Mashad Nemati, Paul Saad, Peter Singerl, and Christian Fager. An RF Carrier Bursting System using Partial Quantization Noise Cancellation. *IEEE Transactions on Circuits and Systems I: Regular Papers*, 59(3):515–528, 2012.

[79] S. Han, C.-l. I, Z. Xu, and C. Rowell. Large-scale Antenna Systems with Hybrid Analog and Digital Beamforming for Millimeter Wave 5G. *IEEE Communications Magazine*, 53(1):186–194, January 2015.

[80] Christoph Haslach, Daniel Markert, Andreas Frotzscher, and Andreas Pascht. New Efficient Architectures for RF Pulse Width Modulators. In *IEEE MTT-S International Microwave Symposium Digest*, pages 1–4, 2013.

[81] Safar Hatami, Mohamed Helaoui, Fadhel M. Ghannouchi, and Massoud Pedram. Single-Bit Pseudoparallel Processing Low-Oversampling Delta–Sigma Modulator Suitable for SDR Wireless Transmitters. *IEEE Transactions on Very Large Scale Integration Systems*, 22(4):922–931, 2014.

[82] Safar Hatami, Mohamed Helaoui, Fadhel M. Ghannouchi, and Massoud Pedram. Single-Bit Pseudoparallel Processing Low-Oversampling Delta–Sigma Modulator Suitable for SDR Wireless Transmitters. *IEEE Transactions on Very Large Scale Integration Systems*, 22(4):922–931, 2014.

[83] Katharina Hausmair, Shuli Chi, Peter Singerl, and Christian Vogel. Aliasing-free Digital Pulse-width Modulation for Burst-mode RF Transmitters. *IEEE Transactions on Circuits and Systems I: Regular Papers*, 60(2):415–427, 2013.

[84] M. Helaoui, S. Hatami, R. Negra, and F. M. Ghannouchi. A Novel Architecture of Delta-Sigma Modulator Enabling All-Digital Multiband Multistandard RF Transmitters Design. *IEEE Transactions on Circuits and Systems*, 55(11):1129–1133, 2008.

[85] M. Helaoui, S. Hatami, R. Negra, and F. M. Ghannouchi. A Novel Architecture of Delta-sigma Modulator Enabling All-digital Multiband Multistandard RF Transmitters Design. *IEEE Trans. Circuits Syst. II, Exp. Briefs*, 55(11):1129–1133, 2008.

[86] M. Helaoui, S. Hatami, R. Negra, and F. M. Ghannouchi. A Novel Architecture of Delta-Sigma Modulator Enabling All-Digital Multiband Multistandard RF Transmitters Design. *IEEE Transactions on Circuits and Systems*, 55(11):1129–1133, 2008.

[87] Luis Hernandez and Eric Gutierrez. Analytical evaluation of vco-adc quantization noise spectrum using pulse frequency modulation. *IEEE Signal Processing Letters*, 22(2):249–253, 2015.

[88] R. E. Hiorns and M. B. Sandler. Power Digital to Analogue Conversion using Pulse Width Modulation and Digital Signal Processing. *IEE Proceedings G (Circuits, Devices and Systems)*, 140(5):329–338, 1993.

[89] Hon Cheong Hor and Liter Siek. Review on vco based adc in modern deep submicron cmos technology. In *2012 IEEE International Symposium on Radio-Frequency Integration Technology (RFIT)*, pages 86–88, 2012.

[90] Shinichi Hori, Andreas Wentzel, Makoto Hayakawa, Wolfgang Heinrich, and Kazuaki Kunihiro. A watt-class digital transmitter with a voltage-mode class-s power amplifier and an envelope delta-sigma modulator for 450 mhz band. In *2012 IEEE Compound Semiconductor Integrated Circuit Symposium (CSICS)*, pages 1–4, 2012.

[91] Tsai-Pi Hung, J. Rode, L. E. Larson, and P. M. Asbeck. Design of H-Bridge Class-D Power Amplifiers for Digital Pulse Modulation Transmitters. *IEEE Transactions on Microwave Theory and Techniques*, 55(12):2845–2855, 2007.

[92] Hi Inose, Y. Yasuda, and J. Murakami. A Telemetering System by Code Modulation - $\Delta\Sigma$ Modulation. *IRE Transactions on Space Electronics and Telemetry*, SET-8(3):204–209, 1962.

[93] Intel. Intel stratix 10 mx (dram system-in-package) product table. https://www.intel.com/content/dam/www/programmable/us/en/pdfs/literature/pt/stratix-10-mx-product-table.pdf, 2020.

[94] M. Iwamura, K. Etemad, M. h. Fong, R. Nory, and R. Love. Carrier Aggregation Framework in 3GPP LTE-advanced [WiMAX/LTE Update]. *IEEE Communications Magazine*, 48(8):60–67, August 2010.

[95] Wonhoon Jang, Rui Cordeiro, Arnaldo Oliveira, and Nuno Borges Carvalho. A Broadband Almost-digital RF Transmitter with an Efficient Power Amplifier. *IEEE Transactions on Microwave Theory and Techniques*, 64(5):1526–1534, 2016.

[96] Erwin Janssen and Arthur Van Roermund. *Look-Ahead Based Sigma-Delta Modulation.* Springer Science & Business Media, 2011.

[97] Jinseong Jeong and Yuanxun Ethan Wang. Pulsed Load Modulation (PLM) Technique for Efficient Power Amplification. *IEEE Transactions on Circuits and Systems II: Express Briefs*, 55(10):1011–1015, 2008.

[98] A. Jerng and C. G. Sodini. A Wideband Delta Sigma Digital-RF Modulator for High Data Rate Transmitters. *Solid-State Circuits, IEEE Journal of*, 42(8):1710–1722, 2007.

[99] Jesper Johansson and Lars Svensson. A Novel Speculative Pseudo-parallel $\Delta\Sigma$ Modulator. In *IEEE NORCHIP*, pages 1–4, 2014.

[100] David A. Johns and Ken Martin. *Analog Integrated Circuit Design.* John Wiley & Sons, 2008.

[101] Thomas Johnson and Shawn P. Stapleton. RF Class-D Amplification with Bandpass Sigma–delta Modulator Drive Signals. *IEEE Transactions on Circuits and Systems I: Regular Papers*, 53(12):2507–2520, 2006.

[102] Thomas Johnson and Shawn P. Stapleton. RF Class-D Amplification with Bandpass Sigma–delta Modulator Drive Signals. *IEEE Transactions on Circuits and Systems I: Regular Papers*, 53(12):2507–2520, 2006.

[103] Thomas Johnson and Shawn P. Stapleton. Comparison of Bandpass Modulator Coding Efficiency with a Periodic Signal Model. *IEEE Transactions on Circuits and Systems I: Regular Papers*, 55(11):3763–3775, Dec. 2008.

[104] Maryam Jouzdani, Mohammad Mojtaba Ebrahimi, Mohamed Helaoui, and Fadhel M. Ghannouchi. Complex Delta-Sigma-Based Transmitter With Enhanced Linearity Performance Using Pulsed Load Modulation Power Amplifier. *IEEE Transactions on Microwave Theory and Techniques*, 2017.

[105] Maryam Jouzdani, Mohamed Helaoui, and Fadhel M. Ghannouchi. Advanced Envelope Delta-sigma Transmitter Architecture with PLM Power Amplifier for Multi-standard Applications. In *IEEE MTT-S International Microwave Symposium Digest*, pages 1–4, 2016.

[106] Holger Karl. An Overview of Energy-Efficiency Techniques for Mobile Communication Systems. Technical report, Technical University Berlin, September 2003.

[107] J. Keyzer, J. Hinrichs, A. Metzger, M. Iwamoto, I. Galton, and P. Asbeck. Digital Generation of RF Signals for Wireless Communications with Band-pass Delta-sigma Modulation. In *Proc. of IEEE MTT-S International Microwave Symposium*, pages 2127–2130. IEEE, 2001.

[108] J. Keyzer, J. Hinrichs, A. Metzger, M. Iwamoto, I. Galton, and P. Asbeck. Digital generation of RF signals for wireless communications with band-pass delta-sigma modulation. In *Microwave Symposium Digest, 2001 IEEE MTT-S International*, volume 3, pages 2127–2130, 2001.

[109] Toru Kitayabu and Hiroyasu Ishikawa. Generalized Architecture of Concurrent Dual-band Transmitter for Spectrum Aggregation System. In *IEEE International Symposium on Personal, Indoor and Mobile Radio Communications*, pages 111–116, 2010.

[110] Mucahit Kozak and Izzet Kale. Novel Topologies for Time-interleaved Delta-sigma Modulators. *IEEE Transactions on Circuits and Systems II: Analog and Digital Signal Processing*, 47(7):639–654, 2000.

[111] Mücahit Kozak and Izzet Kale. *Oversampled Delta-sigma modulators: Analysis, applications and novel topologies.* Springer Science & Business Media, 2003.

[112] Herbert Lee Krauss and Philip Franklin Ordung. Distortion and Band-Width Characteristics of Pulse Modulation. *Transactions of the American Institute of Electrical Engineers*, 66(1):984–988, 1947.

[113] Erik G. Larsson, Ove Edfors, Fredrik Tufvesson, and Thomas L. Marzetta. Massive mimo for next generation wireless systems. *IEEE Communications Magazine*, 52(2):186–195, 2014.

[114] J. Lee, Y. Kim, Y. Kwak, J. Zhang, A. Papasakellariou, T. Novlan, C. Sun, and Y. Li. LTE-advanced in 3GPP Rel -13/14: an evolution toward 5G. *IEEE Communications Magazine*, 54(3):36–42, March 2016.

[115] Ying Loong Lee, Donghong Qin, Li-Chun Wang, and Gek Hong Sim. 6G massive radio access networks: Key applications, requirements and challenges. *IEEE Open Journal of Vehicular Technology*, 2:54–66, 2021.

[116] Haolin Li, Laurens Breyne, Joris Van Kerrebrouck, Michiel Verplaetse, Chia-Yi Wu, Piet Demeester, and Guy Torfs. A 21-gs/s single-bit second-order delta–sigma modulator for fpgas. *IEEE Transactions on Circuits and Systems II: Express Briefs*, 66(3):482–486, 2019.

[117] Haolin Li, Michiel Verplaetse, Jochem Verbist, Joris Van Kerrebrouck, Laurens Breyne, Chia-yi Wu, Laurens Bogaert, Bart Moeneclaey, Xin Yin, Johan Bauwelinck, Piet Demeester, and Guy Torfs. Real-Time 100-GS / s Sigma-Delta Modulator for All-Digital Radio-Over-Fiber Transmission. *Journal of Lightwave Technology*, 38(2):386–393, 2020.

[118] Abel Lorences-Riesgo, Samuel S. Pereira, Daniel C. Dinis, José Vieira, Arnaldo S. R. Oliveira, and Paulo P. Monteiro. Real-Time FPGA-Based Delta-Sigma-Modulation Transmission for 60 GHz Radio-over-Fiber Fronthaul. In *European Conference on Optical Communication Digest*, pages 1–4, Bucharest, Romania, July 2018.

[119] Fa-Long Luo. *Digital front-end in wireless communications and broadcasting: circuits and signal processing.* Cambridge University Press, 2011.

[120] R. G. Lyons. *Understanding Digital Signal Processing.* Pearson Education, 2010.

[121] T. Maehata, K. Totani, S. Kameda, and N. Suematsu. Concurrent Dual-band 1-bit Digital Transmitter using Band-pass Delta-sigma Modulator. In *Proceedings of European Microwave Conference*, pages 1523 –1526, Nuremberg, Germany, Oct. 2013.

[122] Simone Maier, Wolfgang Kuebart, Christoph Haslach, Ulrich Seyfried, Wolfgang Templ, Andreas Frotzscher, Daniel Markert, Ralf Matz, and Andreas Pascht. Class-O Base Station System with RF Pulse-width-modulation in Downlink and Uplink. In *Proceedings of IEEE Asia-Pacific Microwave Conference Proceedings*, pages 1222–1225, 2011.

[123] Simone Maier, Xin Yu, Holger Heimpel, and Andreas Pascht. Wideband base station receiver with analog-digital conversion based on rf pulse width modulation. In *2013 IEEE MTT-S International Microwave Symposium Digest (MTT)*, pages 1–3, 2013.

[124] Răzvan-Cristian Marin, Antoine Frappé, Andreas Kaiser, and Andreia Cathelin. Considerations for High-speed Configurable-bandwidth Time-interleaved Digital Delta-sigma Modulators and Synthesis in 28 nm UTBB FDSOI. In *IEEE Int. New Circ. and Syst. Conf. Digest*, pages 1–4, Grenoble, France, Jun 2015.

[125] Daniel Markert, Christoph Haslach, Holger Heimpel, Andreas Pascht, and Georg Fischer. Phase-modulated DSM-PWM Hybrids with Pulse Length Restriction for Switch-mode Power Amplifiers. In *Proceedings of IEEE European Microwave Conference*, pages 1364–1367, Oct. 2014.

[126] Daniel Markert, Xin Yu, Holger Heimpel, and Georg Fischer. An All-Digital, Single-Bit RF Transmitter for Massive MIMO. *IEEE Transactions on Circuits and Systems I: Regular Papers*, 64(3):696–704, Mar 2017.

[127] Yasuyudi Matsuya, Kuniharu Uchimura, Atsushi Iwata, Tsutomu Kobayashi, Masayuki Ishikawa, and Takeshi Yoshitome. A 16-bit Oversampling A-to-D Conversion Technology using Triple-integration Noise Shaping. *IEEE Journal of Solid-State Circuits*, 22(6):921–929, 1987.

[128] E. Mensink, E. A. M. Klumperink, and B. Nauta. Distortion Cancellation by Polyphase Multipath Circuits. *IEEE Transactions on Circuits and Systems I: Regular Papers*, 52(9):1785–1794, 2005.

[129] III Mitola, J. and Jr. Maguire, G. Q. Cognitive Radio: Making Software Radios More Personal. *IEEE Personal Communications*, 6(4):13–18, 1999.

[130] J. Mitola. The Software Radio Architecture. *IEEE Communications Magazine*, 33(5):26–38, May 1995.

[131] A. Mohammadi and F. M. Ghannouchi. Single RF front-end MIMO transceivers. *IEEE Communications Magazine*, 49(12):104–109, December 2011.

[132] Marc A. Morin. A 1.6 Gb/s Delta-sigma Modulator with Integrated Wideband Mixer for RF applications. In *Proceedings of the Bipolar/BiCMOS Circuits and Technology Meeting*, pages 148–151, 1998.

[133] Hendrik T. Mouton, Brendan McGrath, Donald Grahame Holmes, and Richardt H Wilkinson. One-dimensional Spectral Analysis of Complex PWM Waveforms using Superposition. *IEEE Transactions on Power Electronics*, 29(12):6762–6778, Dec 2014.

[134] Qianli Mu. *Delta-sigma Modulated RF Transmitters.* PhD thesis, University of Colorado at Boulder, 2009.

[135] Khurram Muhammad, Robert B. Staszewski, and Dirk Leipold. Digital RF Processing: toward Low-cost Reconfigurable Radios. *IEEE Communications Magazine*, 43(8):105, 2005.

[136] B. Nauta and A.-J. Annema. Analog/RF Circuit Design Techniques for Nanometer-scale IC Technologies. In *Proceedings of 35th European Solid-State Device Research Conference, ESSDERC 2005*, pages 45–53, 2005.

[137] H. J. Ng, W. Ahmad, and D. Kissinger. Scalable MIMO Radar utilizing Delta-sigma Modulation-based Frequency-division Multiplexing Technique. In *Proceedings of European Radar Conference*, pages 118–121, Oct 2017.

[138] H. J. Ng and D. Kissinger. A Scalable Frequency-division Multiplexing MIMO Radar utilizing Single-sideband Delta-sigma Modulation. In *Proceedings of IEEE Asia Pacific Microwave Conference*, pages 360–363, Nov. 2017.

[139] Michael Nielsen and Torben Larsen. A Transmitter Architecture based on Delta–sigma Modulation and Switch-mode Power Amplification. *IEEE Transactions on Circuits and Systems II: Express Briefs*, 54(8):735–739, 2007.

[140] Michael Nielsen and Torben Larsen. A 2-GHz GaAs HBT RF Pulsewidth Modulator. *IEEE Transactions on Microwave Theory and Techniques*, 56(2):300–304, Feb. 2008.

[141] S. Norsworthy, R. Schreier, and G. Temes. *Delta-Sigma Data Converters (Theory, Design, and Simulation).* Wiley-IEEE Press, 1 edition, 1996.

[142] Pieter A. J. Nuyts, Patrick Reynaert, and Wim Dehaene. *Continuous-Time Digital Front-Ends for Multistandard Wireless Transmission.* Springer, 2014.

[143] Frank W. J. Olver, Daniel W. Lozier, Ronald F. Boisvert, and Charles W. Clark. *NIST Handbook of Mathematical Functions Hardback and CD-ROM.* Cambridge university press, 2010.

[144] P. Ostrovskyy, H. Gustat, Ch Scheytt, and Y. Manoli. A 9 GS/s 2.1.. 2.2 GHz Bandpass Delta-sigma Modulator for Class-S Power Amplifier. In *Proceedings of the IEEE MTT-S International Microwave Symposium*, pages 1129–1132, Boston, MA, USA, Jun. 2009.

[145] Philip Ostrovskyy, Hans Gustat, Maurits Ortmanns, and J. Christoph Scheytt. A 5-Gb/s 2.1–2.2-GHz Bandpass $\Delta\Sigma$ Modulator for Switch-Mode Power Amplifier. *IEEE Transactions on Microwave Theory and Techniques*, 60(8):2524–2531, Aug 2012.

[146] Marouan Othmani, Noureddine Boulejfen, Alberto Brihuega, Fadhel M. Ghannouchi, Markus Allén, and Mikko Valkama. Delta-sigma modulator-embedded digital predistortion for 5g transmitter linearization. *IEEE Transactions on Communications*, 70(8):5558–5571, 2022.

[147] M. Ozen, C. M. Andersson, T. Eriksson, M. Acar, R. Jos, and C. Fager. Linearization study of a highly efficient CMOS-GaN RF pulse width modulation based transmitter. In *42nd European Microwave Conference (EuMC)*, pages 136–139, 2012.

[148] M. Ozen, R. Jos, C. M. Andersson, M. Acar, and C. Fager. High-Efficiency RF Pulsewidth Modulation of Class-E Power Amplifiers. *IEEE Transactions on Microwave Theory and Techniques*, 59(11):2931–2942, 2011.

[149] Mustafa Ozen. *Advanced Transmitter Architectures Based on SMPAs.* PhD thesis, Chalmers University of Technology, 2014.

[150] Bonghyuk Park and Jaeho Jung. A Fully Integrated Pulsewidth Modulator for Class-S System. In *Advanced Communication Technology (ICACT), 2012 14th International Conference on*, pages 274–277, 2012.

[151] T. Parks and J. McClellan. Chebyshev approximation for nonrecursive digital filters with linear phase. *IEEE Transactions on Circuit Theory*, 19(2):189–194, 1972.

[152] S. S. Pereira, L. Almeida, A. S. R. Oliveira, N. B. Carvalho, and P. P. Monteiro. Improving coding efficiency in all-digital transmitters. In *2023 IEEE Radio and Wireless Symposium (RWS)*, pages 115–117, 2023.

[153] S. S. Pereira, Luís Filipe Almeida, D. C. Dinis, Arnaldo S. R. Oliveira, P. P. Monteiro, and N. B. Carvalho. Frequency-agile real-time all-digital transmitter with 1 GHz of bandwidth. *IEEE Transactions on Circuits and Systems II: Express Briefs*, 70(8):2844–2848, 2023.

[154] S. S. Pereira, Luís Filipe Almeida, Daniel C. Dinis, Arnaldo S. R. Oliveira, Paulo P. Monteiro, and Nuno Borges Carvalho. Frequency-agile real-time all-digital transmitter with 1 ghz of bandwidth. *IEEE Transactions on Circuits and Systems II: Express Briefs*, 70(8):2844–2848, 2023.

[155] Samuel S. Pereira, Abel Lorences-Riesgo, Daniel C. Dinis, Arnaldo S. R. Oliveira, José Vieira, and Paulo P. Monteiro. Millimetre-wave Real-time All-Digital Transmitter with Electro-Optical Upconversion. In *International Conf. on Transparent Optical Networks Digest*, pages 1–4, Roma, Italy, Sept. 2018.

[156] Samuel Santos Pereira, Luís Filipe Almeida, Rui F. Cordeiro, Arnaldo S. R. Oliveira, Paulo P. Monteiro, and Nuno Borges Carvalho. Scalable resource optimized LUT-based all-digital transmitter. *IEEE Transactions on Circuits and Systems I: Regular Papers*, 70(8):3212–3220, 2023.

[157] Sérgio C. C. Pires. *Multi-standard Reconfigurable Wireless Transmitter Architectures.* PhD thesis, Universidade de Aveiro, 2014.

[158] Tomasz Podsiadlik, John Dooley, and Ronan Farrell. Improved coding-efficiency two-level source encoder for rf switch-mode power amplifiers. *IEEE Transactions on Circuits and Systems II: Express Briefs*, 59(10):618–622, 2012.

[159] A. S. Y. Poon and M. Taghivand. Supporting and Enabling Circuits for Antenna Arrays in Wireless Communications. *Proceedings of the IEEE*, 100(7):2207–2218, July 2012.

[160] R. Khoini Poorfard and D. A. Johns. Time-interleaved Oversampling Converters. *Electronics Letters*, 29:1673–1674, 1993.

[161] André Prata, Arnaldo S. R. Oliveira, and Nuno B. Carvalho. An Agile and Wideband All-Digital SDR Receiver for 5G Wireless Communications. In *Proceedings of Euromicro Conference on Digital System Design*, pages 146–151, Madeira, Portugal, Aug. 2015.

[162] André Prata, Arnaldo S. R. Oliveira, and Nuno B. Carvalho. All-digital Flexible Uplink Remote Radio Head for C-RAN. In *IEEE MTT-S International Microwave Symposium Digest*, pages 1–3. IEEE, May. 2016.

[163] J. G. Proakis and D. G. Manolakis. *Digital Signal Processing: Principles, Algorithms and Applications.* Prentice Hall, 2001.

[164] Sravan Pulipati and Rui Ma. Design considerations and fpga implementation of a wideband all-digital transmit beamformer with 50% fractional bandwidth. *2020 IEEE/MTT-S International Microwave Symposium (IMS)*, pages 1073–1076, 2020.

[165] Frederick H. Raab, Peter Asbeck, Steve Cripps, Peter B. Kenington, Zoya B. Popovic, Nick Pothecary, John F. Sevic, and Nathan O. Sokal. Power Amplifiers and Transmitters for RF and Microwave. *IEEE Transactions on Microwave Theory and Techniques*, 50(3):814–826, 2002.

[166] Nandana Rajatheva, Italo Atzeni, Emil Bjornson, Andre Bourdoux, Stefano Buzzi, Jean-Baptiste Dore, Serhat Erkucuk, Manuel Fuentes, Ke Guan, Yuzhou Hu, et al. White paper on broadband connectivity in 6g. *arXiv preprint arXiv:2004.14247*, 2020.

[167] Ian Stuart Robinson, Jeffrey Mark Hinrichs, Jasmine Upendra Patel, Paul Charles MacFalda, Reza Dehmohseni, Frederic J. Harris, and Kenneth Weber. Look-up Table Delta-sigma Conversion, March 13 2007. US Patent 7,190,288.

[168] P. J. Roche, J. M. D. Murphy, and M. G. Egan. Reduction of Quantization Noise in Position Servosystems. In *Proc. of International Conference on Industrial Electronics, Control, Instrumentation, and Automation, Power Electronics and Motion Control*, pages 464–469, 1992.

[169] Enrico Roverato, Marko Kosunen, Jerry Lemberg, Kari Stadius, and J. Ryynanen. RX-Band Noise Reduction in All-Digital Transmitters With Configurable Spectral Shaping of Quantization and Mismatch Errors. *IEEE Transactions on Circuits and Systems I: Regular Papers*, 61(11):3256–3265, Nov. 2014.

[170] Enrico Roverato, Marko Kosunen, and Jussi Ryynänen. The Synthesis of Noise Transfer Functions for Bandpass Delta-sigma Modulators with Tunable Center Frequency. In *European Conference on Circuit Theory and Design Digest*, pages 1–4, Trondheim, Norway, Aug. 2015.

[171] H. Ruotsalainen, H. Arthaber, T. I. Laakso, and G. Magerl. Quantization Noise Reduction Techniques for Digital Pulsed RF Signal Generation based on Quadrature Noise Shaped Encoding. *IEEE Transactions on Circuits and Systems I: Regular Papers*, 61(9):2525–2536, 2014.

[172] H. Ruotsalainen, H. Arthaber, and G. Magerl. Quantization Noise Cancelation Scheme for Digital Quadrature RF Pulse Encoding. In *IEEE MTT-S International Microwave Symposium Digest*, pages 1–4, Seattle, WA, USA, Jun. 2013.

[173] Henri Ruotsalainen, Holger Arthaber, and Gottfried Magerl. A New Quadrature PWM Modulator with Tunable Center Frequency for Digital RF Transmitters. *IEEE Transactions on Circuits and Systems II: Express Briefs*, 59(11):756–760, 2012.

[174] M. Schmidt, S. Haug, M. Grozing, and M. Berroth. A Pipelined 3-level Bandpass Delta-Sigma Modulator for Class-S Power Amplifiers. In *IEEE International Symposium on Circuits and Systems (ISCAS)*, pages 2757 –2760, May 2011.

[175] Martin Schmidt, Markus Grozing, Stefan Heck, Ingo Dettmann, Manfred Berroth, Dirk Wiegner, Wolfgang Templ, and Andreas Pascht. A 1.55 GHz to 2.45 GHz Center Frequency Continuous-time Bandpass Delta-sigma Modulator for Frequency Agile

Transmitters. In *Proceedings of IEEE Radio Frequency Integrated Circuits Symposium*, pages 153–156, Boston, MA, USA, Jun 2009.

[176] P. J Schreier and L. L. Scharf. *Statistical Signal Processing of Complex-valued Data: the Theory of Improper and Noncircular Signals.* Cambridge University Press, 2010.

[177] Richard Schreier. MATLAB Delta Sigma Toolbox. Available at `http://www.mathworks.com/matlabcentral/fileexchange/19`.

[178] Richard Schreier. *Noise-shaped Coding.* PhD thesis, University of Toronto, Toronto, Canada, 1991.

[179] Richard Schreier and Gabor C. Temes. *Understanding Delta-sigma Data Converters*, volume 74. IEEE press Piscataway, NJ, 2005.

[180] Richard Schreier, Gabor C. Temes, and Steven R. Norsworthy. *Delta-Sigma Data Converters: Theory, Design, and Simulation.* IEEE, 1997.

[181] Elena Serebryakova. *High-Power Comb-Line Filter Architectures for Switched-Mode RF Power Amplifier Systems.* PhD thesis, PhD Thesis, Fakultät für Elektrotechnik und Informationstechnik, Technischen Universität Ilmenau, 2012.

[182] K. A. Shehata, M. A. Aboul-Dahab, S. H. El Ramly, and K. A. Hamouda. An FPGA Based 1-bit All Digital Transmitter Employing Delta-Sigma Modulation with RF Output for SDR. In *Proceedings of 2nd International Conference on Signals, Circuits and Systems SCS 2008*, pages 1–6, 2008.

[183] N. V. Silva, A. S. R. Oliveira, and N. B. Carvalho. Design and optimization of flexible and coding efficient all-digital RF transmitters. *IEEE Transactions on Microwave Theory and Techniques*, 61(1):625–632, 2013.

[184] N. V. Silva, A. S. R. Oliveira, U. Gustavsson, and N. B. Carvalho. A Novel All-Digital Multichannel Multimode RF Transmitter Using Delta-Sigma Modulation. *IEEE Microwave and Wireless Components Letters*, 22(3):156–158, 2012.

[185] Nelson V. Silva, Arnaldo S. R. Oliveira, and Nuno B. Carvalho. Evaluation of Pulse Modulators for All-digital Agile Transmitters. In *IEEE MTT-S Int. Microw. Symp. Dig.*, pages 1–3, Montreal, QC, Canada, Jun. 2012.

[186] Nelson V. Silva, Arnaldo S. R. Oliveira, and Nuno B. Carvalho. Novel Fine Tunable Multichannel All-digital Transmitter. In *IEEE MTT-S International Microwave Symposium Dig.*, pages 1–3, 2013.

[187] Nelson V. Silva, Arnaldo S. R. Oliveira, and Nuno Borges Carvalho. Evaluation of Pulse Modulators for All-digital Agile Transmitters. In *IEEE MTT-S International Microwave Symposium Digest*, pages 1–3, 2012.

[188] Nelson V. Silva, Arnaldo S. R. Oliveira, and Nuno Borges Carvalho. Evaluation of Pulse Modulators for All-Digital Agile Transmitters. In *IEEE MTT-S International Microwave Symposium Digest (MTT)*, pages 1–3. IEEE, June 2012.

[189] Nelson V. Silva, Arnaldo S. R. Oliveira, and Nuno Borges Carvalho. Design and Optimization of Flexible and Coding Efficient All-Digital RF Transmitters. *IEEE Transactions on Microwave Theory and Techniques*, 61(1):625–632, 2013.

[190] Nelson V. Silva, Arnaldo S. R. Oliveira, and Nuno Borges Carvalho. Novel fine tunable multichannel all-digital transmitter. In *IEEE MTT-S International Microwave Symposium Digest (MTT)*. IEEE, June 2013.

[191] Nelson V Silva, Arnaldo S. R. Oliveira, Ulf Gustavsson, and Nuno Borges Carvalho. A Dynamically Reconfigurable Architecture enabling All-digital Transmission for Cognitive Radios. In *IEEE Radio and Wireless Symposium Dig.*, pages 1–4, 2012.

[192] Nelson V. Silva, Arnaldo S. R. Oliveira, Ulf Gustavsson, and Nuno Borges Carvalho. A Novel All-digital Multichannel Multimode RF Transmitter using Delta-sigma Modulation. *IEEE Microwave and Wireless Components Letters*, 22(3):156–158, 2012.

[193] Nelson V. Silva, Manuel Ventura, Arnaldo S. R. Oliveira, and Nuno Borges Carvalho. Evaluation of an FPGA-based Reconfigurable SoC for All-Digital Flexible RF Transmitters. In *15th Euromicro Conference on Digital System Design*, pages 890–895, September 2012.

[194] F. Sohrabi and W. Yu. Hybrid Digital and Analog Beamforming Design for Large-Scale Antenna Arrays. *IEEE Journal of Selected Topics in Signal Processing*, 10(3):501–513, April 2016.

[195] Hyejeong Song and Ranjit Gharpurey. Digitally Intensive Wireless Transmitter Architecture Employing RF Pulse Width Modulation. In *IEEE Dallas Circ. and Syst. Conf. Dig.*, pages 1–4, Dallas, TX, USA, Oct 2014. IEEE.

[196] Zukui Song and Dilip V. Sarwate. The frequency spectrum of pulse width modulated signals. *Signal Processing*, 83(10):2227–2258, 2003.

[197] SPI, Inc. Stream Processing: Enabling the New Generation of Easy to Use, High-Performance DSPs. White Paper, June 2008.

[198] Luis Suarez, Loutfi Nuaymi, and Jean-Marie Bonnin. An Overview and Classification of Research Approaches in Green Wireless Networks. *EURASIP Journal on Wireless Communications and Networking*, 1:1–18, 2012.

[199] Shahin Mehdizad Taleie, Tino Copani, Bertan Bakkaloglu, and Sayfe Kiaei. A Bandpass $\Delta\Sigma$ RF-DAC with Embedded FIR Reconstruction Filter. In *IEEE International Solid-State Circuits Conference*, 2006.

[200] Masaaki Tanio, Shinichi Hori, Noriaki Tawa, and Kazuaki Kunihiro. An FPGA-based All-Digital Transmitter with 9.6-GHz 2nd Order Time-interleaved Delta-sigma Modulation for 500-MHz Bandwidth. In *IEEE MTT-S International Microwave Symposium Digest*, pages 1–4, Honolulu, HI, USA, Jun. 2017. IEEE.

[201] Masaaki Tanio, Shinichi Hori, Noriaki Tawa, Tomoyuki Yamase, and Kazuaki Kunihiro. An FPGA-based All-digital Transmitter with 28-GHz Time-interleaved Delta-sigma Modulation. In *IEEE MTT-S International Microwave Symposium Digest*, pages 1–4, San Francisco, CA, USA, May 2016. IEEE.

[202] Omer Tanovic, Rui Ma, and Koon H. Teo. Simultaneous Power Encoding and Upconversion for All-Digital Transmitters Using Digital PWM. In *Proceedings of Asia-Pacific Microwave Conference*, pages 1588–1591, 2017.

[203] Omer Tanovic, Rui Ma, and Koon Hoo Teo. Optimal Delta-Sigma Modulation Based Noise Shaping for Truly Aliasing-Free digital PWM. In *Proceedings European Microwave Conference*, 2017.

[204] Bjorn Thorsten Thiel, Alp Ozmert, Junqing Guan, and Renato Negra. Lowpass Delta-Sigma Modulator with Digital Upconversion for Switching-Mode Power Amplifiers. In *Proceedings of IEEE MTT-S International Microwave Symposium Digest (MTT)*, pages 1–4, 2011.

[205] P. P. Vaidyanathan. *Multirate Systems And Filter Banks*. Prentice Hall, 1993.

[206] Gianluca Camillo Valastro, Daniela Panno, and Salvatore Riolo. A SDN/NFV based C-RAN architecture for 5G mobile networks. In *2018 International Conference on Selected Topics in Mobile and Wireless Networking, MoWNeT 2018*. Institute of Electrical and Electronics Engineers Inc., aug 2018.

[207] Joris Van Kerrebrouck, Laurens Breyne, Haolin Li, Johan Bauwelinck, Guy Torfs, Piet Demeester, and Thomas Bohn. Real-time All-digital Radio-over-fiber LTE Transmission. In *Advances in Wireless and Optical Communications*, pages 83–86, 2017.

[208] Jouko Vankka, Jaakko Ketola, Student Member, Johan Sommarek, and Olli Väänänen. A GSM / EDGE / WCDMA Modulator With On-Chip. *Ieee Transactions on Circuits and Systems—Ii: Analog and Digital Signal Processing*, 49(10):645–655, 2002.

[209] S. Bojja Venkatakrishnan, D. K. Papantonis, A. A. Akhiyat, E. A. Alwan, and J. L. Volakis. Experimental Validation of On-Site Coding Digital Beamformer With Ultra-Wideband Antenna Arrays. *IEEE Transactions on Microwave Theory and Techniques*, 65(11):4408–4417, Nov 2017.

[210] Poojan Wagh, Pallab Midya, Pat Rakers, James Caldwell, and Tony Schooler. An All-digital Universal RF Ttransmitter [CMOS RF modulator and PA]. In *Proc. of the IEEE Custom Integrated Circuits Conference*, pages 549–552, 2004.

[211] Poojan Wagh, Pallab Midya, and Patrick Rakers. Distortionless RF Pulse Width Modulation. In *IEEE Midwest Symposium on Circuits and Systems*, pages 11–14, 2002.

[212] Jing Wang, Zhensheng Jia, Luis Alberto Campos, Lin Cheng, Curtis Knittle, and Gee-Kung Chang. Delta-sigma digitization and optical coherent transmission of docsis 3.1 signals in hybrid fiber coax networks. *Journal of Lightwave Technology*, 36(2):568–579, 2018.

[213] Jing Wang, Zhenhua Yu, Kai Ying, Junwen Zhang, Feng Lu, Mu Xu, Lin Cheng, Xiaoli Ma, and Gee-Kung Chang. Digital Mobile Fronthaul Based on Delta–Sigma Modulation for 32 LTE Carrier Aggregation and FBMC Signals. *Journal of Optical Communications and Networking*, 9(2):A233–A244, 2017.

[214] Yuanxun Wang and Jin-Seong Jeong. Pulsed Load Modulation Amplifier and Method, August 28 2012. US Patent 8,254,854.

[215] Andreas Wentzel, Chafik Meliani, and Wolfgang Heinrich. RF Class-S Power Amplifiers: State-of-the-art Results and Potential. In *Proceedings of IEEE MTT-S International Microwave Symposium*, pages 812–815. IEEE, 2010.

[216] Richard M. White, Ben H. Cantrell, John P. McConnell, and James J. Alter. Delta-sigma Waveform Generation for Digital Radars. In *Proceedings of the IEEE Radar Conference, 2004.*, pages 154–156, 2004.

[217] Bernard Widrow, Istvan Kollar, and Ming-Chang Liu. Statistical Theory of Quantization. *IEEE Transactions on Instrumentation and Measurement*, 45(2):353–361, Apr. 1996.

[218] Huibo Wu, Wenhua Chen, Xin Liu, Zhenghe Feng, and Fadhel M. Ghannouchi. A uniform neural network digital predistortion model of rf power amplifiers for scalable applications. *IEEE Transactions on Microwave Theory and Techniques*, 70(11):4885–4899, 2022.

[219] Xilinx. DDS Compiler v6.0 Logicore IP. *Product Guide PG141*, 2015.

[220] Xilinx. UltraScale Architecture GTY Transceivers. *User Guide UG578*, 2017.

[221] Xilinx. 58g pam4 technology. https://www.xilinx.com/products/technology/high-speed-serial/58g.html, 2020.

[222] P. Xingdong, H. Wei, Y. Tianyang, and L. Linsheng. Design and implementation of an active multibeam antenna system with 64 RF channels and 256 antenna elements for massive MIMO application in 5G wireless communications. *China Communications*, 11(11):16–23, Nov 2014.

[223] J. Yang, S. Y. Yang, Z. H. Chen, and X. Y. Zhang. An agile LUT-based all-digital transmitter. *IEEE Transactions on Circuits and Systems I: Regular Papers*, 67(12):5550–5560, 2020.

[224] Si Yuan Yang, Jun Yang, Li Ying Huang, Jin Liang Bai, and Xiu Yin Zhang. A dual-band RF all-digital transmitter based on MPWM encoding. *IEEE Transactions on Microwave Theory and Techniques*, pages 1–1, 2022.

[225] Hao Ye, Geoffrey Ye Li, and Biing-Hwang Juang. Power of deep learning for channel estimation and signal detection in ofdm systems. *IEEE Wireless Communications Letters*, 7(1):114–117, 2017.

[226] Z. Ye, J. Grosspietsch, and G. Memik. An fpga based all-digital transmitter with radio frequency output for software defined radio. In *Proc. Design, Automation & Test in Europe Conference & Exhibition DATE '07*, pages 1–6, 2007.

[227] Yongs Zeng, Qingqing Wu, and Rui Zhang. Accessing from the sky: A tutorial on uav communications for 5g and beyond. *Proceedings of the IEEE*, 107(12):2327–2375, 2019.

[228] Cristian Zet, Cătălin Damian, and Cristian Foşalău. New type adc using pwm intermediary conversion. In *Proceedings of 12th TC4 International Workshop on ADC Modelling and Testing*, pages 113–117, 2007.

[229] Qiuyao Zhu, Rui Ma, Chunjie Duan, Kenji Mukai, Shintaro Shinjo, and Koon Hoo Teo. A 5-level Efficient IFPWM Power Coding Approach Encoding LTE for Class-S Digital-RF Transmitter with Distortion Correction. In *Proceedings of IEEE Radio and Wireless Symposium*, pages 184–186, Jan. 2014.

[230] Qiuyao Zhu, Rui Ma, Chunjie Duan, Koon Hoo Teo, and Kieran Parsons. A 5-level Discrete-time Power Encoder with Measured Coding Efficiency of 70% for 20-MHz LTE Digital Transmitter. In *IEEE MTT-S International Microwave Symposium Digest*, pages 1–3, Tampa Bay, FL, USA, Jun 2014. IEEE.

# Index

For Product Safety Concerns and Information please contact our EU representative GPSR@taylorandfrancis.com Taylor & Francis Verlag GmbH, Kaufingerstraße 24, 80331 München, Germany

**Batch number: 10777154**

Printed by Printforce, the Netherlands